3D QSAR in Drug Design

Recent Advances

QSAR = Three-Dimensional Quantitative Structure Activity Relationships

VOLUME 3

The titles published in this series are listed at the end of this volume.

3D QSAR in Drug Design Volume 3

Recent Advances

Edited by

Hugo Kubinyi
ZHF/G, A30, BASF AG, D-67056 Ludwigshafen, Germany

Gerd Folkers
ETH-Zürich, Department Pharmazie, Winterthurer Strasse 190, CH-8057 Zürich, Switzerland

Yvonne C. Martin
Abbott Laboratories, Pharmaceutical Products Division, 100 Abbott Park Rd., Abbott Park, IL 60064-3500, USA

KLUWER / ESCOM
Dordrecht / Boston / London

A C.I.P. Catalogue record for this book is available from the Library of Congress

ISBN 0–7923–4791–8
ISBN set 0–7923–4792–7

Published, under the KLUWER/ESCOM imprint,
by Kluwer Academic Publishers,
P.O. Box 17, 3300 AA Dordrecht, The Netherlands

Sold and distributed in North, Central and South America
by Kluwer Academic Publishers,
101 Philip Drive, Norwell, MA 02061, U.S.A.

In all other countries, sold and distributed
by Kluwer Academic Publishers,
P.O. Box 322, 3300 AH Dordrecht, The Netherlands.

Printed on acid-free paper

Printed in Great Britain

Contents

Part III. 3D QSAR Applications

Preface

Significant progress has been made in the study of three-dimensional quantitative structure–activity relationships (3D QSAR) since the first publication by Richard Cramer in 1988 and the first volume in the series, *3D QSAR in Drug Design. Theory, Methods and Applications*, published in 1993. The aim of that early book was to contribute to the understanding and the further application of CoMFA and related approaches and to facilitate the appropriate use of these methods.

Since then, hundreds of papers have appeared using the quickly developing techniques of both 3D QSAR and computational sciences to study a broad variety of biological problems. Again the editor(s) felt that the time had come to solicit reviews on published and new viewpoints to document the state of the art of 3D QSAR in its broadest definition and to provide visions of where new techniques will emerge or new applications may be found. The intention is not only to highlight new ideas but also to show the shortcomings, inaccuracies, and abuses of the methods. We hope this book will enable others to separate trivial from visionary approaches and me-too methodology from innovative techniques. These concerns guided our choice of contributors. To our delight, our call for papers elicited a great many manuscripts. These articles are collected in two bound volumes, which are each published simultaneously in two related series: they form Volumes 2 and 3 of the *3D QSAR in Drug Design* series which correspond to volumes 9–11 and 12–14, respectively, in *Perspectives in Drug Discovery and Design*. Indeed, the field is growing so rapidly that we solicited additional chapters even as the early chapters were being finished. Ultimately it will be the scientific community who will decide if the collective biases of the editors have furthered development in the field.

The challenge of the quantitative prediction of the biological potency of a new molecule has not yet been met. However, in the four years since the publication of the first volume, there have been major advances in our understanding of ligand–receptor interactions, molecular similarity, pharmacophores, and macromolecular structures. Although currently we are well prepared computationally to describe ligand–receptor interactions, the thorny problem lies in the complex physical chemistry of intermolecular interactions. Structural biologists, whether experimental or theoretical in approach, continue to struggle with the field's limited quantitative understanding of the enthalpic and entropic contributions to the overall free energy of binding of a ligand to a protein. With very few exceptions, we do not have experimental data on the thermodynamics of intermolecular interactions. The recent explosion of 3D protein structures helps us to refine our understanding of the geometry of ligand–protein complexes. However, as traditionally practiced, both crystallographic and NMR methods yield static pictures and relatively coarse results considering that an attraction between two non-bonded atoms may change to repulsion within a tenth of an Ångstrom. This is well below the typical accuracy of either method. Additionally, neither provides information about the energetics of the transfer of the ligand from solvent to the binding site.

With these challenges in mind, one aim of these volumes is to provide an overview of the current state of the quantitative description of ligand–receptor interactions. To aid this understanding, quantum chemical methods, molecular dynamics simulations and the important aspects of molecular similarity of protein ligands are treated in detail in Volume 2. In the first part 'Ligand–Protein Interactions,' seven chapters examine the problem from very different points of view. Rule- and group-contribution-based approaches as well as force-field methods are included. The second part 'Quantum Chemical Models and Molecular Dynamics Simulations' highlights the recent extensions of *ab initio* and semi-empirical quantum chemical methods to ligand–protein complexes. An additional chapter illustrates the advantages of molecular dynamics simulations for the understanding of such complexes. The third part 'Pharmacophore Modelling and Molecular Similarity' discusses bioisosterism, pharmacophores and molecular similarity, as related to both medicinal and computational chemistry. These chapters present new techniques, software tools and parameters for the quantitative description of molecular similarity.

Volume 3 describes recent advances in Comparative Molecular Field Analysis and related methods. In the first part '3D QSAR Methodology. CoMFA and Related Approaches', two overviews on the current state, scope and limitations, and recent progress in CoMFA and related techniques are given. The next four chapters describe improvements of the classical CoMFA approach as well as the CoMSIA method, an alternative to CoMFA. The last chapter of this part presents recent progress in Partial Least Squares (PLS) analysis. The part 'Receptor Models and Other 3D QSAR Approaches' describes 3D QSAR methods that are not directly related to CoMFA, i.e., Receptor Surface Models, Pseudo-receptor Modelling and Genetically Evolved Receptor Models. The last two chapters describe alignment-free 3D QSAR methods. The part '3D QSAR Applications' completes Volume 3. It gives a comprehensive overview of recent applications but also of some problems in CoMFA studies. The first chapter should give a warning to all computational chemists. Its conclusion is that all investigations on the classic corticosteroid-binding globulin dataset suffer from serious errors in the chemical structures of several steroids, in the affinity data and/or in their results. Different authors made different mistakes and sometimes the structures used in the investigations are different from the published structures. Accordingly it is not possible to make any exact comparison of the reported results! The next three chapters should be of great value to both 3D QSAR practitioners and to medicinal chemists, as they provide overviews on CoMFA applications in different fields, together with a detailed evaluation of many important CoMFA publications. Two chapters by Ki Kim and his comprehensive list of 1993–1997 CoMFA papers are a highly valuable source of information.

These volumes are written not only for QSAR and modelling scientists. Because of their broad coverage of ligand binding, molecular similarity, and pharmacophore and receptor modelling, they will help synthetic chemists to design and optimize new leads, especially to a protein whose 3D structure is known. Medicinal chemists as well as agricultural chemists, toxicologists and environmental scientists will benefit from the description of so many different approaches that are suited to correlating structure–activity relationships in cases where the biological targets, or at least their 3D structures, are still unknown.

This project would not have been realized without the ongoing enthusiasm of Mrs. Elizabeth Schram, founder and former owner of ESCOM Science Publishers, who initiated and strongly supported the idea of publishing further volumes on 3D QSAR in Drug Design. Special thanks belong also to Professor Robert Pearlman, University of Texas, Austin, Texas, who was involved in the first planning and gave additional support and input. Although during the preparation of the chapters Kluwer Academic Publishers acquired ESCOM, the project continued without any break or delay in the work. Thus, the Editors would also like to thank the new publisher, especially Ms. Maaike Oosting and Dr. John Martin, for their interest and open-mindedness, which helped to finish this project in time.

Lastly, the Editors are grateful to all the authors. They made it possible for these volumes to be published only 16 months after the very first author was contacted. It is the authors' diligence that has made these volumes as complete and timely as was Volume 1 on its publication in 1993.

Hugo Kubinyi, BASF AG, Ludwigshafen, Germany
Gerd Folkers, ETH Zürich, Switzerland
Yvonne C. Martin, Abbott Laboratories, Abbott Park, IL, USA

October 1997

Part I

3D QSAR Methodology CoMFA and Related Approaches

3D QSAR: Current State, Scope, and Limitations

Yvonne Connolly Martin
D-47E/AP10-2, Pharmaceutical Products Division, Abbott Laboratories, 100 Abbott Park Rd, Abbott Park, IL 60064-3500, U.S.A.

3D QSAR continues to be a vigorous field as evidenced by the 363 CoMFA models reported in this volume [1] and the number of alternative strategies for 3D QSAR suggested recently [2–11]. This chapter will examine some of the factors that make 3D QSAR such an attractive discipline and those limitations that are fundamental to the approaches, as well as those that might be overcome with improved methodology. Indeed, it is this author's opinion that, in spite of challenges, there are opportunities for improving its generality, precision of forecasts, and ease of use and interpretation.

Any 3D QSAR method wouldn't be tried for a dataset unless the experimenter expects that the study will provide useful three-dimensional structure–activity insights. Since scientists know that it is the 3D properties of molecules that govern their biological properties, it is especially gratifying to see a 3D summary of how changes in structure change biological properties. Methods that do not provide such a graphical result are often less attractive to the scientific community.

A major factor in the continuing enthusiasm for 3D QSAR comes from the proven ability of several of the methods to forecast correctly the potency of compounds not used in their derivation [1,12,13]. For example, CoMFA forecasts the potencies of 297 compounds in 25 datasets with a root mean square error of 0.70 logs or 0.98 kcal/mol [12]. Validation by forecasting compounds not used in the derivation is usually included in 3D QSAR reports, a difference from traditional QSAR methods. This ability to forecast affinity is gaining new respect as scientists realize that we are far away from the hoped-for fast and accurate forecast of affinity from the structure of a protein–ligand complex [14,15].

A final factor in the enthusiasm for 3D QSAR is that the software and hardware for performing 3D QSAR are accessible to laboratory scientists. The commercial software is easy to use and gaining access to the requisite computer power is no longer difficult, at least partly because of more efficient algorithms for model development [16]. Thus scientists whose primary focus is laboratory work can use the computer to gain 3D insights into the structure–activity relationships of their compounds.

1. Scientific Roots of 3D QSAR

Even before computers, medicinal chemists knew that a set of molecules will typically display an understandable structure–activity relationship [17]. Usually this is manifest in the observation that the smaller the change in the structure of the molecule, the less likely is there to be a change in its biological properties. The similarity principle is another way to say the same thing: compounds with similar chemical and physical properties also have similar biological properties [18]. In QSAR the similarity principle is considered to apply within a series or structural class only [19], although the

H. Kubinyi et al. (eds.), 3D QSAR in Drug Design, Volume 3. 3–23.

pharmacophore hypothesis generalizes the similarity to 3D properties independent of the underlying structure diagrams of the compounds [20,21]. Another important observation is that the effect on biological activity of changing a substituent at one position of a molecule is often independent of the effect of changing a substituent at a second position, quantified in the early Free–Wilson QSAR method [22]. Supplanting these qualitative insights by 3D quantitative structure–activity relationships was accomplished by the conscious or unconscious incorporation of insights from many different disciplines.

Structural chemistry provides valuable insights into why changing a substituent on a molecule might change its biological activity. For decades scientists have realized that the three-dimensional arrangement of dispersion, electrostatic and hydrophobic interactions, as well as hydrogen-bonds, determines the strength of intermolecular interactions [23]. Small-molecule crystallography has contributed greatly to our knowledge of the structural aspects of intermolecular interactions [24–27]. However, only recently have we had the requisite macromolecular structural information, theoretical models and computer power to attempt to forecast macromolecular structure and binding affinity [14,15,28]. 3D QSAR capitalizes on these developments and insights of structural and physical biochemistry.

Quantum chemistry changes focus from the nuclei of the atoms, the traditional structure, to the electrons of molecule. Today's computers have changed this discipline from one practiced by only devoted experts [29] to one that laboratory chemists can practice or at least set up on their desk-top computer. Although *ab initio* methods remain the benchmark method, semiempirical quantum mechanical methods allow one to calculate fairly accurately the molecular structure and electronic properties of almost any organic molecule — one doesn't need numerous parameters to do so [30–33]. Recently developed solvation models [34–37] expand the scope of problems that one can tackle.

Although physical organic chemistry traditionally focuses on the rate and equilibrium constants of organic reactions [38], it has provided both a precedent and an understanding that has been critical to the development of 3D methods. First, it has provided methods for the quantitation of the electronic, steric and hydrophobic effects of substituents on the reaction center. Second, it demonstrated that multivariate statistical analysis can suggest the physical basis of biological structure–activity relationships, QSAR [39–41]. It provided the jump-start to combine molecular modelling and statics into 3D QSAR.

Molecular modelling in the form of molecular mechanics [42] of small molecules grew from the early hand-held molecular models so useful in conformation analysis. The computer allows the incorporation of electrostatic effects as well as steric ones; the generation and comparison of many conformers of the same molecule; and comparison of the 3D structures of different molecules. Kier pioneered comparing the 3D structures of bioactive molecules to discovering the pharmacophore, the 3D requirements, for a particular biological activity [20] which Marshall later developed into the active analog approach [43].

Lastly, the development of computer graphics provided the platform with which scientists would interact with their structure–activity data [44,45]. Molecular graphics provides visual insight into 3D structures with color used to distinguish atoms types and color-coded dot surfaces showing the surface distribution of molecular properties such as electrostatic or hydrophobic potential [46]. It also allows one to easily compare, by superimposing, different molecules. Most 3D QSAR methods provide some 3D graphics as part of their output.

Since 3D QSAR uses insights from so many scientific disciplines, different implementations differ in the concepts and strategies employed. In a perfect world, we would have the requisite understanding to develop a perfect method. In the current world, our scientific understanding is primitive and often qualitative and we continually strive to approximate the truth more closely. Part of the enthusiasm for continued development of 3D QSAR methods is that researchers recognize that each approach has deficiencies in either theoretical background or implementation. This recognition provides the incentive for continuing attempts to improve the methods.

2. 3D QSAR versus Traditional 2D QSAR

As noted in the previous section, computer analysis in the form of linear free energy relationships allowed scientists for the first time to quantitate the relationship between the change in structure of molecules with the change in their biological activity [39]. Traditional QSAR, also known as Hansch-Fujita or 2D QSAR [39,47], accurately forecasts the potency of additional compounds and has led to the development of several commercial drugs and pesticides [41,48–50]. Statistical analysis distinguishes between steric, hydrophobic and electrostatic effects of substituents on biological activity. This strategy identifies which few of these are the dominant features behind the change in biological properties. When only the statistically important features are considered, a larger number of substituents will be predicted to have the same effect on biological activity. For example, if the QSAR indicates that increasing hydrophobicity leads to increased potency, then both electron-donating and electron-withdrawing substituents can increase potency if they are hydrophobic, and neither will if they are hydrophilic. This is true provided, of course, that the original QSAR was derived from a dataset that included both electron-donating and electron-withdrawing substituents. 3D QSAR methods generalize further to hypothesize that the critical factor is the 3D spatial arrangement of these chemical and physical properties.

There are those who conjecture that its structure diagram encodes all the information about the chemical, physical and biological properties of a molecule [51]. In fact, our own studies demonstrated that simple substructure keys are more successful in grouping diverse active compounds together than are more elaborate keys based on 3D structures [52]. Indeed, we found the same trend for the prediction of octanol–water and cyclohexane–water logP, pK_a, surface area and a number of other physical properties [53]. Although we have found more sophisticated 3D descriptors that separate actives from inactives more effectively [54], the impressive performance of simple descriptors must not be ignored.

A key difference between traditional and 3D QSAR is the form of the output. Although both provide statistical evidence for the validity of the proposed relationships, the result of a 3D QSAR analysis is typically supplied as a 3D graphics image superimposed on a molecule of the dataset. This visualization of the results increases the fidelity of the communication between the QSAR modeler and collaborators, such as the synthetic chemists who are interested to see why or if certain molecules are suggested by the model.

Another key difference between traditional and 3D QSAR lies in the source of the numerical descriptors of the molecules. In traditional QSAR, one relies on the observed correlation between the effect of a particular substituent on the rate or equilibrium constant for one reaction with the effect of the same substituent on the rate or equilibrium constant for another reaction. Since substituents affect the electronic, steric and hydrophobic properties of molecules, independent parameters are used for each of these properties. The substituent constants themselves are derived from measured effects in model reactions or equilibria. Accordingly, to derive a traditional QSAR equation the scientist or the computer looks up in a table the values of such parameters for each substituent. In contrast, in 3D QSAR one calculates the properties of the molecules of interest. Usually these properties are calculated in such a way that their 3D distribution is retained in the final model.

Although they are appealing because they are measured and not estimated by calculation, a fundamental problem with using measured substituent constants is that the model reactions used to define substituent constants are often themselves only postulated to represent the named feature. This is particularly true of the long-standing argument whether Taft Es values are purely steric, as originally proposed, or whether the measured rate is also influenced by electronic effects [41,55]. Moreover, recent studies of solvation properties of molecules emphasize that the relative octanol–water partition coefficients of molecules depend on their hydrogen-bonding character, as well as their 'innate' hydrophobicities [56]. Thus, the traditional logP is a composite measure of the hydrophobic and hydrogen-bonding properties of the compounds.

A practical handicap to using traditional QSAR can be the unavailability of substituent constants for the compounds of interest. Should one then omit those compounds, or guess at the values? Another problem arises when the molecules do not represent a series that can be described by substituent constants. In some cases, overall molecular properties, such as octanol–water logP and calculated pK_a, will provide a useful equation. However, this is not always true. Of course, the solution to the difficulty of finding tabulated parameters is to use calculated properties since the definitions are clear and usually all the compounds can be included. However, since this usually involves calculations on the 3D structures of the molecules, why not move directly to 3D QSAR? One must also ask if the calculations are accurate enough to represent such measured properties, a question answered affirmatively by several workers [1]. A final limitation of traditional QSAR, and a reason why 3D QSAR is considered so attractive by contrast, is that the equations discovered by traditional QSAR do not directly suggest new compounds to synthesize. Rather, one must be experienced with the values of the substituent constants in order to imagine which molecules will have the desired properties.

In spite of these limitations, traditional QSAR has contributed greatly to computer-assisted molecular design. Many other types of descriptors have been suggested: often these can be directly calculated from the structure diagram of the compounds [57–59]. Equally important, workers in this field have introduced a wide variety of methods for the quantitative analysis of structure–property relationships. These supplement or replace the traditional multiple regression analysis with statistically based methods such as discriminant analysis, principal components and partial least squares; neural networks; genetic algorithms; and artificial intelligence strategies [60]. Important also is the early recognition that, in order to derive a satisfactory QSAR, one must design the set of compounds carefully [61–64]: this presages the current interest in diversity analysis and selection of subsets of compound collections [65–67].

Two early 3D QSAR methods used traditional QSAR descriptors for electronic and hydrophobic effects of substituents, but generate a single steric descriptor by comparing the 3D structures of the molecules with references [68,69]. Although these methods include 3D properties, they suffer from difficulties in choosing the appropriate reference for the calculation and from ambiguities in how to handle both positive and negative steric influences on potency. An alternative early 3D QSAR method describes the properties of the molecules by their calculated interaction energies with a model of the binding site [70]. Although this method has led to interesting results and enhancements, it was too complex and ambiguous to be adapted for general use.

3D QSAR, as we know it, started with CoMFA. It was invented when Cramer and colleagues recognized that (i) they could describe, as had others before or simultaneously with them, the 3D distribution of electrostatic and steric properties of molecules by calculating interaction energies on a 3D lattice surrounding the molecules [71–73]; (ii) they could use partial least squares to extract the relationships between biological potency and these fields [74]; and (iii) they could produce a visual summary of the QSAR by contouring of the influence of each lattice point to potency [75]. In the literature up to 1993, CoMFA models reported from 90 biological datasets show the range of R^2_{fit} to be 0.73–1.00, of $RMSE_{fit}$ to be 0.034–0.91 and of $RMSE_{cv}$ to be 0.32–1.52 [12]. Although CoMFA overcomes some of the deficiencies of traditional QSAR, new difficulties arise; these will be discussed below. We showed that CoMFA reproduces traditional QSAR descriptors; that is, that a traditional QSAR and a CoMFA analysis provide the same information [76,77].

Whether traditional or 3D QSAR, only the structure–activity relationships of the ligands contribute to the statistical comparisons. They require no knowledge or hypothesis of the 3D structure or chemical nature of the complementary macromolecule. The comparisons may imply something about this macromolecule, but the implication is by correlation and not direct structural evidence. Although it is not necessary for deriving models, both traditional and 3D QSAR models are usually interpreted as if the common portions of all molecules interact in the same way with the target biomolecule.

3. 3D QSAR versus Protein-based Affinity Prediction Methods

The revolution in structural biology means that today the computational chemist often has the 3D structure of the macromolecular binding site with which the ligands of interest interact. Increasing numbers of protein and nucleic acid structures are being solved. As well as being directly useful, these structures supply the basis for homology models of related proteins. Does this make 3D QSAR useless, or do the two approaches complement each other?

Knowing the 3D structure of the target makes it easier to perform a 3D QSAR analysis. Many 3D QSAR methods base their property calculation on some absolute orientation of the molecules in space. Usually this means that either the user or the computer program selects the conformation of each molecule to use and how to compare each molecule to the others. Obviously if one has the 3D structure of the macromolecular target, particularly if one also has the structure of at least one ligand of each series bound to the protein, then it will be easier to propose a bioactive conformation and superposition rule [78,79]. The location of key binding sites should help suggest an orientation for the other molecules of interest. One could also directly observe the structure of the complex crystallographically [80], or optimize a model to provide a bioactive conformation [79].

Is 3D QSAR necessary if one has a 3D structure of the protein on which to base predictions [14]? Much attention has been paid recently to perturbation free energy method of predicting protein–ligand affinity [81]. Although this method is based on solid theoretical foundations, in practice such calculations involve days to weeks of computer time per pair of ligands and are limited to calculating affinity differences resulting from rather modest differences in structure. Their accuracy is probably limited by the approximations used in the force fields and electrostatic calculations: greater computer power and deeper insight into the biophysics of macromolecular structure may result in improved precision of calculations [15,82,83].

A more recent method, Linear Interaction Energy calculations, combines features of perturbation free energy calculations and QSAR to produce simple equations in steric and electronic energy using only three to four compounds [28,84,85]. The calculation on each ligand requires less than a day of computer time. In one report, four compounds were used to determine a regression equation that predicted the affinity of seven structurally different compounds with a mean error of 0.55 kcal/mol [86]. Clearly, this method deserves watching: it currently would be useful for predicting the potency of a handful of compounds, more if several computers were available and as computer speeds increase. However, its limitations are also becoming known: both errors in prediction [87] and correct predictions of affinity based on the wrong structure of the complex [88].

Another approach to using protein structures to predict binding affinity involves deriving generalized QSAR equations that predict the strength of any protein–ligand complex [89–94]. They are used mainly in the computer *de novo* design and docking of ligands. The descriptors for each ligand are calculated from an experimental 3D structure of a complex. Typically they include features such as the number and quality of the intermolecular hydrogen-bonds, as well as electrostatic, dispersion and hydrophobic interactions and an estimate of the ligand entropy lost on binding. A universal model is derived by regression or PLS analysis of dissociation constants of a variety of protein–ligand complexes using many different proteins. Once a model is derived, it can be used quickly to predict the affinities of any ligand interacting with any protein. Forecasts from these empirical equations are less precise than from perturbation or linear interaction energy analysis, typically of the order of 1.3 log units. A problem with these approaches is that steric misfit is not explicitly included since such molecules will bind in another configuration. In contrast, all QSAR methods include explicit terms that reflect steric misfit.

In yet another approach to using the structure of a protein–ligand complex as a basis of a QSAR analysis, several groups have used molecular descriptors derived from energy minimization of docked ligands with a target protein [7,8,95–98]. Either the calculated interaction energy or separated components of the interaction energy are correlated with affinity. Sometimes other properties, such as estimates of the relative entropy cost of binding the ligand, are added to the prediction equation [97]. Interestingly, the cross-validation statistics suggest that these equations are approximately of the same precision as typical equations derived without knowledge of the protein structure. One problem with this approach may be that since the force fields are parameterized to reproduce the structure and dynamics of a single compound, they may be deficient in the treatment of solvation energy. This varies more dramatically between compounds than between different conformations of the same compound. Additionally, the parameter values for the types of atoms of the ligands may not have been as carefully established: it appears that especially assigning values for the partial atomic charges may present a problem [8].

An emerging method to predict binding energy is based on the observed preferences of certain types of atoms to be near each other in macromolecular complexes [99–101]. The accuracy appears to be approximately the same as the generalized QSAR equations. The main limitation of this approach, at the moment, is the limited numbers of better than 2.0 Å resolution protein–ligand complexes available compared to the number of atom types present in drug molecules and the number of examples of each that would be needed to derive a preference score.

This survey suggests that 3D QSAR methods are an important complement to structure-based affinity prediction methods. If one already has a series of molecules and their corresponding binding affinities, then a 3D QSAR equation may provide a valuable method to forecast affinity of further analogs. Knowledge of the structure of the binding site would guide the molecular modelling and should prevent unwarranted extrapolation of such equations. At the moment, the observed structure–activity relationships of ligands provide a more sensitive measure of ligand–receptor affinity than do computational methods. On the other hand, structure-based calculations of affinity can be done, even if one has no or limited structure–activity and if the suggested compounds are very different from any known ligands.

4. Limitations, Challenges, Opportunities for the Future Application of 3D QSAR

4.1. Choosing the bioactive conformation and alignment

Many of the 3D QSAR methods discussed in this volume require that the chosen conformations of the molecules be aligned before the software develops the quantitative model; other methods select a conformation and an alignment as part of the development of the model. Usually one assumes that the conformation used should be the best assessment of the bioactive conformation and, furthermore, that the alignment represents how the different molecules bind to the target macromolecule. In fact, a 3D QSAR model simply provides a summary of how changes in the structure of the ligand affect its affinity for a target molecule. Furthermore, in many cases, either multiple binding modes of the same compound or closely related compounds have been observed crystallographically [88,102,103] and could be expected for many of the series studied by 3D QSAR. Consider a 3D QSAR model that suggests that increased affinity results from added steric bulk (or electronegative group) at a certain position with respect to the groups used for the alignment. A simple explanation would be a hydrophobic (or electropositive) pocket accessible in the given alignment, whereas the true one might be that this steric bulk (or electronegative group) leads to favored binding in an alternative orientation.

Although one would expect that alignment of ligands based on minimizing the structures of the corresponding ligand–macromolecule complexes would produce the most robust 3D QSAR models, several groups have found this not to be the case [104–106]. This is probably a reflection of the uncertainties in the structure minimization programs [15]. However, as noted above, the structure of the macromolecular binding site does provide a starting point for choosing the bioactive conformation and alignment.

If one has no structure of the macromolecular target but yet has decided to use a method that needs at least a starting orientation and conformation of every molecule, then either manual molecular modelling or automated pharmacophore mapping tools will be needed; along with advances in 3D QSAR, recent years have produced advances in these techniques as well [21]. However, no computer program can substitute for good structure–activity data. A pharmacophore mapping exercise can be expected to be successful if there is one relatively rigid active compound or several somewhat rigid compounds that collectively restrict the common distances between key recognition atoms or site points. A truly complete study would involve synthesis and testing of such molecules before a pharmacophore and a 3D QSAR study was undertaken [107–109].

There have been a number of interesting suggestions of ways to improve the alignment of molecules. Usually these are applied once one has chosen the bioactive conformation or a preliminary model [3,11,104,106,110–112]. The downside of these strategies that modify alignment or conformation to improve fit or predicted activity is that one must become increasingly alert to the possibility of deriving a chance model [112]. With the receptor surface strategy, it is suggested to optimize the structures of the less potent compounds within the model receptor surface generated from the three or four most potent compounds [3]. This could lead to very distorted structures of molecules that in a CoMFA analysis penetrate into negative steric regions. Investigating alternative alignment strategies should certainly be an area of active research; hopefully, more analysis of the reliability of the forecasts that result from different strategies will provide definitive guidelines for future work.

CoMMA [10], EVA [4] or the WHIM [9] descriptors promise an advantage because they provide 3D descriptors that are independent of the orientation of the molecules in space; they do not have to be aligned. However, the reader is reminded that the CoMMA inertial, dipole, and quadrapole moments are sensitive to conformation, as are most of the WHIM descriptors. The best way to find corresponding conformations in a set of molecules is to align them with each other, so one does not totally escape the alignment problem. However, the CoMMA and WHIM descriptors are less sensitive to exact conformation than are lattice-based energy values used in CoMFA and related methods. The EVA descriptors appear to be even less sensitive to conformation. This is somewhat adjustable within a run; sometimes the lack of sensitivity to conformation occurs at the expense of statistical quality of the model [4]. A philosophical issue arises: if a method is insensitive to the 3D structure, the conformation, of a molecule, is it really a 3D QSAR method? Clearly, there are opportunities to continue to explore the role these and other alignment-free methods will play in QSAR analyses.

4.2. Choosing the type of descriptors

Many workers have investigated alternative molecular descriptors for 3D QSAR. For lattice-based methods, there is now evidence that hydrophobic fields do not generally increase the statistical quality of the model, that steric fields can profitably be replaced with somewhat softer functions and that electrostatic fields based on semiempirical electrostatic potentials are superior to empirical schemes [112,113]. The CoMSIA descriptors appear to contain the same information as those of traditional CoMFA but produce contour plots that are easier to transform mentally into molecules to synthesize [5].

Several groups have proposed 3D QSAR methods that are not based on properties calculated at a lattice. The GERM [6], COMPASS [11] and receptor surface [3] methods rely on properties calculated at discrete locations in the space at or near the union surface of the active molecules, presumably a model of the macromolecular binding site. If all molecules of the set do bind in a manner that doesn't distort the binding site too much, this can be a reasonable strategy as evidenced by the fact that these methods have led to the development of reasonable models. However, in series for which there is a large positive contribution of steric energy at certain points, as in the case of our D1 dopaminergic agonists [108], this type of descriptor might not be able to detect that the *absence* of steric bulk at a certain point leads to a decrease in potency. Both of these methods base their 3D QSAR on interaction energies with the hypothetical receptor and, hence, are subject to all the limitations of such interaction energies, even when the structure of the target macromolecule is known (see section 3; above). The positive feature of these two methods is that the model is presented as a 3D display of properties of the receptor in space.

The EVA, CoMMA and WHIM descriptors differ from the lattice- or surface-based descriptors, in that they do not consider properties at locations in space, but rather 3D properties of the molecules themselves. Hence, it is not possible to provide a 3D display of the resulting models.

4.3. Designing the series and choosing the training set

Within the CoMFA paradigm, some attention has been paid to the design of series for 3D QSAR analysis [112–117]. For example, one might generate a number of principal components from the steric and electrostatic fields of the aligned molecules and cluster the molecules based on these descriptors. Alternatively, one might choose to use steric field descriptors suited to substituents [59,118]. However, today most models are derived from datasets that were not designed for 3D QSAR analysis. A particular concern is that, in poorly designed series, electrostatic and steric properties are not varied independently, nor are they varied continuously. Although good statistical models may result, their predictivity may be low if the new compounds break the correlations in the training set. The use of 3D QSAR or related descriptors in series planning represents an opportunity to help the medicinal chemist synthesize fewer and better distributed compounds for the derivation of the first QSAR model, or to select substituents for combinatorial libraries.

Sometimes it happens that there are too few active compounds to derive a CoMFA model, even one based on active versus inactive sets. In that case, simply designing compounds that are similar to the active ones but different from the known inactives in one or more dimensions might lead to the identification of more active compounds.

There is also evidence that one can derive 3D QSAR models of equivalent or better quality by considering a carefully selected subset of the compounds in the dataset [112,116] and that such models are more robust and provide more accurate forecasts of affinity [113,117]. Some even suggest that one constructs many models from subsets of the data [119]. Accordingly, for retrospective analyses, it appears advantageous to select a training subset of all compounds tested and to use the remaining compounds as a biased test set.

4.4. Selecting variables for the model

CoMFA requires that one considers thousands of 3D descriptors rather than the small number used in traditional QSAR. Even after discarding descriptors that do not vary significantly in the data set, there are often thousands remaining. Additionally there is the conflict between using many lattice points to produce more accurate energy values (smaller lattice spacing) and the notion of keeping the number of variables low (larger lattice spacing) to reduce the noise in the models. Since PLS is very sensitive to noise in the descriptors [120], more predictive models should result if we could eliminate unnecessary descriptors.

Experiences with HASL [121] and genetic PLS [122] suggest that for typical CoMFA models the energy at only a very few points explains most of the variance in biological potency. Models derived with the steroid dataset using different approaches reinforces this point since several of the methods use very few descriptors to provide the same level of statistical quality [123]. Similarly, traditional QSAR provides equations in very few variables.

However, in spite of the promise of cross-validated R^2-guided region selection [124] and GOLPE-guided region selection [125], it is too early to tell if variable reduction based on preliminary QSARs lead to models with better ability to forecast the potency of new compounds [113]. The same problem might apply to genetic selection based on cross-validation [122,126]. Again, it is to be expected that variable selection for 3D QSAR will continue to be an area of active research just as it is currently in traditional QSAR and other lower-dimensional problems [127–135].

4.5. Deriving the model

For those methods that use only a few descriptors or that calculate a single interaction energy to be correlated with biological potency [6,136,137], multiple linear regression is a suitable method. However, if several variables are considered for possible inclusion in the model, it is all too easy to overfit a regression equation [138], suggesting a preference for partial least squares, PLS, modelling instead [74]. Although the simplicity of PLS is a positive attribute, its modelling power decreases when noise is mixed with the relevant descriptors. Additionally, a PLS model is linear in the descriptors [139], although quadratic PLS identifies certain nonlinear relationships [139]. Hence, there is considerable interest in finding new methods to establish the relationship between (selected) 3D descriptors and biological potency. However, one should be aware that the deficiencies of PLS may be more noticed only because so much more attention has been devoted to PLS, and that alternative methods may suffer from the same problems.

Nonlinear relationships can be detected by the PLS analysis of a transformation of the original data matrix into a matrix of the distances between each pair of observations as measured in the original property space [140–142]. A problem with using this approach with CoMFA fields is that there is no obvious way to display the nonlinear relationship on the CoMFA lattice. Another problem is that including irrelevant descriptors in the distance calculation can weaken the nonlinear signal.

Several chapters in this volume report modelling with neural networks [3,11]. This is another area that deserves more attention to establish the conditions for reliable 3D QSAR model development [132,143,144].

4.6. Validating the model

The primary test of any model is how well it forecasts the potency of compounds not used in its derivation, typically a test set reserved for this purpose [113]. Less common, but to be recommended, is to repeat the model derivation on different subsets of the data to test for the consistency of the models produced [112]. Despite all the caution one uses, it is all too easy to overfit the training set data [112,113,145]. Hence, it is becoming common to scramble the biological data, often many times, and repeat the variable selection and model generation procedure [4,7,112,113,146]. This randomization procedure preserves the correlations between the predictor variables and the distribution of the potency while breaking any true relationship between them.

It is becoming clear that the cross-validated R^2 is not a good measure of the quality of a 3D QSAR method, particularly if variable- or alignment-selection strategies have been used [112,113]. A further complication with this statistic is that it is sensitive to the composition of the dataset: if there are many near-duplicates, then the cross-validation will indicate a robust model, whereas it will indicate no or a poor model if the dataset has been consciously designed to include no similar compounds. Larger datasets, usually preferred by QSAR modelers, have a larger chance of containing many near-duplicates.

If the 3D structures of the target macromolecule becomes available after the QSAR determination, then one can compare it with the 3D QSAR model. Of course, such comparisons are fraught with the complexities discussed in section 4.1, with choosing, and the molecular alignment of the molecules.

4.7. Forecasting potency

Most forecasts of potency from 3D QSAR models are simply a value with no estimate of reliability, except the cross-validated root mean square error. However, it is important to know if the test compound is very different from every molecule in the training set and, hence, that its potency forecast is much less accurate than one for which a very similar molecule is in the training set. The use of molecular similarity to align molecules for potency forecasts [112] suggests that all 3D QSAR forecasts should also include how similar the test molecule is to one in the dataset. The similarity should be calculated over all the properties considered for the model, rather than for those properties that were found important for the model, since if a new compound changes a property that was not previously changed, then no QSAR model can be expected to give reliable forecasts.

There is no perfect way to summarize the accuracy of potency forecasts, because each method depends on the distribution of potency in the test set. Typically, authors report either the R^2_{pred} or the mean of the absolute error of prediction. Consider two QSAR methods: the first predicts only fairly accurately but consistently under-predicts potent compounds and over-predicts less active ones, whereas the second method predicts each compound more closely but has no such bias. For datasets that contain most compounds at the extremes of activity, the former will have a higher R^2_{pred}, even though the slope between observed and forecast is not 1.0. On the other hand, for datasets in which all compounds have potency near the mean, the mean unsigned error of prediction would favor the latter method. The common use of plots of observed versus forecast affinities, on the same figure or at least the same scale as a similar figure for the training set, provides a more detailed picture of the quality of the forecasts.

4.8. Comparing 3D QSAR methods

A serious problem in comparing methods is that often the only information provided by the authors concerns the relative precision of models derived from the same dataset with different methods, whereas what one wants to know is how well the different methods forecast the affinity of new compounds. In particular, the comparison of methods must deal with the perception that at least some variable-selection methods provide optimistic cross-validation estimates of model accuracy [113] and that feedback neural networks may overfit a model [143,144]. Compounds to consider for true potency forecasting may be hard to find, and it is tempting to include all known molecules in the development of a model or when statistically selecting those to include and those to predict.

Although most new methods provide a result on a reference set of compounds, errors of many sorts can confound these comparisons [123]. Furthermore, it is possible that some methods are unintentionally tuned to the test datasets and will perform less well with other data. Until benchmark studies are done, how does one choose which method to use? Frequently, the choice depends on the software available. However, if no satisfactory quantitative relationship is found, one must decide if another method will be successful.

5. Role of 3D QSAR in Combinatorial Chemistry and High-throughput Screening

5.1. Generating 3D QSARs and forecasts quickly

The modern pharmaceutical industry has embraced two strategies that were just emerging a decade ago, when CoMFA was devised: mass or high-throughput screening hundreds of thousands of compounds in a particular assay and synthesis and testing of mixtures of compounds. In view of its success in small sets of compounds, it would be an important contribution if 3D QSAR could contribute to the success of these ventures.

In industry today, computational chemists often participate in the design of targeted combinatorial libraries that can include any of millions of compounds. A QSAR method that could efficiently forecast the potency of so many compounds would be very attractive, even if it were less accurate than more time-consuming methods. Yet another challenge is to develop QSAR models based on high-throughput screening of thousands of compounds with associated errors in structure.

The first challenge to basing a 3D QSAR model on high-throughput screening or screening of combinatorial libraries will be to establish the validity of the structures actually tested. Typically, the success of the chemistry to produce combinatorial libraries is measured only in rehearsal runs and on compounds identified as active. Similarly, the identity of the structures of the compounds in collections is often assessed only when activity has been identified. In both cases, the modeler cannot be assured that certain compounds are not active because there is a small chance that they have not been tested. This ambiguity suggests that methods that tolerate ambiguity might find application in this context.

The second challenge to developing a QSAR based on high-throughput screening is that often the biological activities are simple active versus inactive. Hence, the PLS variant of discriminant analysis [147] or a neural network method might be useful. Since there are usually 10–1000 times more inactive compounds than active ones, a clever strategy to select only a subset of the inactive compounds for model development will conserve considerable time.

A third challenge is for the computer to be fast enough to complement high-throughput screening methods or SAR by NMR [148] for the identification of novel existing compounds to fit a target of known 3D structure.

A final challenge is that the QSAR modelling must be done quickly. Often, not only must a QSAR be derived, but new compounds for combinatorial synthesis must be designed within a matter of a week or two. This challenge means that any QSAR method used must be robust without human valuation of the results. The positive aspect is that the QSAR need not be especially reliable since any enrichment of active compounds in a second library will improve the efficiency of the search for new compounds. It is an open question whether a traditional [149] or 3D QSAR approach will be more useful in this context.

5.2. Designing, diverse combinatorial libraries

The success of 3D QSAR in predicting the affinity of new compounds suggests that this type of descriptor has relevance to biological properties of molecules. Accordingly, some have based their selection of substituents for combinatorial libraries on 3D fields [118]. A positive aspect of combinatorial library synthesis is that often there are more potential compounds that can be made than will actually be made. The result is that the computational chemist can influence the decision of which compounds to make and design a set that should lead to an interpretable QSAR.

6. Conclusion

All evidence suggests that 3D QSAR techniques will continue to make a valuable contribution to the computer-assisted analysis of structure–bioactivity relationships. The search for new descriptors of 3D properties of ligands and innovative strategies to investigate the relationships between these properties and bioactivity continues to be a fruitful research enterprise. Increasing information from structural biology will provide valuable feedback to the hypotheses that form the basis of 3D QSAR methods.

3D QSAR methods complement traditional QSAR based on physical properties. They offer the advantage that it is easy to calculate descriptors for most molecules, and the disadvantage that one must select a conformation and usually a superposition rule as part of the analysis.

Because of their speed and accuracy, 3D QSAR methods complement calculations based on the structure of the ligand–macromolecular complex. Whereas the structure of at least one complex aids in the selection of the bioactive conformation and the alignment of the molecules for 3D QSAR, a QSAR model can be derived much more quickly than calculations based on the complex. Frequently, it is just as predictive. Knowledge of the structure of the complex can also prevent unwarranted extrapolation from a QSAR model.

It is expected that concepts from 3D QSAR will continue to impact the analysis of high-throughput screening structure–activity data and the diversity of compound collections and combinatorial libraries.

References

1. Kim, K.H., Greco, G. and Novellino, E., *A critical review of recent CoMFA applications*, In Kubinyi, H., Folkers, G., and Martin, Y.C., (Eds.) 3D QSAR in drug design: Vol. 3, Kluwer Academic Publishers, Dordrecht, The Netherlands, 1998, pp. 257–316.
2. Dunn III, W.J. and Hopfinger, A.J., *3D QSAR of flexible molecules using tensor representation*, In Kubinyi, H., Folkers, G. and Martin, Y.C. (Eds.) 3D QSAR in drug design: Vol. 3, Kluwer Academic Publishers, Dordrecht, The Netherlands, 1998, pp. 167–182.

3. Hahn, M. and Rogers, D., *Receptor surface models*, in Kubinyi, H., Folkers, G. and Martin, Y.C. (Eds.) 3D QSAR in drug design: Vol. 3, Kluwer Academic Publishers, Dordrecht, The Netherlands, 1998, pp. 117–134.
4. Heritage, T.W., Ferguson, A.M., Turner, D.B. and Willett, P., *EVA — a novel theoretical descriptor for QSAR studies*, In Kubinyi, H., Folkers, G. and Martin, Y.C. (Eds.) 3D QSAR in drug design: Vol. 2, Kluwer Academic Publishers, Dordrecht, The Netherlands, 1998, pp. 381–398.
5. Klebe, G., *Comparative molecular similarity indices analysis — CoMSIA*, In Kubinyi, H., Folkers, G. and Martin, Y.C. (Eds.) 3D QSAR in drug design: Vol. 3, Kluwer Academic Publishers, Dordrecht, The Netherlands, 1998, pp. 87–104.
6. Walters, D.E., *Genetically evolved receptor models (GERM) as a 3D QSAR tool*, In Kubinyi, H., Folkers, G. and Martin, Y.C. (Eds.) 3D QSAR in drug design: Vol. 3, Kluwer Academic Publishers, Dordrecht, The Netherlands, 1998, pp. 159–166.
7. Wade, R.C., Ortiz, A.R. and Gago, F., *Comparative binding energy analysis*, In Kubinyi, H., Folkers, G. and Martin, Y.C. (Eds.) 3D QSAR in drug design: Vol. 2, Kluwer Academic Publishers, Dordrecht, The Netherlands, 1998, pp. 19–34.
8. Holloway, M.K., *A priori predicition of ligand affinity by energy minimization*, In Kubinyi, H., Folkers, G. and Martin, Y.C. (Eds.) 3D QSAR in drug design: Vol. 2, Kluwer Academic Publishers, Dordrecht, The Netherlands, 1998, pp. 63–84.
9. Todeschini, R. and Gramatica, P., *New 3D molecular descriptors: The WHIM theory and QSAR applications*, In Kubinyi, H., Folkers, G. and Martin, Y.C. (Eds.) 3D QSAR in drug design: Vol. 2, Kluwer Academic Publishers, Dordrecht, The Netherlands, 1998, pp. 355–380.
10. Silverman, B.D., Platt, D.E., Pitman, M. and Rigoutsos, I., *Comparative molecular moment analysis (COMMA)*, in Kubinyi, H., Folkers, G. and Martin, Y.C. (Eds.) 3D QSAR in drug design: Vol. 3, Kluwer Academic Publishers, Dordrecht, The Netherlands, 1998, pp. 183–196.
11. Jain, A.N., Koile, K. and Chapman, D., *Compass: Predicting biological activities from molecular surface properties — performance comparisons on a steroid benchmark*, J. Med. Chem., 37 (1994) 2315–2327.
12. Martin, Y.C., Kim, K.-H. and Lin, C.T., *Comparative molecular field analysis: CoMFA*, In Charton, M. (Ed.) Advances in quantitative structure property relationships, JAI Press, Greenwich, CT, 1996, pp. 1–52.
13. Greco, G., Novellino, E. and Martin, Y.C., *Approaches to 3D-QSAR*, In Martin, Y.C. and Willett, P. (Eds.) Designing bioactive molecules: Three-dimensional techniques and applications, America Chemical Society, Washington, DC, 1997 (in press).
14. Ajay and Murcko, M.A., *Computational methods to predict binding free-energy in ligand–receptor complexes*, J. Med. Chem., 38 (1995) 4953–4967.
15. Kollman, P.A., *Advances and continuing challenges in achieving realistic and predictive simulations of the properties of organic and biological molecules*, Acc. Chem. Res., 29 (1996) 461–469.
16. Bush, B.L. and Nachbar Jr., R.B., *Sample-distance partial least-squares — PLS optimized for many variables, with application to CoMFA*, J. Comput.-Aided Mol. Design, 7 (1993) 587–619.
17. Burger, A., *Medical chemistry — the first century*, Med. Chem. Res., 4 (1994) 3–15.
18. Willett, P., *Similarity and clustering techniques in chemical information systems*, Research Studies Press, Letchworth, 1987.
19. Hodgkin, E.E. and Richards, W.G., *Molecular similarity based on electrostatic potential and electric field*, Int. J. Quantum Chem., 14 (1987) 105–110.
20. Kier, L.B., *Molecular orbital theory in drug research*, Academic Press, New York, 1971, p. 258.
21. Martin, Y.C., *Pharmacophore mapping*, In Martin, Y.C. and Willett, P. (Eds.) Designing bioactive molecules: Three-dimensional techniques and applications, American Chemical Society, Washington, DC, 1997 (in press).
22. Free, S.M. and Wilson, J., *A mathematical contribution to structure–activity studies*, J. Med. Chem., 7 (1964) 395–399.
23. Pauling, L., Campbell, D.H. and Pressman, D., *The nature of the forces between antigen and antibody and of the precipitation reaction*, Physiol. Rev., 23 (1943) 203–219.
24. Allen, F.H., Kennard, O. and Taylor, R., *Systematic analysis of structural data as a research tool in organic chemistry*, Acc. Chem. Res., 16 (1983) 146–153.

25. Bürgi, H.-B. and Dunitz, J.D., Structure Correlation, 1st Ed., VCH Verlagsgesellschaft mbH, Weinheim, Germany, 1994, Vols. 1 and 2, pp. 900.
26. Allen, F.H., Bird, C.M., Rowland, R.S., Harris, S.E. and Schwalbe, C.H., *Correlation of the hydrogen-bond acceptor properties of nitrogen with the geometry of the Nsp(2)-Nsp(3) transition in R(1)(X=)C-NR(2)R(3) substructures — Reaction pathway for the protonation of nitrogen*, Acta Crystallogr., Sec. B, 51 (1995) 1068–108.
27. Mills, J. and Dean, P.M., *3-Dimensional hydrogen-bond geometry and probability information from a crystal survey*, J. Comput.-Aided Mol. Design, 10 (1996) 607–622.
28. Åqvist, J., Medina, C. and Samulesson, J.-E., *A new method for predicting binding affinity in computer-aided drug design*, Protein Eng., 7 (1994) 385–391.
29. Dirac, P.A.M., Proc. R. Soc. London, Ser. A, 123 (1929) 714.
30. Dewar, M.J.S., Zoebish, E.G., Healy, E.F. and Stewart, J.J.P., *AM1: A new general purpose quantum mechanical molecular model*, J. Am. Chem. Soc., 107 (1985) 3902–3909.
31. Clark, T., *A handbook of computational chemistry: A practical guide to chemical structure and energy calculations*, Wiley, New York, 1985, pp. 332.
32. Stewart, J.P., *Semiempirical molecular orbital methods*, In Lipkowitz, K.B. and Boyd, D.B. (Eds.) Reviews in computational chemistry, VCH, Weiheim, Germany, 1990, pp. 45–81.
33. Kroemer, R.T., Hecht, P. and Liedl, K.R., *Different electrostatic descriptors in comparative molecular-field analysis: A comparison of molecular electrostatic and Coulomb potentials*, J. Comput. Chem., 17 (1996) 1296–1308.
34. Cramer, C.J. and Truhlar, D.G., *AM1-SM2 and PM3-SM3 parameterized SCF solvation models for free energies in aqueous solution*, J. Comput.-Aided Mol. Design, 6 (1992) 629–666.
35. Klamt, A. and Schuurmann, G., *COSMO: A new approach to dielectric screening in solvents with explicit expressions for the screening energy and its gradient* J. Chem. Soc., Perkin Trans. 2, (1993) 799–805.
36. Giesen, D.J., Chambers, C.C., Cramer, C.J. and Truhlar, D.G., *Solvation model for chloroform based on class-IV atomic charges*, J. Phys. Chem. B, 101 (1997) 2061–2069.
37. Richardson, W.H., Peng, C., Bashford, D., Noodleman, L. and Case, D.A., *Incorporating solvation effects into density-functional theory: Calculation of absolute acidities*, Int. J. Quantum Chem., 61 (1997) 207–217.
38. Hammett, L., *Physical organic chemistry*, McGraw-Hill, New York, 1970.
39. Hansch, C. and Fujita, T., *Rho Sigma pi analysis: A method for the correlation of biological activity and chemical structure*, J. Am. Chem. soc., 86 (1964) 1616–1626.
40. Hansch, C. and Leo, A., *Exploring QSAR: Fundamentals and applications in chemistry and biology*, American Chemical Society, Washington, DC, 1995, pp. 557.
41. Hansch, C., Leo, A. and Hoekman, D., *Exploring QSAR: Hydrophobic, electronic, and steric constants*, American Chemical Society, Washington, DC, 1995, pp. 348.
42. Burkert, U. and Allinger, N.L., *Molecular mechanics*, American Chemical Society, Washington, DC, 1982, pp. 339.
43. Marshall, G.R., Barry, C.D., Bosshard, H.E., Dammkoehler, R.A. and Dunn, D.A., *The conformation parameter in drug design: The active analog approach*, In Olson, E.C. and Christoffersen, R.E. (Eds.) Computer-assisted drug design, American Chemical Society, Washington, DC, 1979, pp. 205–226.
44. Langridge, R., Ferrin, T.E., Kuntz, I.D. and Connolly, M.L., *Real-time color graphics in studies of molecular interactions*, Science, 211 (1981) 661–667.
45. Blaney, J.M., Jorgensen, E.C., Connolly, M.L., Ferrin, T.E., Langridge, R., Oatley, S.J., Burridge, J.M. and Blake, C.C.F., *Computer graphics in drug design: Molecular modeling of thyroid hormone–prealbumin interactions*, J. Med. Chem., 25 (1982) 785- 790.
46. Weiner, P.K., Langridge, R., Blaney, J.M., Schaefer, R. and Kollman, P.A., *Electrostatic potential molecular-surfaces*, Proc. Natl. Acad. Sci. U.S.A., 79 (1982) 3754–3758.
47. Martin, Y.C., *Quantitative drug design*, Dekker, New York, 1978, pp. 425.
48. Fujita, T., *The role of QSAR in drug design*, In Jolles, G. and Wolldridge, K.R.H. (Eds.) Drug design: Fact or fantasy?, Academic Press, London, 1984, pp. 19–33.
49. Boyd, D.B., *Successes of computer-assisted molecular design*, In Lipkowitz, K.B. and Boyd, D.B. (Eds.) Reviews in computational chemistry, VCH, New York, 1990, pp. 355–371.

50. Hansch, C., and Fujita, T., (Ed.), *Classical and three-dimensional QSAR in agrochemistry*, American Chemical Society, Washington, DC, 1995, 342 pp.
51. Weiniger, D., *A Note on the sense and nonsense of searching 3-D databases for pharmaceutical leads*, Network Science, (1995). www.awod.com/netsci/Science/Cheminform/feature 04.html.
52. Brown, R.D. and Martin, Y.C., *Use of structure–activity data to compare structure-based clustering methods and descriptors for use in compound selection*, J. Chem. Inf. Comput. Sci., 36 (1996) 572–584.
53. Brown, R.D. and Martin, Y.C., *The information content of 2D and 3D structural descriptors relevant to ligand–receptor binding*, J. Chem. Inf. Comput. Sci., 37 (1997) 1–9.
54. Brown, R.D., Danaher, E., Lico, I. and Martin, Y.C., unpublished observations.
55. Kim, K.H. and Martin, Y.C., *Evaluation of electrostatic and steric descriptors of 3D-QSAR: The H^+ and CH_3 probes using comparative molecular field analysis (CoMFA) and the modified partial least squares method*, In Silipo, C. and Vittoria, A. (Eds.) QSAR: Rational approaches to the design of bioactive compounds, Elsevier Science Publishers, Amsterdam, The Netherlands, 1991, pp. 151–54.
56. Kamlet, M., Doherty, R., Fiserova-Bergerova, V., Carr, P., Abraham, M. and Taft, R., *Solubility properties in biological media: 9. Prediction of solubility and partition of organic nonelectrolytes in blood and tissues from solvatochronic parameters.*, J. Pharm. Sci., 76 (1987) 14–17.
57. Klopman, G., *Artificial intelligence approach to structure–activity studies: Computer automated structure evaluation of biological activity of organic molecules*, J. Am. Chem. Soc., 106 (1984) 7315–7321.
58. Hall, L.H. and Kier, L.B., *The molecular connectivity chi indexes and kappa shape indexes in structure–property modeling*, In Lipkowitz, K.B. and Boyd, D.B. (Eds.) Reviews in computational chemistry, VCH, New York, 1991, pp. 367–422.
59. Van de Waterbeemd, H., Clementi, S., Costantino, G., Carrupt, P.-A. and Testa, B., *CoMFA-derived substituent descriptors for structure–property correlations*, In Kubinyi, H. (Ed.) 3D QSAR in drug design: Theory, methods and applications, ESCOM, Leiden, The Netherlands, 1993, pp. 697–707.
60. van de Waterbeemd, H. (Ed.), *Chemometric methods in molecular design*, VCH, Weinheim, Germany, 1995, 359 pp.
61. Hansch, C., Unger, S.H. and Forsythe, A.B., *Strategy in drug design: Cluster analysis as an aid in the selection of substituents*, J. Med. Chem., 16 (1973) 1212–1222.
62. Wootton, R., Cranfield, R., Sheppey, G.C. and Goodford, P.J., *Physicophemical–activity relationships in practice: 2. Rational selection of benzenoid substituents*, J. Med. Chem., 18 (1975) 607–613.
63. Martin, Y.C. and Panas, H.N., *Mathematical considerations in series design*, J. Med. Chem., 22 (1979) 784–791.
64. Austel, V., *Experimental design in synthesisis planning and structure–property correlations*, In van de Waterbeemd, H. (Ed.) Chemometric methods in molecular design, VCH, Weinheim, Germany, 1995, pp. 49–62.
65. Downs, G.M. and Willett, P., *Clustering in chemical-structure databases for compound selection*, In van der Waterbeemd, H. (Ed.) Chemometric methods in molecular design, VCH, Weinheim, Germany, 1994, pp. 111–30.
66. Martin, Y.C., Brown, R.D. and Bures, M.G., *Quantifying diversity*, In Kerwin, J.F. and Gordon, E.M. (Eds.) Combinatorial chemistry and molecular diversity, Wiley, New York, 1997 (in press).
67. Turner, D.B., Tyrrell, S.M. and Willett, P., *Rapid quantification of molecular diversity for selective database acquisition*, J. Chem. Inf. Comput. Sci., 37 (1997) 18–22.
68. Simon, Z., Dragomir, N., Plauchitiu, M.G., Holban, S., Glatt, H. and Kerek, F., *Receptor site mapping for cardiotoxic aglicones by the minimal steric difference method*, Eur. J. Med. Chem., 15 (1980) 521–527.
69. Hopfinger, A.J., *A QSAR investigation of dihydrofolate reductase inhibition by Baker triazines based upon molecular shape analysis*, J. Am. Chem. Soc., 102 (1980) 7196–7206.
70. Höltje, H.-D. and Kier, L.B., *Sweet taste receptor studies using model interaction energy calculations*, J. Pharm. Sci., 63 (1974) 1722–1725.
71. Goodford, P.J., *A computational procedure for determining energetically favorable binding sites on biologically important macromolecules*, J. Med. Chem., 28 (1985) 849–857.
72. Kato, Y., Itai, A. and Iitaka, Y., *A novel method for superimposing molecules and receptor mapping*, Tetrahedron, 43 (1987) 5229–5234.

73. Doweyko, A.M., *The hypothetical active site lattice: An approach to modeling active sites from data on inhibitor molecules*, J. Med. Chem., 31 (1988) 1396–1406.
74. Wold, S., Ruhe, A., Wold, H. and Dunn, W.J., *The collinearity problem in linear regression: The partial least square (PLS) approach to generalized inverses*, Siam J. Sci. Stat. Comput., 5 (1984) 735–743.
75. Cramer III, R.D., Patterson, D.E. and Bunce, J.D., *Comparative molecular field analysis (CoMFA): 1. Effect of shape on binding of steroids to carrier proteins*, J. Am. Chem. Soc., 110 (1988) 5959–5967.
76. Kim, K.H. and Martin, Y.C., *Direct prediction of dissociation constants (pK_a's) of clonidine-like imidazolines, 2-substituted imidazoles, and 1-methyl-2-substituted-imidazoles from 3D structures using a comparative molecular field analysis (CoMFA) approach*, J. Med. Chem., 34 (1991) 2056–2060.
77. Kim, K.H., *Comparison of classical and 3D QSAR*, In Kubinyi, H. (Ed.) 3D QSAR in drug design: Theory methods and applications, ESCOM, Leiden, The Netherlands, 1993, pp. 619–642.
78. Waller, C.L., Oprea, T.I., Giolitti, A. and Marshall, G.R., *Three-dimensional QSAR of human immunodeficiency virus (I) protease inhibitors: 1. A CoMFA study employing experimentally-determined alignment rules*, J. Med. Chem., 36 (1993) 4152–4160.
79. Klebe, G. and Abraham, U., *On the prediction of binding properties of drug molecules by comparative molecular field analysis*, J. Med. Chem., 36 (1993) 70–80.
80. Watson, K.A., Mitchell, E.P., Johnson, L.N., Cruciani, G., Son, J.C., Bichard, C.J.F., Fleet, G.W.J., Oikonomakos, N.G., Kontou, M. and Zographos, S.E., *Glucose analog inhibitors of glycogen-phosphorylase — from crystallographic analysis to drug prediction using grid force-field and GOLPE bariable selection*, Acta Crystallogr., Sec. D, 51 (1995) 458–472.
81. Jorgensen, W.L. and Tiradorives, J., *Free-energies of hydration for organic-molecules from Monte Carlo Simulations*, Persp. Drug Discov. Design, 3 (1995) 123–138.
82. Marrone, T.J., Gilson, M.K. and McCammon, J.A., *Comparison of continuum and explicit models of solvation — potentials of mean force for allanine dipeptide*, J. Phys. Chem., 100 (1996) 1439–1441.
83. Madura, J.D., Nakajima, Y., Hamilton, R.M., Wierzbicki, A. and Warshel, A., *Calculations of the electrostatic free-energy contributions to the binding free-energy of sulfonamides to carbonic-anhydrase*, Struct. Chem. 7 (1996) 131–138.
84. Åqvist, J. and Mowbray, S.L., *Sugar recognition by a glucose/galactose receptor: Evaluation of binding energetics from molecular dynamics simulations*, J. Biol. Chem., 270 (1995) 9978–9981.
85. Hansson, T. and Åqvist, J., *Estimation of binding free-energies for HIV proteinase-inhibitors by molecular-dynamics simulations*, Protein Eng., 8 (1995) 1137–1144.
86. Paulsen, M.D. and Ornstein, R.L., *Binding free-energy calculations for P450cam–substrate complexes*, Protein Eng., 9 (1996) 567–571.
87. Hulten, J., Bonham, N.M., Nillroth, U., Hansson, T., Zuccarello, G., Bouzide, A., Åqvist, J., Classon, B., Danielson, U.H., Karlen, A., Kvarnstrom, I., Samuelsson, B. and Hallberg, A., *Cyclic HIV-1 protease inhibitors derived from mannitol: synthesis, inhibitory potencies, and computational predictions of binding affinities*, J. Med. Chem., 40 (1997) 885–897.
88. Backbro, K., Lowgren, S., Osterlund, K., Atepo, J., Unge, T., Hulten, J., Bonham, N.M., Schaal, W., Karlen, A. and Hallberg, A., *Unexpected binding mode of a cyclic sulfamide HIV-1 protease inhibitor*, J. Med. Chem., 40 (1997) 898–902.
89. Blaney, J.M. and Dixon, J.S., *A good ligand is hard to find: Automated docking methods*, Persp. Drug Discovery Design, 1 (1993) 301–319.
90. Böhm, H.-J., *Ligand design*, In H. Kubinyi (Ed.) 3D QSAR in drug design: theory, methods and applications, ESCOM, Leiden, The Netherlands, 1993, pp. 386–405.
91. Böhm, H.-J., *The development of a simple empirical scoring function to estimate the binding constant for a protein–ligand complex of known three-dimensional structure*, J. Comput.-Aided Mol. Design, 8 (1994) 243–256.
92. Head, R.D., Smythe, M.L., Oprea, T.I., Waller, C.L., Green, S.M. and Marshall, G.R., *VALIDATE: A new method for the receptor-based prediction of binding affinities of novel ligands*, J. Am. Chem. Soc., 118 (1996) 3959–3969.
93. Jain, A.N., *Scoring noncovalent protein–ligand interactions: a continuous differentiable function tuned to compute binding affinities*, J. Comput.-Aided Mol. Design, 10 (1996) 427–440.

94. Dixon, S. and Blaney, J., *Docking*, In Martin, Y.C. and Willett, P. (Eds.) Designing bioactive molecules: Three-dimensional techniques and applications, American Chemical Society, Washington, DC, 1997 (in press).
95. Holloway, M.K., Wai, J.M., Halgren, T.A., Fitzgerald, P.M.D., Vacca, J.P., Dorsey, B.D., Levin, R.B., Thompson, W.J., Chen, L.J., deSolms, S.J., Gaffin, N., Ghosh, A.K., Giuliani, E.A., Graham, S.L., Guare, J.P., Hungate, R.W., Lyle, T.A., Sanders, W.M., Tucker, T.J., Wiggins, M., Wiscount, C.M., Woltersdorf, O.W., Young, S.D., Darke, P.L. and Zugay, J.A., *A priori prediction of activity for HIV-1 protease inhibitors employing energy minimization in the active site*, J. Med. Chem., 38 (1995) 305–317.
96. Ortiz, A.R., Pisabarro, M.T., Gago, F. and Wade, R.C., *Prediction of drug binding affinities by comparative binding energy analysis*, J. Med. Chem., 38 (1995) 2681–2691.
97. Reddy, B.V.B., Gopal, V. and Chatterji, D., *Recognition of promoter DNA by subdomain-2 in-4.2 of Escherichia-Coli-sigma(70): A knowledge-based model of -35-hexamer interaction with 4.2-helix-turn-helix motif*, J. Biomol. Struct. Dynamics, 14 (1997) 407–419.
98. Weber, I.T. and Harrison, R.W., *Molecular mechanics calculations on protein–ligand complexes*, In Kubinyi, H., Folkers, G. and Martin, Y.C. (Eds.) 3D QSAR in drug design: Vol. 2, Kluwer Academic Publishers, Dordrecht, The Netherlands, 1998, pp. 115–127.
99. Wallqvist, A., Jeering, R.L. and Coeval, D.G., *A preference-based free-energy parameterization of enzyme–inhibitor binding: Applications to HIV-1-protease inhibitor design*, Protein Science, 4 (1995) 1881–1903.
100. Wallqvist, A. and Covell, D.G., *Docking enzyme–inhibitor complexes using a preference-based free-energy surface, Proteins: Struct. Funk. Genet., 25 (1996) 403–411.*
101. Dewitt, R.S. and Shakhnovich, E.I., *Smog — de novo design method based on simple, fast, and accurate free-energy estimates: 1. Methodology and supporting evidence*, J. Am. Chem. Soc., 118 (1996) 11733–11744.
102. Mattos, C., and Ringe, D., *Multiple binding modes*, In Kubinyi, H. (Ed.) 3D QSAR in drug design: Theory, methods and applications, ESCOM, Leiden, The Netherlands, 1993, pp. 226–254.
103. Meyer, E.F., Boots, I., Scapozza, L. and Zhang, D., *Backward binding and other structural surprises,* Persp. Drug Discov. Design, 3 (1996) 168–195.
104. Klebe, G., Mietzner, T., and Weber, F., *Different approaches toward an automatic structural alignment of drug molecules: Applications to sterol mimics, thrombin and thermolysin inhibitors*, J. Comput.-Aided Mol. Design, 8 (1994) 751–778.
105. Oprea, T.I., Waller, C.L. and Marshall, G.R., *Three dimensional quantitative structure–activity relationship of human immunodeficiency virus (I) protease Inhibitors: 2. Predictive power using limited exploration of alternate binding modes*, J. Med. Chem., 37 (1994) 2206–2215.
106. DePriest, S.A., Mayer, D., Naylor, C.B. and Marshall, G.R., *3D-QSAR of angiotensin-converting enzyme and thermolysin inhibitors: A comparison of CoMFA models based on deduced and experimentally determined active-site geometries*, J. Am. Chem. Soc., 115 (1993) 5372–5384.
107. Schoenleber, R., Martin, Y.C., Wilson, M., DiDomenico, S., Mackenzie, R.G., Artman, L.D., Ackerman, M.S., DeBernardis, J.F., Meyer, M.D., De, B., Hsiao, C.W. and Kebabian, J.W., American Chemical Society Meeting, August, New York, 1991.
108. Martin, Y.C., Kebabian, J.W., MacKenzie, R. and Schoenleber, R., *Molecular Modeling-based Design of Novel, Selective, Potent D1 Dopamine Agonists*, In Silipo, C. and Vittoria, A. (Eds.) QSAR: Rational approaches on the design of bioactive compounds, Elsevier, Amsterdam, The Netherlands, 1991, pp. 469–482.
109. Glen, R., Martin, G., Hill, A., Hyde, R., Woollard, P., Salmon, J., Buckingham, J. and Robertson, A., *Computer-aided-design and synthesis of 5-substituted tryptamines and their pharmacology at the 5-HT1D receptor — discovery of compounds with potential antimigraine properties*, J. Med. Chem., 38 (1995) 3566–3580.
110. Waller, C.L. and Marshall, G.R., *Three-dimensional quantitative structure–activity relationship of angiotensin-converting enzyme and thermolysin inhibitors: 2. A comparison of CoMFA models incorporating molecular-orbital fields and desolvation free-energies based on active-analog and complementary-receptor field alignment rules.*, J. Med. Chem., 36 (1993) 2390–2403.

111. Klebe, G., *Structural alignment of molecules*, In Kubinyi, H. (Ed.) 3D QSAR in drug design: Theory, methods and applications, ESCOM, Leiden, The Netherlands, 1993, pp. 173–99.
112. Kroemer, R.T., Hecht, P., Guessregen, S. and Liedl, K.R., *Improving the predictive quality of CoMFA models*, In Kubinyi, H., Folkers, G. and Martin, Y.C. (Eds.) 3D QSAR in drug design: Vol. 3, Kluwer Academic Publishers, Dordrecht, The Netherlands, 1998, pp. 41–56.
113. Norinder, U., *Recent progress in CoMFA methodology and related techniques*, In Kubinyi, H., Folkers, G. and Martin, Y.C. (Eds.) 3D QSAR in drug design: Vol. 3, Kluwer Academic Publishers, Dordrecht, The Netherlands, 1998, pp. 25–39.
114. Lin, C.T., Pavlik, P.A. and Martin, Y.C., *Use of molecular fields to compare series of potentially bioactive molecules designed by scientists or by computer*, Tetrahedron Comput. Method., 3 (1990) 723–738.
115. Norinder, U., *Experimental design based 3-D QSAR analysis of steroid–protein interactions: Application to human CBG complexes*, J. Comput.-Aided Mol. Design, 4 (1990) 381–389.
116. Caliendo, G., Greco, G., Novellino, E., Perissutti, E. and Santagada, V., *Combined use of factorial design and comparative molecular field analysis (CoMFA): A case study*, Quant. Struct.-Act. Relat., 13 (1994) 249–261.
117. Mabilia, M., Belvisi, L., Bravi, G., Catalano, G. and Scolastico, C., *A PCA/PLS analysis on nonpeptide angiotensin II receptor antagonists*, In Sanz, F., Giraldo, J. and Manaut, F. (Eds.) QSAR and molecular modeling: Concepts, computational tools and biological applications, Proceedings of the 10th European Symposium on Structure–Activity Relationships: QSAR and Molecular Modeling, Barcelona, 4–9 September 1994, Prous, Barcelona, 1995, pp. 456–60.
118. Cramer III, R.D., Clark, R.D., Patterson, D.E. and Ferguson, A.M., *Bioisosterism as a molecular diversity descriptor — steric fields of single topomeric conformers*, J. Med. Chem., 39 (1996) 3060–3069.
119. Mager, P.P., *A random number experiment to simulate resample model evaluations*, J. Chemometrics, 10 (1996) 221–240.
120. Clark, M. and Cramer III, R.D., *The probability of chance correlation using partial least squares (PLS)*, Quant. Struct.-Act. Relat., 12 (1993) 137–145.
121. Doweyko, A.M., *Three-dimensional pharmacophores from binding data*, J. Med. Chem., 37 (1994) 1769–1778.
122. Dunn III, W.J. and Rogers, D., *Genetic partial least squares in QSAR*, In Devillers, J. (Ed.) Genetic algorithms in molecular modeling, Academic Press, London, 1996, pp. 109–130.
123. Coats, E.A., *The CoMFA steroids as a benchmark data set for development of 3D QSAR methods*, In Kubinyi, H., Folkers, G. and Martin, Y.C. (Ed.) 3D QSAR in drug design: Vol. 3, Kluwer Academic Publishers, Dordrecht, The Netherlands, 1998, pp. 199–214.
124. Tropsha, A. and Cho, S.J., *Cross-validated region selection for CoMFA studies*, In Kubinyi, H., Folkers, G. and Martin, Y.C. (Eds.) 3D QSAR in drug design: Vol. 3, Kluwer Academic Publishers, Dordrecht, The Netherlands, 1998, pp. 57–69.
125. Cruciani, G., Clementi, S. and Pastor, M., *GOLPE-Guided Region Selection*, In Kubinyi, H., Folkers, G. and Martin, Y. (Ed.) 3D QSAR in drug design: Vol. 3, Kluwer Academic Publishers, Dordrecht, The Netherlands, 1998, pp. 71–86.
126. Dunn III, W.J. and Rogers, D., *Genetic partial least-squares in QSAR*, In J. Devillers (Ed.) Genetic algorithms in molecular modeling, Academic Press, London, 1996, p. 109–30.
127. Wikel, J.H. W.J. and Dow, E.R., *The use of neural networks for variable selection in QSAR*, Bioorg. Medic. Chem. Lett., 3 (1993) 645–651.
128. Kubinyi, H., *Variable selection in QSAR Studies: 1. An Evolutionary Algorithm*, Quant. Struct.-Act. Relat., 13 (1994) 285–294.
129. Kubinyi, H., *Variable selection in QSAR studies: 2. A highly efficient combination of systematic search and evolution*, Quant. Struct.-Act. Relat., 13 (1994) 393–401.
130. Rogers, D. and Hopfinger, A.J., *Application of genetic function approximation to quantitative structure–activity relationships and quantitative structure–property relationships*, J. Chem. Inf. Comput. Sci., 34 (1994) 854–866.
131. Lingren, F., Geladi, P., Berglund, A., Sjostrum, M. and Wold, S., *Interactive variable selection (IVS) for PLS: 2. Chemical applications*, J. Chemometrics, 9 (1995) 331–342.

132. Tetko, I.V., Villa, A. and Livingstone, D.J., *Neural-network studies: 2. Variable selection*, J. Chem. Inf. Comput. Sci., 36 (1996) 794–803.
133. Baldovin, A., Wu, W., Centner, V., Jouanrimbaud, D., Massart, D.L., Favretto, L. and Turello, A., *Feature-selection for the discrimination between pollution types with partial least-squares modeling*, Analyst, 121 (1996) 1603–1608.
134. Centner, V., Massart, D.L., Denoord, O.E., Dejong, S., Vandeginste, B.M. and Sterna, C., *Elimination of uninformative variables for multivariate calibration*, Anal. Chem., 68 (1996) 3851–3858.
135. Hasegawa, K., Miyashita, Y. and Funatsu, K., *GA strategy for variable selection in QSAR studies: GA-based PLS analysis of calcium-channel antagonists*, J. Chem. Inf. Comput. Sci., 37 (1997) 306–310.
136. Höltje, H.-D., Anzali, S., Dall, N. and Höltje, M., *Binding Site Models*, In Kubinyi, H. (Ed.) 3D QSAR in drug design: Theory, methods and applications, ESCOM, Leiden, The Netherlands, 1993, pp. 320–335.
137. Vedani, A., Zbinden, P., Snyder, J.P. and Greenidge, P.A., *Pseudoreceptor modeling: The construction of three-dimensional receptor surrogates*, J. Am. Chem. Soc., 117 (1995) 4987–4994.
138. Topliss, J.G. and Edwards, R.P., *Chance factors in studies of quantitative structure–activity relationships*, J. Med. Chem., 22 (1979) 1238–1244.
139. Hoskuldsson, A., *Quadratic PLS regression*, J. Chemometrics, 6 (1992) 307–334.
140. Benigni, R. and Guiliani, A., *Analysis of distance matrices for studying data structures and separating classes*, Quant. Struct.-Act. Relat., 12 (1993) 397–401.
141. Kubinyi, H., *QSAR: Hansch analysis and related approaches*, VCH, Weinheim, Germany, 1993, Vol. 1, pp. 240.
142. Martin, Y.C., Lin, C.T., Hetti, C. and DeLazzer, J., *PLS analysis of distance matrices detects non-linear relationships between biological potency and molecular properties*, J. Med. Chem., 38 (1995) 3009–3015.
143. Livingstone, D. and Manallack, D.T., *Statics using neural networks: Chance effects*, J. Med. Chem., 36 (1993) 1295–1297.
144. Tetko, I.V., Livingstone, D.J. and Luik, A.I., *Neural-network studies: 1. Comparison of overfitting and overtraining*, J. Chem. Inf. Comput. Sci., 35 (1995) 826–833.
145. Devries, S. and Terbraak, C., *Prediction error in partial least-squares regression: A critique on the deviation used in the unscrambler*, Chemometrics Intelligent Lab. systems, 30 (1995) 239–245.
146. Jonathan, P., Mccarthy, W.V. and Roberts, A., *Discriminant-analysis with singular covariance matrices: A method incorporating cross-validation and efficient randomized permutation tests*, J. Chemometrics, 10 (1996) 189–213.
147. Kemsley, E.K., *Discriminant-analysis of high-dimensional data: A comparison of principal components-analysis and partial least-squares data reduction methods*, Chemometrics Intelligent Lab. Systems, 33 (1996) 47–61.
148. Shuker, S., Hajduk, P., Meadows, R. and Fesik, S., *Discovering high-affinity ligands for proteins: SAR by NMR*, Science, 274 (1996) 1531–1534.
149. Sheridan, R.P. and Kearsley, S.K., *Using a genetic algorithm to suggest combinatorial libraries*, J. Chem. Inf. Comput. Sci., 35 (1995) 310–320.

Recent Progress in CoMFA Methodology and Related Techniques

Ulf Norinder
Astra Pain Control AB, S-151 85 Södertälje, Sweden

1. Introduction

Since the advent of 3D QSAR techniques, such as the hypothetical active site lattice (HASL) method [1], receptor modelling from the three-dimensional structure and physico-chemical properties of the ligand molecules (REMOTEDISC) [2] and Comparative Molecular Field Analysis (CoMFA) related methods [3–5] in the late 1980s, a large number of investigations have been described in the literature. The development and application of 3D QSAR methods up to 1993 have been compiled in the book *3D QSAR in Drug Design* [6]. After 1993, more than 340 articles have been published in the 3D QSAR area (For a list of published articles 1993–1996, see the final chapter in this volume by Ki H. Kim). The vast majority of these publications are applications using CoMFA.

The advances with respect to technological development, in the area of CoMFA-related methods since 1993, can be divided into four main areas:

1. Protocols for the alignments of compounds.
2. Introduction of new fields.
3. Variable selection techniques.
4. Statistical developments.

Significant progress has also been made in other types of 3D QSAR methods where new mathematical/statistical tools for deriving consistent and predictive QSAR models, such as *neural networks* [7–9] and *genetic/evolutionary algorithms* [10], have been introduced. In one of these approaches, which is discussed in more detail in section 3.2, the Comparative Molecular Moment Analysis (CoMMA) [11], the alignment problem is eliminated. Several methods [12,13] have also been developed in the ligand–receptor-based direction due to the rapidly increasing number of crystal structures of ligand–macromolecule complexes of good quality that have become available in recent years.

2. CoMFA-related Methods

2.1. Approaches to find relevant alignment rules

Several investigations have tried to use alignments based on crystallographic data. One of the first investigations of this kind was that of Klebe and Abraham [14], where they compared datasets related to *human rhinovirus14* (HRV14) and thermolysin with alignments obtained from multiple-fit and field-fit procedures. For the HRV14 dataset, they found that both types of alignment resulted in predictions of moderate quality. For the

H. Kubinyi et al. (eds.), 3D QSAR in Drug Design, Volume 3. 25–39.

thermolysin dataset, however, the fitted hypothetical alignments gave substantially better predictions than those based on experimental data. DePriest et al. [15] have investigated some ACE and thermolysin inhibitors using alignment rules determined from a systematic *conformational search* (ACE dataset) and experimentally determined *active site alignments* (thermolysin). They also found that the ACE models showed significantly better predictivity for an external test set ($r^2 = 0.53$) compared to the predictivity of the thermolysin model ($r^2 = 0.29$). It may, at first, seem somewhat strange that experimental geometries result in inferior models compared with those models based on a more simplistic scheme. However, the fundamental basis of any good predictive QSAR model is a consistent description of the structures under investigation. By using experimental geometries, that are more or less perturbed from one another, an orientation-related element is introduced in all variables which is different for each structure. Thus, the grid-points do not contain an altogether consistent structural description anymore, which makes it difficult to derive a predictive 3D QSAR model. The situation is further complicated by the use, in most CoMFA investigations, of the 6–12 type potential functions for calculation of the non-bonded interactions, which have a very steep repulsion component and are, consequently, sensitive to orientational distortions of the investigated structure (see reference [16] for a more complete discussion on this topic and section 2.2 regarding new fields).

Waller et al. [17] have also used experimentally determined alignments in an investigation of *HIV-1 protease inhibitors*. They also found that the alignments based on field-fit minimizations gave statistically better and more predictive 3D QSAR models than those based on crystallographic data from ligand–receptor complexes. However, the difference in predictivity between the two types of alignments on an external test set (18 compounds) was not very large.

Waller and Marshall [18] have further investigated the use of alignments based on knowledge of the active-site of the receptor using a '*complementary receptor field' technique* for the same thermolysin inhibitors as previously analyzed by DePriest et al. [15] with promising results. The 'complementary receptor field' method improved the predictions of the 11 test set inhibitors from that of $r^2 = 0.29$ calculated by DePriest et al. to $r^2 = 0.40$. Waller and Marshall also used considerably fewer PLS components (3) than previously used (11) in the study by DePriest et al. [15].

An additional step in the 'active site' direction (i.e. the use of a known active site geometry) was taken by Oprea et al. [19]. They devised a semiautomated procedure called NewPred with which they analyzed the predictivity for a series of 30 HIV-1 protease inhibitors from on a model based on 59 inhibitors. NewPred uses a limited exploration of *alternative binding modes* and several conformers for each compound which are individually relaxed in the binding site. The predictivity for the same test set, as earlier studied by Waller and Marshall [18], did not change significantly using neutral (uncharged) ligands. Both studies showed $r^2 \approx 0.67$ for the test set. However, the predictivity of the test set from models based on charged ligands improved from $r^2 = 0.327$ to $r^2 = 0.662$ when using NewPred. Thus, a more consistent protocol for alignments seems to result from using NewPred. NewPred can also be used in the absence of a known active site geometry. In this case, the conformers of each molecule are minimized and aligned in the average CoMFA fields.

Additional examples of the use of *X-ray structure information* for the alignment of compounds include that of Brandt et al. [20] in a CoMFA study of some artificial peptide inhibitors of the serine protease *thermitase*, and Kroemer et al. [21] in an investigation of some HIV-1 protease inhibitors of statine type. In both of these examples, the investigated inhibitors were fitted to a reference structure in a crystallized complex exhibiting high structural similarity with the studied compounds. In the latter study, a large number of compounds were divided into a training set (100 compounds) and a test set (75 compounds) and predictive models, as determined by internal validation, but more importantly by the predictivity of the test set ($r^2 = 0.552 - 0.569$), were derived. The resulting CoMFA maps were compared with the surface of the active site of the receptor and a high degree of consistency was found. This fact, also noted by Cruciani and Watson [22], is encouraging from a methodological point of view since it, in favorable cases, allows a better understanding of the binding process, as well as the fact that it may aid the design of new potent compounds in a better manner.

An interesting and promising technique was recently published by Gamper et al. [23], where they studied the binding of 27 *haptens* to the *monoclonal antibody* IgE (Lb4) using the automated docking program AUTODOCK [24]. A small starting set of 9 ligands was used that had either two or three distinct orientations. The alignments that resulted in the best cross-validated r^2 value were further used in the study. A small set of 3 sulphur-containing haptens was used as a test set with good predictivity. However, a more balanced selection of training set and test set would have been desirable in this study in order to estimate the consistency of the technique since a 'tuning' procedure is employed by the authors in order to establish relevant alignments.

The same situation prevails in a study by Cho et al. [25] of some *AChE inhibitors* using structure-based alignments combined with a region variable selection technique. CoMFA models with high cross-validated r^2 values result, as can be expected from variable selection procedures (see section 2.3 for a more detailed presentation), but no external evidence of the predictivity — i.e. using an external test set — or the stability with respect to randomization of the biological activities are presented by the authors. Since the dataset contained 56 compounds, the division of these inhibitors into a balanced training set and test set, respectively, seems possible which would have made the investigation more valuable from a methodological development point of view.

A different approach for improving the predictivity of CoMFA models has been adapted by Kroemer and Hecht [26]. They used a scheme of fixed translations and rotations for the underpredicted ligands of the training set to maximize their respective predicted activities. The dataset studied was a set of *DHFR triazine inhibitors* where they used 80 compounds as a training set and 70 molecules as a test set. The construction of the CoMFA model is straightforward using the scheme mentioned above. However, the predictions of new molecules (e.g. the test set) is somewhat more complex. Kroemer and Hecht devised two similar schemes for that purpose based on the highest similarity, determined by the molecular CoMFA fields (for a more extensive description of the method see reference [27]), between each test molecule and an arbitrarily chosen number of training set compounds (6 in their study). Thus, the predicted activity of a test compound is weighted according the 6 highest similarity scores to 6 training set compounds. The difference between the two schemes is that in the more 'complex' one the inaccuracy of the CoMFA model is also taken into account by introducing the residuals of the template (training set) molecules into the prediction scheme. Predictive models (r^2_{test} = 0.484 – 0.645) resulted. However, the authors of the study also brought to light one of the potential problems with this kind of 'tuning' operation, namely, that random models with an initially negative (!!) cross-validated r^2 value may be taken into what may seem to be consistent CoMFA models with high positive cross-validated r^2 values! This dangerous fact will be further discussed in conjunction with variable selection techniques in section 2.3. Fortunately, the use of a test set, which still resulted in negative r^2 values, shows the poor quality of these 'refined' random models. This study further emphasizes the necessity of an external test set to be able to assess the quality of the derived models as pointed out by Kroemer and Hecht in their article. In the investigation by Kroemer and Hecht, the compounds were only allowed, by choice, to be translated a maximum of 0.3 Å in any direction and rotated a maximum of 3.0° around any axis. Is this enough to obtain a consistent model?

Another investigation toward the same objective — i.e. to create 'consistent' 3D QSAR models of CoMFA type with improved predictivity — is the TDQ (Three-dimensional QSAR) approach of Norinder [28]. Two data sets, the Tripos steroids and some *tyrosine kinase inhibitors*, were studied using a COMPASS-related approach [29] implemented in a CoMFA-like framework. A conformational analysis of Catalyst [30] type was initially performed for every compound. A starting conformer and alignment was selected for each compound belonging to the training set. The conformer and orientation, using a series of rigid-body translations and rotations of each compound, with the highest predicted activity were selected to update the model. This iterative scheme was pursued until self-consistency of the model was achieved. Predictions of test set compounds were performed with an analogous scheme. The conformer and orientation with the highest predicted activity were chosen to represent the activity of the test compound. Two different schemes, a traditional one using non-bonded and charge–charge interactions, as well as a COMPASS-like description using squared distances between atoms and grid-points, were used to represent the fields in the study. Predictive models were derived for both datasets. However, models based on the *distance representation* had a wider range of structural predictivity compared to the traditional description. Again, this observation points to the limitations and problems associated with using a functional form of 6–12 type to represent the non-bonded interactions (for further discussions on this topic see reference [17] and section 2.3). No randomization experiments were performed in the study by Norinder; thus, no conclusions with respect to determining the robustness of the method can be drawn.

A somewhat different approach for arriving at reasonable alignments to be used in 3D QSAR studies has been investigated by Norinder [31], Palomer et al. [32] and Hoffmann and Langer [33]. They all used the Catalyst [30] software to determine the alignments of investigated compounds. These orientations of the structures were subsequently used to derive 3D QSAR models of CoMFA type. The use of the program SEAL for obtaining reasonable alignments has been reported by Klebe and co-workers [34–35].

2.2. New fields in CoMFA applications

Apart from perhaps the largest problem in 3D QSAR investigations, namely inadequate alignment of structures, other reasons for not obtaining good models, which show predictivity and robustness, certainly include an insufficient representation of the investigated structures. To handle this problem, a number of new fields and other parameters have been introduced into CoMFA.

The hydropathic interaction (HINT) technique of Kellogg et al. [36] has been used in CoMFA applications for a number of years now (see reference 37 for a more thorough description of the HINT method).

The GRID program [38,39] has been used by a number of authors [40,41] as an alternative to the original CoMFA method for calculating the interaction fields in molecular field analysis (MFA). An advantage of using GRID in MFA investigations, apart from the large number of different probes available, is the use of a 6-4 potential function, which is smoother than the 6-12 form of Lennard-Jones type, for computing the interaction energies at the grid lattice points.

An interesting dataset of some *glycogen phosphorylase* b *inhibitors* has been analyzed by Cruciani and Watson [22] using the GRID force field in conjunction with GOLPE (see section 2.3 for further details on the GOLPE procedure). The particularly interesting aspect of this dataset is that the three-dimensional X-ray structures of all ligands complexed to glycogen phosphorylase b are known. This allows many opportunities to investigate the dataset using new and different methodological ideas to further the development of 3D QSAR techniques, as well as to relate the results of such studies back to ligand–receptor complexes for analysis.

Kim et al. [40,42,43] have introduced a *hydrogen-bonding field* into 3D QSAR. This was useful for some benzodiazepines where the GRID H_2O probe successfully described the hydrophobic effects not adequately described by the standard CH_3 (sp^3) probe used in most CoMFA studies.

Kenny has investigated the use of electrostatic properties to predict hydrogen-bonding and their implications for CoMFA [44]. He found that the electrostatic potential is not sampled closely enough to hydrogen-bonding atoms with the typically used standard CoMFA probe and grid spacing of 1.5 Å. He also noted that at greater distances from atoms capable of hydrogen bonding a more effective descriptor of hydrogen bonding is the electric field strength. Thus, a combination of electrostatic potentials and electric fields may provide a better-defined CoMFA field for describing electrostatic interactions including hydrogen-bond contributions.

Development of new fields in recent years, which consists of adding lipophilic information to CoMFA analysis, are centered on the use of *molecular lipophilicity potentials* (*MLP*s) [45]. Testa and co-workers have published a number of articles using MLPs based on atomistic hydrophobicity parameters [46]. They have studied 5-HT1A receptor ligands [47], indeno[1,2-c]pyridazines [48] and some isoquinolines [49]. However, the incorporation of the MLP field did not improve the statistical quality of the models and their predictivity, as measured by external tests, to any significant extent. Masuda et al. [50] have used a similar MLP field in a CoMFA study of glycine conjugation of some aromatic and aliphatic carboxylic acids. They used a Fuchère-type [45] MLP equation previously used by Norinder [51] in a 3D QSAR study. The predictivity of the resulting model, albeit only using internal cross-validation, improved somewhat using the MLP field in conjunction with the traditional CoMFA fields of non-bonded and electrostatic nature as compared to using only the two latter fields.

However, the greatest benefit from adding an MLP field to 3D QSAR models seems at the present time, in view of the results obtained so far, not to be that of improving the statistical quality, but rather to add interpretability to CoMFA/3D QSAR models in physico-chemical terms. This is an important aspect, not to be forgotten or obscured by only focusing on the statistical parameters of the derived model, since the interpretation of the resulting CoMFA maps is sometimes quite difficult to understand and utilise in drug development.

The incorporation of *molecular orbital fields* into CoMFA has attracted interest. Waller and Marshall [18] have used a *HOMO field* in order to refine a CoMFA study on some *ACE inhibitors* previously investigated by DePriest et al. [15] using traditional field representations — i.e. non-bonded and electrostatic interactions. The main advantage of using an orbital field in the Waller-Marshall study was to describe the interactions between the ligands and a zinc metal present in the system in better detail. The HOMO field in this (and other) studies was incorporated into the model as the electron density at the respective grid positions of the defined CoMFA region.

Poso et al. [52] have used a *LUMO field* in a study of mutagenicity of some 16 *MX compounds* (furanones) related to TA100 *mutagenicity*. The use of a LUMO field did improve the internal predictivity of the model significantly. The two best models, as judged by their cross-validated r^2 values, were based on steric/LUMO and steric/electrostatic/LUMO fields that showed values of 0.903 (!) and 0.910 (!), respectively. However, the exact numbers of PLS components (less than 10) used in the models were not mentioned in the article, nor was an external test set deployed to verify the predictivity of the models. Navajas et al. [53] have studied the same set of compounds. In their study, they concluded that the AM1 and PM3 methods for calculating electronic characteristics were superior to MNDO but, more interestingly, derived models based on 3 PLS components which showed cross-validated r^2 values of 0.733–0.742 that seem somewhat more realistic from a non-over-fitting-the-model point of view.

Kim et al. have in earlier studies investigated the quality of electrostatic descriptors calculated at different levels of approximation — e.g. semi empirical AM1, GRID and *ab initio* STO-3G — used in the CoMFA method and found that the use of semi empirical calculated charges is a reasonable computational level on which to operate in 3D QSAR studies [54,55].

Kroemer et al. [56] have also investigated the quality of electrostatic descriptors used in the CoMFA method. They studied some 37 ligands of the *benzodiazepine receptor* inverse agonist–antagonist site. The methods deployed for calculating *electrostatic potentials* and charges included that of Gasteiger-Marsili [57], semiempirical (MNDO, AM1 and PM3) and *ab initio* (HF/STO-3G, HF/3-21G* and HF/6-31G*). Atomistic charges were derived both from Mulliken population analysis (MPA) or from fitting the charges to the molecular electrostatic potentials (MEP) (ESPFIT), as well as using MEPs from *ab initio* calculations directly mapped onto the CoMFA grid points. Kroemer et al. concluded that ESPFIT charges were superior to MPA-derived charges and that semiempirical ESPFIT charges were of comparable quality to those computed with *ab initio* methods. MEPs mapped directly onto the grid-points did not prove to be superior to ESPFIT potentials. The results of Kroemer et al. further support the use of semiempirical calculated charges as a reasonable computational level on which to operate in 3D QSAR studies. This is especially valuable keeping the *combinatorial chemistry* implications at hand — i.e. the possibility to run virtual libraries of compounds through a developed CoMFA/3D QSAR model in order to determine a synthetic combinatorial strategy for a particular drug development programme.

Another promising method for the addition of electrostatic information to CoMFA-related methods (and other techniques as well) is the use of *electrotopological state (E-state) fields*. Recently, Kellogg et al. [58] have applied the E-state formalism of Kier and Hall [59] to develop an E-state (non-hydrogen atoms) and a *hydrogen electrotopological state (HE-state) field* suitable for incorporation into 3D QSAR investigations. Kellogg et al. studied the classical CoMFA steroid dataset and investigated the influence of grid size, as well as various functional forms for computing the new fields. The best model in their study resulted from the combined use of the E- and HE-state fields alone. The use of the E- and/or HE-state fields in combination with other fields (steric, electrostatic and hydropathic) gave models with improved statistics as compared with the traditional representation (steric and electrostatic) where the (H)E-state fields provided a significant contribution. Unfortunately, the study was only conducted and evaluated using the training set of 21 steroids. Thus, the 'true' predictivity and potential of the new fields based on the evaluation of an external test set — e.g. the 10 steroids included in the original paper by Cramer et al. [3] — cannot be assessed at this point in time.

Desolvation energy fields computed by the Delphi technique [60,61] have been used in a CoMFA study by Waller and Marshall [18] on some ACE and thermolysin inhibitors. The inclusion of a desolvation energy field did not improve the statistical quality of models and the desolvation energy field was found to be rather colinear with the electrostatic field [62].

The problems associated with the functional form of the Lennard-Jones 6-12 potential used to compute the non-bonded (steric) interactions in most CoMFA studies have attracted the attention of Kroemer and Hecht [63]. They suggest that the steric descriptors are replaced by *indicator variables* representing the presence of an atom in a predefined volume element within the CoMFA region of the aligned molecules. Kroemer and Hecht found a significant improvement of the derived models, as indicated by both the cross-validated r^2 values for the training sets and the predictive r^2 values for the test sets, using five randomly selected training sets (80 compounds each) and test sets (60 compounds each) of some DHFR inhibitors, with the indicator-based description of the steric field. A similar result with respect to changing the computation of the steric field from the Lennard-Jones type potential into a distance-based representation has also been noted by Norinder [28] (see section 2.1 for a more detailed description of the method). Klebe et al. have developed molecular similarity fields (see section 3.1 for further details) to address similar issues related to the use of Lennard-Jones type potentials in CoMFA related methods [35]. For a recent mini-review on adding new fields to CoMFA/3D QSAR models, see reference [62].

2.3. *Variable selection techniques*

The creation and incorporation of new fields have introduced another problem into 3D QSAR techniques with respect to the statistical analysis, namely the rapidly decreasing signal-to-noise ratio in the descriptor matrix. Although the introduction of additional variables is advantageous from a molecular representation point of view, as they (at best) allow a better and more comprehensive description of the investigated structures. These variables make it increasingly difficult for multivariate projection methods, such as PLS [64], to distinguish the useful information contained in the descriptor matrix from that of less quality or noise.

Thus, methods for selecting the 'useful' variables, defined by some criteria, from the less useful ones were needed. A chemometric tool called GOLPE (Generating Optimal Linear PLS Estimations) was developed by Baroni et al. [65] to achieve the objective of improving the consistency and predictivity of QSAR models in general, and 3D QSAR models in particular, by means of variable selection. In the earlier versions of the GOLPE protocol, a preselection of variables, by means of *D-optimal design*, was performed. This step was later abandoned, as computational capacity has increased considerably; and because it introduced unnecessary bias into the final selection procedure and, hence, the final model. The predictivity of the analyzed variables was determined by the use of a fractional factorial design (FFD) protocol where a large number of 3D QSAR models were evaluated. The predictivity of each model was determined by SDEP (Standard Deviation of Error of Prediction). After the completion of an FFD protocol, each variable was evaluated and classified into one of three categories: positive (helpful for predictivity), negative (detrimental for predictivity) or uncertain. Also in the earlier versions of the GOLPE procedure, a number of FFD cycles were performed until very few (or no) uncertain variables remained. This repetitive procedure was later abandoned since it has a strong tendency to result in models which are over-fitted. Today only one cycle of an FFD evaluation is used.

However, there are several problems associated with variable selection techniques on single variables in 3D QSAR applications. One problem is the tendency to result in improved models for the training set without improved predictivity on an external test set [66]. The models may also show quite non-contiguous CoMFA maps, which does not aid the interpretation of these maps. Furthermore, by using single variable selection procedures of GOLPE type in an inappropriate manner — e.g. starting with a model having a negative cross-validated r^2 value (!!) — it is possible to achieve what may seem to be a consistent and good 3D QSAR model as determined by internal validation. This was nicely demonstrated by Nordén et al. [67], where a set of *randomly aligned structures* resulted in a 'good' CoMFA-type model using internal validation and single variable selection!

To circumvent these problems and to obtain more contiguous coefficient maps, *region or domain variable selection* procedures have been developed by Cho and Tropsha [68], Norinder [66] and Cruciani et al. [69,70]. The method of Cho and Tropsha, called cross-validated R^2-Guided Region Selection (q^2-GRS), divides the original CoMFA region into smaller regular boxes (regions). A CoMFA analysis, using a leave-one-out (LOO) procedure, is then performed on each of the small regions. Regions with a cross-validated r^2 value greater than a specified cutoff value are selected for further use. Finally, a CoMFA analysis is performed using all variables belonging to the selected regions. In the first work, Cho and Tropsha [68] analyzed 3 datasets of reasonable size (20 5-HT1A receptor ligands, 59 HIV-1 inhibitors and the 21 steroids of the classic Tripos data set). They derived q^2-GRS selected models with higher cross-validated r^2-values than the corresponding conventional CoMFA procedure as can be expected using variable selection. However, no external test sets were used in that study to evaluate the increase in predictivity, as a result of variable selection, in a more unbiased manner than through internal cross-validation using a LOO approach. A favorable result from that study was that the q^2-GRS routine resulted in orientation-independent models with respect to translations/rotations of all structures. This is otherwise a potential problem using the conventional CoMFA protocol. The q^2-GRS procedure has been further developed to incorporate different types of probe atoms reported in a study by Cho et al. [71] on some 101 antitumor agents of 4′-O-demethyl-epiodophyllotoxin type. In that investigation, they used a training set of 59 compounds and a test set of 41 compounds. The cross-validated r^2 values for the training set increased from 0.34 (conventional CoMFA-type procedure) to 0.58 using the q^2-GRS method. However, the predictivity of the test set by the latter model was rather poor ($r^2 = 0.24$).

Similar results with respect to poor predictivity of external test sets have been reported by Norinder [66] using a GOLPE-like protocol and small domains (boxes) of similar type as used in the q^2-GRS method. Norinder studied 3 steroid datasets (the 31 steroids of the classic Tripos dataset and 49 steroids with affinity for the progesterone and glucocorticoid steroid receptors) but found no improvements on predictivity for the test sets using variable selection. The performance on the training sets increased as a result of variable selection. This is, however, to be expected since variable selection methods of this kind (as well as the q^2-GRS procedure) has changed the role of the cross-validation procedure from an internal validation technique into an object function which is to be maximized. Thus, other tools, such as the use of balanced training sets and test sets as well as randomization trials, quality criteria and monitoring methods are needed to measure the performance of variable selection procedures. The use of internal validation only in conjunction with 'tuning' operations, such as variable selection and geometry realignments (see section 2.1), says very little about the 'true' performance, stability and consistency of the derived 3D QSAR models. An interesting method, in this respect, has been deployed by Sutter et al. [72] in property estimations using neural networks, which are known for their tendency towards being over-trained, where the investigated set of compounds has been divided into three parts: a training set, an internal test set with which the predictivity of the model is monitored and an external test set with which the predictivity of the final model is determined. The SDEP parameter developed by Baroni et al. [65] is similar in nature to the technique used by Sutter et al., in that a number of training sets are automatically created and employed during the variable selection process to determine which parameters or regions are useful or detrimental, respectively, for improving the predictivity of the model.

Cruciani et al. [69,70] have developed a slightly different form of region selection. Initially, a number of seeds are placed in the CoMFA/3D QSAR region defined by the investigated compounds. The seeds exhibit a representative distribution in variable space. Each variable is then assigned to the nearest seed, thus forming a number of polyhedra. The polyhedra are then collapsed into larger regions if the polyhedra are close in space and contain the same information — i.e. they are correlated to a high degree. Application of this approach to some glycose phosphorylase b inhibitors resulted in better predictivity for an external test set compared to the region and domain variable selection techniques of Cho et al. [25,68,71] and Norinder [66], respectively.

2.4. *Statistical developments*

Through the introduction of new fields and by the subsequent need for variable selection, many rounds of statistical analysis, most often using the PLS method [64], are needed today as compared to one or few analyses required by the original CoMFA protocol.

In order to speed up the computational process 'kernel'-like PLS algorithms have been developed by Rännar et al. [73,74], and by Bush and Nachbar [75] (the SAMPLS method). These methods work by using the equivalent of a covariance matrix instead of the whole descriptor matrix [76]. Thus, instead of having to handle an N × M matrix (N objects, M variables ; M >> N), the methods only compute on a N × N matrix (the so-called kernel and association matrices). An impressive computational 'speed-up' has been reported by Bush and Nachbar [75] for the classic Tripos steroid dataset using SAMPLS.

An interesting development using an N-way PLS method with emphasis on the *3-way PLS* version has recently been described by Bro [77]. Application of this algorithm to 3D QSAR investigations seems attractive since the unfolding step of the original 3D matrix into a 2D matrix is avoided. So far, only a few applications of the 3-way PLS method to 3D QSAR problems have been presented [78,79]. According to the authors of the presentations [80], the method seems to give more robust and consistent PLS models, especially with respect to the optimum number of PLS components (ONC) to be used in a particular model. This is of great importance for 3D QSAR methods since the present procedures (methodologies) often suggest different ONCs that should be used depending on the protocol employed — e.g. the deployed statistical significance tests. A similar statistical approach has recently been presented by Dunn et al. [81] in conjunction with molecular shape analysis.

3. Other CoMFA-Related Techniques

3.1. *Comparative Molecular Similarity Indices Analysis (CoMSIA)*

Due to the problems associated with the fields presently used in most CoMFA-related methods (see section 2.2 for further discussions on the subject), Klebe et al. [35] have developed a similarity indices-based CoMFA-related method (CoMSIA) using Gaussian-type functions. Three different indices related to steric, electrostatic and hydrophobic potentials were used in the study of the classic Tripos steroid dataset and some thermolysin inhibitors previously studied by DePriest et al. [15]. Models of comparable statistical significance with respect to internal cross-validation of the training sets, as well as predictivities of the test sets, were obtained using CoMSIA as compared with traditional CoMFA analysis. The clear advantage of CoMSIA lies in the functions used to describe the compounds under investigation, as well as the resulting contour maps. The CoMSIA approach produces contour maps that are more contiguous compared to maps resulting from the traditional CoMFA method, which makes the CoMSIA maps easier to interpret. The CoMSIA approach also avoids the cutoff values used in CoMFA to restrict the potential functions from assuming unacceptably large values.

3.2. Comparative Molecular Moment Analysis (CoMMA)

The most crucial and difficult step in a CoMFA-related analysis is how to align the investigated compounds in a 'correct' manner (see section 2.1 for further discussions on this topic). A development of the CoMFA method to possibly avoid the 'alignment problem' has recently been described by Silverman and Platt [11]. The method requires no superposition step and use descriptors that characterize shape and charge distribution such as the *principal moments* of inertia and properties derived from dipole and quadropole moments, respectively. Silverman and Platt analyzed a number of datasets, which included the classic Tripos steroids, and obtained models with good consistency, as determined by an internal LOO–CV procedure. Analysis of the 21 steroids gave cross-validated $r^2 = 0.67 - 0.83$ with respect to CBG binding. Unfortunately, although used in a study with all 31 steroids as training set, the authors do not report the predictivity of the steroid models, or any other models for that matter, using the available external test set. The study would have been more informative had such external predictions been reported which would have allowed comparisons with other 3D QSAR investigations — e.g. CoMFA [3], CoMSIA [35], COMPASS [29] and TDQ [28] — which have used the Tripos steroid dataset and reported external predictions for the test set.

References

1. Doweyko, A.M., *The hypothetical active site lattice: An approach to modeling sites from data on inhibitor molecules*, J. Med. Chem., 31 (1988) 1396–1406.
2. Ghose, A., Crippen, G., Revankar, G., McKernan, P., Smee, D. and Robbins, R., *Analysis of the in vitro activity of certain ribonucleosides against parainfluenza virus using a novel computer-aided molecular modeling procedure*, J. Med. Chem., 32 (1989) 746–756.
3. Cramer, R.D., Patterson, D.E. and Bunce, J.C., *Comparative molecular field analysis (CoMFA): 1. Effect of shape on binding of steroids to carrier proteins*, J. Am. Chem. Soc., 110 (1988) 5959–5967.
4. Norinder, U., *A PLS QSAR analysis using 3D generated aromatic descriptors of principal property type: Application to some dopamine D_2 benzamide antagonists*, J. Comput.-Aided Mol. Design, 7 (1993) 671–682.
5. Floersheim, P., Nozulak, J. and Weber, J., *Experience with molecular field analysis*, In Wermuth, C.G. (Ed.) Trends in QSAR and molecular modeling 92: Proceedings of the 9th European Symposium on Structure–Activity Relationships — QSAR and Molecular Modeling, ESCOM, Leiden, The Netherlands, 1993, pp. 227–232.
6. Kubinyi, H. (Ed.), *3D QSAR in drug design: Theory, methods and applications*, ESCOM, Leiden, The Netherlands, 1993.

7. Jain, A.N., Harris, N.L. and Park, J.Y., *Quantitative binding site model generation: Compass applied to multiple chemotypes targeting the 5-HT_{1A} receptor*, J. Med. Chem., 38 (1995) 1295–1308.
8. Head., R.D., Smythe, M.L., Oprea, T.I., Waller, C.L., Green, S.M. and Marshall, G.R., *VALIDATE: A new method for the receptor-based prediction of binding affinities of novel ligands*, J. Am. Chem. Soc., 118 (1996) 3959–3969.
9. Anzali, S., Barnickel, G., Krug, M., Sadowski, J., Wagener, M., Gastaiger, J. and Polanski, J., *The comparison of geometric and electronic properties of molecular surfaces by neural networks: Application to the analysis of corticosteroid-binding globulin activity of steroids*, J. Comput.-Aided Mol. Des., 10 (1996) 521–534.
10. Rogers, D.R. and Hopfinger, A.J., *Application of genetic function approximation to quantitative structure–activity relationships and quantitative structure–property relationships*, J. Chem. Inf. Comput. Sci., 34 (1994) 854–866.
11. Silverman, B.D. and Platt, D.E., *Comparative molecular moment analysis (CoMMA): 3D-QSAR without molecular superposition*, J. Med. Chem., 39 (1996) 2129–2140.
12. Ortiz, A.R., Pisabarro, M.T., Gago, F. and Wade, R., *Prediction of drug binding affinities by comparative binding energy analysis*, J. Med. Chem., 38 (1995) 2681–2691.
13. Gusso, R., Pattabiraman, N., Zaharevitz, D.W., Kellogg, G.E., Topol, I.A., Rice, W.G., Schaeffer, C.A., Erickson, J.W. and Burt, S.K., *All-atom models for the non-nucleoside binding site of HIV-1 reverse transcriptase complexed with inhibitors: A 3D QSAR approach*, J. Med. Chem., 39 (1996) 1645–1650.
14. Klebe, G. and Abraham, U., *On the prediction of binding properties of drug molecules by comparative molecular field analysis*, J. Med. Chem., 36 (1993) 70–80.
15. DePriest, S.A., Mayer, D., Naylor, C.B. and Marshall, G.R., *3D-QSAR of angiotensin-converting enzyme and thermolysin inhibitors: A comparison of CoMFA models based on deduced and experimentally determined active site geometries*, J. Am. Chem. Soc., 115 (1993) 5372–5384.
16. Folkers, G., Merz, A. and Rognan, D., *CoMFA: Scope and limitations*, In Kubinyi, H. (Ed.) 3D QSAR in drug design: Theory, methods and applications, ESCOM, Leiden, The Netherlands, 1993, pp. 583–618.
17. Waller, C.L., Oprea, T.I., Giolitti, A. and Marshall, G.R., *Three-dimensional QSAR of human immunodeficiency virus (I) protease inhibitors: 1. A CoMFA study employing experimentally-determined alignment rules*, J. Med. Chem., 36 (1993) 4152–4160.
18. Waller, C.L. and Marshall, G.R., *Three-dimensional quantitative structure–activity relationship of angiotensin-converting enzyme and thermolysin inhibitors: 2. A comparison of CoMFA models incorporating molecular orbital fields and desolvation free energies based on active-analog and complementary-receptor-field alignment rules*, J. Med. Chem., 36 (1993) 2390–2403.
19. Oprea, T.I., Waller, C.L. and Marshall, G.R., *Three-dimensional quantitative structure–activity relationship of human immunodeficiency virus (I) protease inhibitors: 2. Predictive power using limited exploration of alternative binding modes*, J. Med. Chem., 37 (1994) 2206–2215.
20. Brandt, W., Lehmann, T., Willkomm, C., Fittkau, S. and Barth, A., *CoMFA investigation on two series of artificial peptide inhibitors of the serine protease thermitase*, Int. J. Peptide Protein Res., 46 (1995) 73–78.
21. Kroemer, R.T., Ettmayer, P. and Hecht, P., *3D-quantitative structure–activity relationships of human immunodeficiency virus type-1 protease inhibitors: Comparative molecular field analysis of 2-hetero-substituted statine derivatives — implications for the design of novel inhibitors*, J. Med. Chem., 38 (1995) 4917–4928.
22. Cruciani, G. and Watson, K.A., *Comparative molecular field analysis using GRID force-field and GOLPE variable selection methods in a study of inhibitors of glycogen phosphorylase b*, J. Med. Chem., 37 (1994) 2589–2601.
23. Gamper, A.M., Winger, R.H., Liedl, K.R., Sotriffer, C.A., Varga, J.M., Kroemer, R.T. and Rode, B.M., *Comparative molecular field analysis of haptens docked to the multispecific antibody IgE (Lb4)*, J. Med. Chem., 39 (1996) 3882–3888.
24. Goodsell, D.S. and Olson, A.J., *Automated docking of substrates to proteins by simulated annealing*, Proteins: Struct. Funct. Genet., 8 (1990) 195–202.
25. Cho, J.-C., Garsia, M.L.S., Bier, J. and Tropsha, A., *Structure-based alignments and comparative molecular field analysis of acetylcholinesterase inhibitors*, J. Med. Chem., 39 (1996) 5064–5071.

26. Kroemer, R.T. and Hecht, P., *A new procedure for improving the predictiveness of CoMFA models and its application to a set of dihydrofolate reductase inhibitors*, J. Comput.-Aided Mol. Design, 9 (1995) 396–406.
27. Kroemer, R.T., Hecht, P., Guessregen, S. and Liedl, K.R., *Improving the predictive quality of CoMFA models*, In Kubinyi, H., Folkers, G. and Martin, Y.C. (Eds.) 3D QSAR in drug design: Vol. 3, Kluwer Academic Publishers, Dordrecht, The Netherlands, 1998, pp. 41–56.
28. Norinder, U., *3D-QSAR investigation of the tripos benchmark steroids and some protein-tyrosine kinase inhibitors of styrene type using the TDQ approach*, J. Chemometrics, 10 (1996) 533–545.
29. Jain, A.N., Koile, K. and Chapman, D., *Compass: Predicting biological activities from molecular surface properties. Performance comparisons on a steroid benchmark*, J. Med. Chem., 37 (1994) 2315–2327.
30. Catalyst, Molecular Simulations Inc., San Diego, CA, U.S.A.
31. Norinder, U., *The alignment problem in 3D-QSAR: A combined approach using catalyst and a 3D-QSAR technique*, In Sanz, F., Giraldo, J. and Manaut, F. (Eds.) QSAR and molecular modeling: Concepts, computational tools and biological applications, Prous Science Publishers, Barcelona, Spain, 1995, pp. 433–438.
32. Palomer, A., Giolitti, A., Garcia, M.L., Cabre, F., Mauleon, D. and Carganico, G., *Molecular modeling and CoMFA investigations on LTD4 receptor antagonists*, In Sanz, F., Giraldo, J. and Manaut, F. (Eds.) QSAR and molecular modeling: Concepts, computational tools and biological applications, Prous Science Publishers, Barcelona, Spain, 1995, pp. 444–450.
33. Hoffmann, R.D. and Langer, T., *Use of the Catalyst program as a new alignment tool for 3D-QSAR*, In Sanz, F., Giraldo, J. and Manaut, F. (Eds.) QSAR and molecular modeling: concepts, computational tools and biological applications, Prous Science Publishers, Barcelona, Spain, 1995, pp. 466–469.
34. For a review of methods of alignments of molecules see Klebe, G., *Structural alignment of molecules*, In Kubinyi, H. (Ed.) 3D QSAR in drug design: Theory, methods and applications, ESCOM, Leiden, The Netherlands, 1993, pp. 173–199.
35. Klebe, G., Abraham, U. and Mietzner, T., *Molecular similarity indices in a comparative analysis (CoMSIA) of drug molecules to correlate and predict their biological activity*, J. Med. Chem. 37 (1994) 4130–4146.
36. Kellogg, G.E., Semus, S.F. and Abraham, D.J., *HINT: A new method of empirical field calculation of CoMFA*, J. Comput.-Aided Mol. Design, 5 (1991) 545–552.
37. Kellogg, G.E. and Abraham, D.J., *Hydrophobic fields*, In Kubinyi, H. (Ed.) 3D QSAR in drug design: Theory, methods and applications, ESCOM, Leiden, The Netherlands, 1993, pp. 506–522.
38. Goodford, P.J., *A Computational procedure for determining energetically favorable binding sites on biologically important macromolecules*, J. Med. Chem., 28 (1985) 849–857.
39. Wade, R.C., *Molecular interaction fields*, In Kubinyi, H. (Ed.) 3D QSAR in drug design: Theory, methods and applications, ESCOM, Leiden, The Netherlands, 1993, pp. 486–505.
40. Kim, K.H., Greco, G., Novellino, E., Silipo, C. and Vittoria, A., *Use of the hydrogen bond potential function in a comparative molecular field analysis (CoMFA) on a set of benzodiazepines*, J. Comput.-Aided Mol. Design, 7 (1993) 263–280.
41. Davis, A.M., Gensmantel N.P., Johansson, E. and Marriott, D.P., *The use of the GRID program in the 3D QSAR analysis of a series of calcium-channel agonists*, J. Med. Chem., 37 (1994) 963–972.
42. Kim, K.H., *A novel method of describing hydrophobic effects directly from 3D structures in 3D-quantitative structure–activity relationships study*, Med. Chem. Res., 1 (1991) 259–264.
43. Kim, K.H., *3D-Quantitative structure–activity relationships: Describing hydrophobic interactions directly from 3D structures using a comparative molecular field analysis (CoMFA) approach*, Quant. Struct.-Act. Relat., 12 (1993) 232–238.
44. Kenny, P.W., *Prediction of hydrogen bond basicity from computed molecular electrostatic properties: Implications for comparative molecular field analysis*, J. Chem. Soc. Perkin Trans., 2 (1994) 199–202.
45. Fuchère, J.L., Quarendon, P. and Kaetterer, L.J., *Estimating and representing hydrophobicity potential*, J. Mol. Graph., 8 (1988) 202–206.
46. For a recent review see Testa, B., Carrupt, P.A., Gaillard, P., Billois, F. and Weber, P., *Lipophilicity in molecular modeling*, Pharm. Res., 13 (1996) 335–343.
47. Gaillard, P., Carrupt, P.A., Testa, B. and Schambel, P., *Binding of arylpiperazines, (aryloxy) propanolamines and tetrahydropyridyl-indoles to the 5-HT$_{1A}$ receptor: Contribution of the molecular*

lipophilicity potential to three-dimensional quantitative structure–activity relationship models, J. Med. Chem., 39 (1996) 126–134.
48. Kneubühler, S., Thull, U., Altomare, C., Carta, V., Gaillard, P., Carrupt, P.A., Carotti, A. and Testa, B., *Inhibition of monoamine oxidase-B by 5H-indeno[1,2-c]pyridazine derivatives: Biological activities, quantitative structure–activity relationships (QSARs) and 3D-QSARs*, J. Med. Chem., 38 (1995) 3874–3883.
49. Thull, U., Kneubühler, S., Gaillard, P., Carrupt, P.A., Testa, B., Altomare, C., Carotti, A., Jenner, P. and McNaught, K.S.P., *Inhibition of monoamine oxidase by isoquinoline derivatives: Qualitative and 3D-quantitative structure–activity relationships*, Biochem. Pharmacol., 50 (1995) 869–877.
50. Masuda, T., Nakamura, K., Jikihara, T., Kasuya, F., Igarashi, K., Fukui, M., Takagi, T. and Fujiwara, H., *3D-quantitative structure–activity relationships for hydrophobic interactions: Comparative molecular field analysis (CoMFA) including molecular lipophilicity potentials as applied to the glycine conjugation of aromatic as well as aliphatic carboxylic acids*, Quant. Struct.-Act. Relat., 15 (1996) 194–200.
51. Norinder, U., *Experimental design based 3-D QSAR analysis of steroid–protein interactions: Application to human CBG complexes*, J. Comput.-Aided Mol. Design, 4 (1990) 381–389.
52. Poso, A., Tuppurainen, K. and Gynther, J., *Modeling of molecular mutagenicity with comparative molecular field analysis (CoMFA): Structural and electronic properties of MX compounds related to TA100 mutagenicity*, J. Mol. Struc. (Theochem), 304 (1994) 255–260.
53. Navajas, C., Poso, A., Tuppurainen, K. and Gynther, J. *Comparative molecular field analysis (CoMFA) of MX compounds using different semi-empirical methods: LUMO field and its correlation with mutagenic activity*, Quant. Struct.-Act. Relat., 15 (1996) 189–193.
54. Kim, K.H. and Martin, Y.C., *Direct prediction of linear free energy substituent effects from 3D structures using comparative molecular field analysis: 1. Electronic effects of substituted benzoic acids*, J. Org. Chem., 56 (1991) 2723–2729.
55. Kim, K.H. and Martin, Y.C., *Direct prediction of dissociation constants (pK_a's) of clonidine-line imidazolines, 2-substituted imidazoles, and 1-methy-2-substituted-imidazoles from 3D structures using a comparative molecular field analysis (CoMFA) approach*, J. Med. Chem., 34 (1991) 2056–2060.
56. Kroemer, R.T., Hecht, P. and Liedl, K.R., *Different electrostatic descriptors in Comparative molecular field analysis: A comparison of molecular electrostatic and coulumb potentials*, J. Comput. Chem., 17 (1996) 1296–1308.
57. Gasteiger, J. and Marsili, M., *Iterative partial equalization of orbital electronegativity — a rapid access to atomic charges*, Tetrahedron, 36 (1980) 3219–3288.
58. Kellogg, G.E., Kier, L.B., Gaillard, P. and Hall, L.H., *E-state fields: Applications to 3D QSAR*, J. Comput.-Aided Mol. Design, 10 (1996) 513–520.
59. Hall, L.H. and Kier, L.H., *Binding of salicylamides: QSAR analysis with electrotopological state indices*, Med. Chem. Res., 2 (1992) 497–502.
60. Delphi, Molecular simulations, San Diego, CA, U.S.A.
61. Gilson, M.K. and Honig, B.H., *Calculations of electrostatic potentials in an active site*, Nature, 330 (1987) 84–86.
62. Waller, C.L. and Kellogg, G.E., *Adding chemical information to CoMFA models with alternative 3D QSAR fields*, NetSci, January 1996: http://www.awod.com/netsci/Science/Compchem/feature10.html.
63. Kroemer, R.T. and Hecht, P., *Replacement of steric 6-12 potential-derived interaction energies by atom-based indicator variables in CoMFA leads to models of higher consistency*, J. Comput.-Aided Mol. Design, 9 (1995) 205–212.
64. Wold, S., Johansson, E. and Cocchi, M., *PLS — partial least-squares projections to latent structures*, In Kubinyi, H. (Ed.) 3D QSAR in drug design: Theory, methods and applications, ESCOM, Leiden, The Netherlands, 1993, pp. 523–550.
65. Baroni, M., Constantino, G., Cruciani, G., Riganelli, D., Valigi, R. and Clementi, S., *Generating optimal linear PLS estimations (GOLPE): An advanced chemometric tool for handling 3D QSAR problems*, Quant. Struct.-Act. Relat., 12 (1993) 9–20.
66. Norinder, U., *Single and domain mode variable selection in 3D QSAR applications*, J. Chemometrics, 10 (1996) 95–105.

67. Nordén, B., Svensson, P. and Carter, R.E., oral presentation at the 10th European Symposium on Structure–Activity Relationships, Barcelona, 1994.
68. Cho, S.-J. and Tropsha, A., *Cross-validated R^2-guided region selection for comparative molecular field analysis: A simple method to achieve consistent results*, J. Med. Chem., 38 (1995) 1060–1066.
69. Cruciani, G., Pastor, M. and Clementi, S., *Region selection in 3D QSAR*, In van der Waterbeemd, H. (Ed.) Computer lead finding and optimization: Proceedings of the 11th European Symposium on Structure–Activity Relationships, Wiley-VCH, Basel, Switzerland, 1977, pp. 379–395.
70. Pastor, M., Cruciani, G. and Clementi, S., *Smart Region Definition SRD: A new way to improve the predictive ability and interpretability of three-dimensional quantitative structure–activity relationships*, J. Med. Chem., 40 (1997) 1455–1464.
71. Cho, S.-J., Tropsha, A., Suffness, M., Cheng Y.-C. and Lee, K.-H., *Antitumor agents: 163. Three-dimensional quantitative structure–activity relationship study of 4′-O-demethylepipodophyllotoxin analogs using the modified CoMFA/q^2*-GRS approach, J. Med. Chem., 39 (1996) 1383–1395.
72. Sutter, J.M., Dixon, S.L. and Jurs, P.C., *Automated descriptor selection for quantitative structure–activity relationships using generalized simulated annealing*, J. Chem. Inf. Comput. Sci., 35 (1995) 77–84.
73. Rännar, S., Lindgren, F., Geladi, P. and Wold, S., *A PLS kernel algorithm for data sets with many variables and fewer objects: Part 1. Theory and algorithm*, J. Chemometrics, 8 (1994) 111–125.
74. Rännar, S., Geladi, P., Lindgren, F. and Wold, S., *A PLS kernel algorithm for data sets with many variables and fewer objects: Part 2. Cross-validation, missing data and examples*, J. Chemometrics, 9 (1995) 459–470.
75. Bush, B.L. and Nachbar, Jr., R.B., *Sample-distance partial least squares: PLS optimized for many variables, with application to CoMFA*, J. Comput.-Aided Mol. Design, 7 (1993) 587–619.
76. See the chapter by F. Lindgren and S. Rännar in this volume, pp. 105–113, for a more detailed presentation of kernel PLS methods.
77. Bro, R., *Multiway calibration: Multilinear PLS*, J. Chemometrics, 10 (1996) 47–61.
78. Nilsson, J., Bro, R., Wikström, H. and Smilde, A., *A comparison between multi-way PLS and GOLPE utilised as variable selection tools, applied on GRID-parameters from a set of compounds with affinity for the dopamine D_3 receptor subtype*. Poster presentation at the 11th European symposium on Structure–Activity Relationships, Lausanne, 1996.
79. Nilsson, J. and Smilde, A., *Multiway calibration in 3D QSAR*, J. Chemometrics (in press).
80. Nilsson, J., personal communication.
81. Dunn III, W.J., Hopfinger, A.J., Catana, C. and Duraiswami, C., *Solution of the conformation and alignment tensors for the binding of trimethoprim and its analogs to dihydrofolate reductase: 3D-quantitative structure–activity relationship study using molecular shape analysis, 3-way partial least squares regression, and 3-way factor analysis*, J. Med. Chem. 39 (1996) 4825–4832.

Improving the Predictive Quality of CoMFA Models

Romano T. Kroemer[a*], Peter Hecht[b*], Stefan Guessregen[b] and Klaus R. Liedl[c]

[a] *Physical and Theoretical Chemistry Laboratory, University of Oxford, South Parks Road, Oxford OX1 3QZ, U.K.*

[b] *Tripos GmbH, Martin-Kollar-Str. 15, D-81829 Munich, Germany*

[c] *Department of General, Inorganic and Theoretical Chemistry, University of Innsbruck, Innrain 52a, A-6020 Innsbruck, Austria*

1. Introduction

Comparative molecular field analysis (CoMFA) [1] has proven a very useful QSAR technique in the field of medicinal chemistry, as indicated by many publications over the past years. At the time of introduction, its two cornerstones were probably not novel *per se*, but their combination certainly was. Molecules are described by three-dimensional (3D) fields evaluated over a grid of points, and only steric and electrostatic fields were used initially. This description leads to over-squared matrices containing the corresponding field values. Therefore, in order to correlate these data with some target properties (such as biological activities), a statistical method was applied which is referred to as partial least squares (PLS) [2–4]. PLS is able to extract linear equations from over-squared matrices by applying a latent model technique. This statistical technique was combined with cross-validation (CV) in order to evaluate the predictive quality of the resulting method, using the training set as an internal test set [5–7].

Despite its enormous success, various attempts have been made to further improve the predictive quality of CoMFA. Related to these topics are two major points: (i) how can the degree of predictive quality for a given model be analyzed?; and (ii) is it possible to improve the predictive quality of a CoMFA without losing general applicability, in particular the ability to predict the activities of novel molecules?

2. Analysis of the Predictive Quality of a Given Model

The first CoMFA studies were performed on rather small datasets (smaller than 50 molecules) [8]. Normally, in order to assess the internal predictive quality (consistency), cross-validation with the leave-one-out (LOO) method has been applied. This implies that each compound is excluded once from the dataset and predicted by the sub-model generated from the remaining molecules. In other words, each compound serves once as an internal test set. Of course, this method has the advantage of being reproducible, as opposed to the random selection of internal training and test sets. However, large datasets have a higher probability of considerable pairwise similarity of compounds.

---

*To whom correspondence should be addressed.

H. Kubinyi et al. (eds.), 3D QSAR in Drug Design, Volume 3. 41–56.

Therefore, the LOO method could lead to overfitting of the data in these cases, depending on the similarity distribution of the training set, and it might be necessary to employ other cross-validation strategies.

3. Improvement of Predictive Quality without Loss of General Applicability

There are several points where the predictive quality of CoMFA might be improved. One problem associated with PLS is its noise-sensitivity [9], which might have an impact on the predictive quality of the model. Also, very basic descriptors — i.e. the Lennard-Jones 6-12 potential and the Coulomb potential — are normally used in CoMFA. Another point is that CoMFA is very dependent on the alignment rule. Furthermore, one might have to deal with an intrapolation versus extrapolation problem. Having an analysis which is internally consistent does guarantee good predictions within the data space covered by the training set (intrapolation), but does not guarantee good predictions for compounds outside the data space of the training set (extrapolation).

3.1. Description of molecules

Usually two different descriptor types, the steric and electrostatic fields, have been used in CoMFA. The steric interaction energy between the probe and the molecules is described by a Lennard-Jones 6-12 potential. This potential is characterized by a very steep slope of the function in the repulsive part (i.e. near the molecules). The electrostatic descriptors calculated are dependent on partial charges assigned to the atoms of the molecules under investigation.

3.2. Alignment of molecules

Probably the most crucial point for performing a successful CoMFA study is the alignment of the molecules, as it determines the field values calculated. The basic idea is to superimpose the molecules in the orientation that they are thought to bind to the (putative) receptor. However, a strict alignment rule cannot account for the receptor flexibility and, in some cases, there is no unique alignment rule.

3.3. Analysis of molecules/descriptors

Another question with respect to CoMFA is: are there ways to overcome the noise sensitivity of PLS? Noise, in this context, means that parts of the molecules are included in the description which are not relevant for biological activity. In some cases, this noise might even overwhelm the field values important for a proper description of the target property. Therefore, it is desirable to focus only on the relevant parts of the molecules.

3.4. Reliability of the predictions

As mentioned above, one might have the problem of internal consistency versus general predictive quality, the intrapolation versus extrapolation problem. Intrapolations and their assessment can be handled by the cross-validation approach. With respect to extrapolations, one needs to consider how dissimilar a compound is to the training set. The higher the degree of dissimilarity, the more uncertain the prediction will become.

In the following, we will focus on the topics introduced above and describe some of the attempts made in this context. However, we would like to point out at this stage that ideally any method aiming at an improvement of predictive quality in CoMFA should not focus only on the training set, the method should improve the predictive quality for test compounds as well. In order to avoid subjective interference, one might envisage incorporation of the method in an automated process.

4. Results

4.1. Analysis/assessment of predictive quality

The potential problems with cross-validation of large datasets and an analysis of the predictive quality have been illustrated by a recent study of HIV-protease inhibitors [10]; in this study, 100 compounds served as a training set. Using the LOO method fairly high cross-validated r^2 (r^2_{cv}) values between 0.572 and 0.593 were achieved using different field types and grid spacings.

However, the LOO method might lead to high r^2_{cv} values which do not necessarily reflect a general predictive quality of the underlying model [5–7]. Therefore, analyses with two cross-validation groups were performed: each of the respective sub-models consisted of 50% of the compounds (randomly selected) and the remaining ones were predicted. As the random formation of cross-validation groups might have an impact on the results, this kind of analysis was repeated 100 times for the analyses mentioned above with an identical set of cross-validation groups, respectively (Table 1). The mean r^2_{cv} for each of the 100 runs was slightly lower compared to the values obtained with the LOO method, and the standard deviation for these values was rather low. Nevertheless, in all three cases a few analyses with a rather poor r^2_{cv} could be obtained indicating a certain degree of inconsistency in the underlying dataset. On the other hand, a few higher r^2_{cv} values were obtained, too. These 'extrema' were found with identical cross-validation groups within the different analyses.

Table 1 *Summary of the CoMFA study on HIV-protease inhibitors, reference [10]*

			r^2_{cv} [a], $cv = 2$, 100 runs			
Analysis[b]	LOO r^2_{cv}	q^2	Mean	Std dev.[c]	High[d]	Low[e]
1	0.593	0.569	0.452	0.094	0.629	0.022
2	0.574	0.555	0.433	0.108	0.659	–0.063
3	0.572	0.552	0.436	0.104	0.643	–0.034

[a] Highest r^2_{cv} found within the first five components extracted.
[b] Denoted as analyses 1, 6, and 9 in the original paper.
[c] Standard deviation.
[d] Highest value found.
[e] Lowest value found.

An interesting conclusion from this study can be drawn by comparing the averaged r^2_{cv} values with the predictive r^2 (q^2) values for the test set. While the r^2_{cv} values obtained with the LOO method are higher, the averaged r^2_{cv} gives a conservative estimate of the q^2 to be expected, verified in this case by test sets. This indicates that the averaged r^2_{cv} values are, indeed, a better measure of the predictive quality of the CoMFA model, even without confirmation by the prediction of a suitable test set. Furthermore, the spread of the r^2_{cv} values gives an indication of the internal data structure of the set investigated.

4.2. Methods to improve predictive quality: description of molecules

The most common field types used in CoMFA are the steric and electrostatic fields. However, other field types have also been introduced such as hydrophobic fields [11]. In the following, we concentrate on the steric and electrostatic fields and their manipulation in order to improve the results.

4.2.1. Steric descriptors

As the steep increase of the Lennard-Jones 6-12 potential might lead to high variances in energy values at grid-points near the molecules, several attempts have been made to deal with this problem. So it has been suggested to truncate the probe-ligand steric energies to 4.0 or 5.0 kcal/mol, as opposed to the 30.0 kcal/mol standard cutoff in SYBYL-CoMFA [12–14]. A different method was the generation of 'shape potentials' in combination with PLS by Floersheim et al. [15]. Here the values of either 1 or 0 were assigned to grid-points, depending on whether the grid-point is within, or not within, the van der Waals radius of any atom of the molecule in a predefined grid (distance of the lattice intersections: 2.0 Å) [16].

In another approach the Lennard-Jones potentials were replaced by variables indicating the presence of an atom in predefined volume elements (cubes) within the region enclosing the ensemble of superimposed molecules [17]. The resulting 'atom indicator vectors' were used as steric fields in the subsequent PLS analyses (Fig. 1).

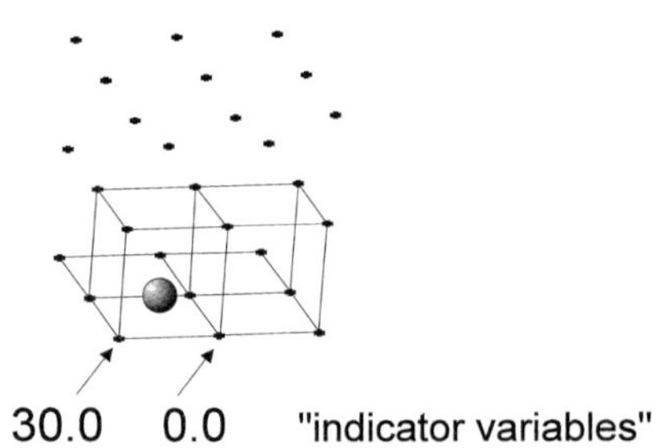

Fig. 1. Schematic representation of the atom mapping procedure. The sphere represents an atom centre. If an atom (i.e. its center) is found within one of the cubes, a value of 30.0 is assigned to the corresponding grid-point.

Five training sets (80 compounds each) and five test sets (60 compounds each), randomly selected from an ensemble of 256 dihydrofolate reductase inhibitors, were investigated. Two different grid positions and four different grid spacings (2.0, 1.0, 0.75 and 0.57 Å) were used and compared to the standard fields at these positions, also applying different cutoffs. The analyses were performed with and without the inclusion of standard electrostatic fields.

The trends derived from this study (Table 2) can be summarized as follows. (i) In the CoMFAs with the standard 6–12 potentials a reduction of the grid spacing did not lead to an improvement of the statistical parameters (r^2_{cv} and predictive r^2). This result was, in fact, no surprise, as it is known that a reduction of the lattice spacing does not improve r^2_{cv} [18–21]; most of the associated increase in field information is noise in so far as a PLS correlation is concerned. (ii) In contrast, for the analyses using indicator fields, narrower lattice spacings resulted in a significant increase of the r^2_{cv} and predictive r^2 values. (iii) The attempt to improve the standard CoMFAs by truncating the probe-ligand steric energies at a value lower than the default setting (5.0 instead of 30.0) did not yield significant improvements. (iv) Comparison of the results obtained with the two different steric field types after inclusion of electrostatic descriptors indicated that the analyses with the indicator fields were still superior. (v) The analyses with indicator fields showed, in some cases, a significant dependency on the grid position used. However, at both positions investigated they were superior to those using Lennard-Jones derived fields.

On average, for the analyses using indicator fields, the grid spacing of 0.75 Å gave the best results. In many cases, at a narrower distance of the lattice intersections (0.57 Å), a decrease of the statistical parameters became apparent. This phenomenon may be interpreted as a compromise of two contrary developments: on the one hand, the shape of the structures should be described exactly; and on the other hand, the degree of differentiation should not be too high. Atoms of different molecules which are located at almost identical positions in space should be described as being equal. A very fine grid will differentiate such atoms and puts the corresponding indicator values into different columns of the descriptor matrix, thus describing these two atoms as not

Table 2 *Comparison of two different steric descriptor types in a CoMFA, reference [17]*

	Lennard-Jones fields				Indicator fields			
	Best[a]		Worst[a]		Best[a]		Worst[a]	
Set	r^2_{cv}	q^2	r^2_{cv}	q^2	r^2_{cv}	q^2	r^2_{cv}	q^2
1	0.160	0.450	0.007	0.282	0.334	0.595	0.150	0.669
2	0.534	0.214	0.378	0.063	0.594	0.517	0.422	0.214
3	0.331	0.386	0.207	−0.054	0.543	0.466	0.251	0.343
4	0.359	0.535	0.113	0.151	0.574	0.665	0.308	0.650
5	0.633	0.506	0.451	0.171	0.668	0.498	0.445	0.455

[a] q^2 values listed are of the corresponding analyses.

superimposable. But this was not the intention of the method, since it was intended to level out high differences in the descriptors for 'similar' atoms. Therefore, the grid spacing of 0.75 Å appeared to be the best compromise between exactness of shape description and inaccuracy in differentiation of atoms.

4.2.2. Electrostatic description

The other field type normally used in CoMFA contains the Coulomb potential between the probe and the molecules bearing atom centered point charges. However, the assignment of atomic electron populations has been a subject of intensive discussion for two reasons: first, it is *per se* problematic to represent the electrostatic properties of molecules by atomic charges, thus exaggerating an ionic character of the bonds; and second, the charge calculation methods themselves have been discussed very often, in particular because of the partitioning schemes which are applied.

Due to the wide variety of charge calculation methods available and the fundamental differences in their algorithms, the electrostatic fields derived from them also show significant differences. Therefore, a variety of charge calculation methods was applied to a dataset consisting of 37 ligands of the benzodiazepine receptor inverse agonist/antagonist active site [22,23], and a CoMFA study was performed [24]. The charge calculation methods included Gasteiger-Marsilli [25], semiempirical (MNDO [26], AM1 [27] and PM3 [28]) and *ab initio* (HF/STO-3G, HF/3-21G* and HF/6-31G*) charges. Semiempirical and also *ab initio* electron populations, were derived both from the Mulliken Population Analysis (MPA) [29] and from fitting the charges to the molecular electrostatic potential (ESPFIT charges) [30–33]. In addition, the molecular electrostatic potentials (MEPs) resulting from *ab initio* calculations were mapped directly onto the CoMFA grid. In order to estimate to what extent the results were affected by variations in the statistical parameters, two different column filters and scaling options were applied.

The results obtained in this study can be summarized as follows. With regard to the r^2_{cv} values of the resulting QSAR models, the ESPFIT-derived potentials yielded generally higher r^2_{cv} values than those resulting from MPA charges. For example, at the HF/3-21G* level the r^2_{cv} rose from 0.61 (MPA-derived potentials) to 0.76 (ESPFIT fields). The MEPs mapped directly onto the CoMFA grid were not superior to the corresponding ESPFIT-derived potentials. Semiempirical ESPFIT charges appeared to be of similar quality compared with *ab initio* ESPFIT electron populations in the CoMFAs.

Another important result was the fact that the electrostatic coefficient contour map of the QSAR might be significantly influenced by the charge-calculation method applied. For example, a comparison of the coefficient contour map of an analysis derived from HF/6-31G*/MEP descriptors with the one generated using HF/3-21G*/MPA charges showed remarkable differences. Despite a low correlation coefficient of 0.66, reversal of the sign of the contours within a certain region was also found (Fig. 2). This is certainly a result which must be kept in mind when interpreting the contour maps of a CoMFA study.

Also of interest was the finding that when no scaling between steric and electrostatic descriptors was applied, the analyses were significantly affected, in particular with

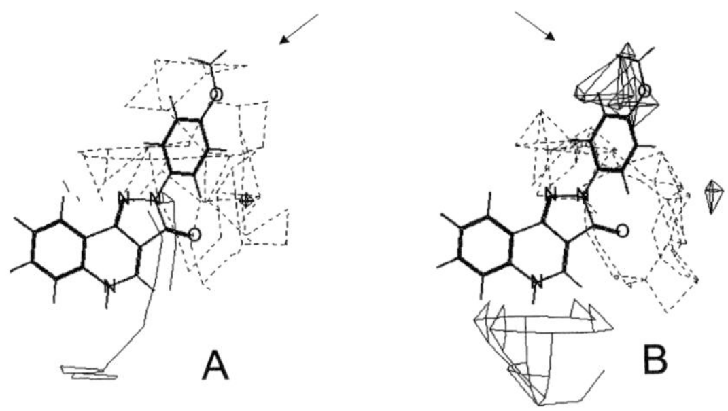

Fig. 2. CoMFA coefficient fields resulting from analyses including steric and electrostatic potentials, the latter being HF/3-21G*/MPA Coulomb derived (A) and HF/6-31G*/MEP (B) fields. Negative contours (values < -0.003) are represented by broken lines, positive contours (values $> +0.003$) are shown with solid lines. The arrows indicate the region where significant differences between the contour maps can be found.

respect to the contributions of the electrostatic fields. In this case, a direct correlation between magnitude of electrostatic field values and contribution of these descriptors was observed. When discussing the problem of calculating partial atomic charges, one may distinguish between two aspects: on the one hand, the 'quality' of the charges — i.e. their sign (whether they are positive or negative) and their relative magnitudes; and on the other hand, the 'quantity' of the charges — i.e. their absolute values, or the scaling factor between different calculation methods. By scaling the steric and electrostatic descriptor matrices relative to each other in CoMFA, the actual physico-chemical relevance (e.g. the binding enthalpy of the molecules to a putative receptor) gets lost. However, since it is difficult to decide what is the 'correct' magnitude of partial charges, it is justified to apply such a scaling procedure (which is, in fact, usually done), especially when application of scaling leads to more consistent results.

4.3. Methods to improve predictive quality: alignment of molecules

Certainly the crucial problem in CoMFA is to generate a proper alignment of the molecules investigated [1]. In many cases, the datasets contain fairly similar molecules [34–37] where an atom-based alignment or methods like the 'active analog approach' are sufficient for obtaining good correlations. However, different methods or considerations are, in some cases, necessary in order to perform a successful study or to improve the predictive quality.

4.3.1. Alignment via automated pharmacophore analysis

In a recent study, a set of uncompetitive N-methyl-D-aspartate (NMDA) receptor antagonists was investigated applying CoMFA [38]. The dataset comprised a number of structurally very diverse compounds (Fig. 3). Therefore, the molecules were subjected first to a pharmacophore analysis using the DISCO method [39]. One of the features of this method is that putative receptor residues interacting with the molecules are taken into account as well. This analysis does not only yield a pharmacophore model, but also generates an alignment which can be used for a subsequent CoMFA study.

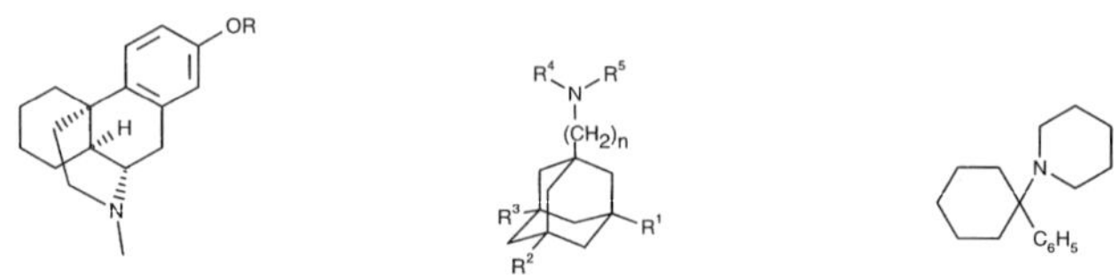

Dextrorphan, R = H   Adamantan amine   Phencyclidine (PCP)

Fig. 3. Examples for the structural variety of the NMDA *inhibitor set investigated.*

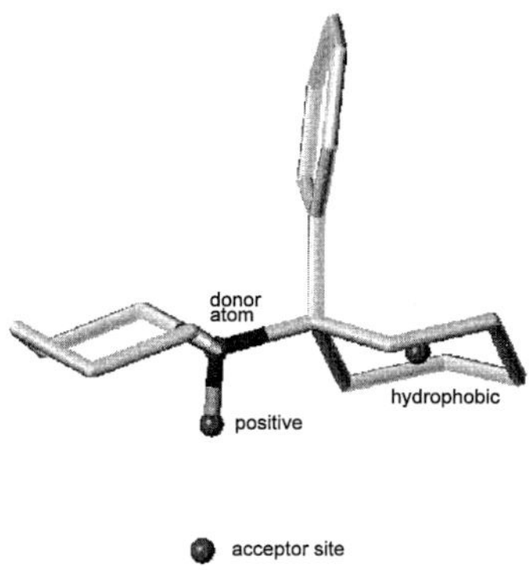

Fig. 4. Representation of the 4-point pharmacopore used as a basis for the alignment in the subsequent CoMFA.

The resulting QSAR proved to be highly consistent, as indicated by r^2_{cv} value of 0.72. This was not only important with respect to inner consistency and predictive quality, but also supported the validity of the pharmacophore model it was based upon (Fig. 4). Furthermore, the CoMFA proved not only to be self-consistent, but could also be used to predict the activities of several other molecules with good accuracy. Noteworthy also in this context is the fact that the predicted molecules were unique in some aspects compared to the training set. Apparently, their alignment via the pharmacophore model generated was good enough for a successful prediction. In general, this study indicated the usefulness of an automated pharmacophore analysis for generating an alignment as a basis for a consistent CoMFA.

4.3.2. Alignment via automated docking to a receptor

Lately, a very different strategy has been applied in order to generate an alignment for a CoMFA study [40]. In this case, structurally very diverse antigens were docked to the receptor structure of IgE(Lb4) using the automated docking program AUTODOCK [41].

The antigens investigated covered a very large property space, ranging from DNP-substituted amino acids to diaspirin, and from negatively charged molecules such as hemimellitic acid to double positive prolonium iodide (Fig. 5). Initial trials to superimpose these diverse molecules applying systematic conformational searches (using dis-

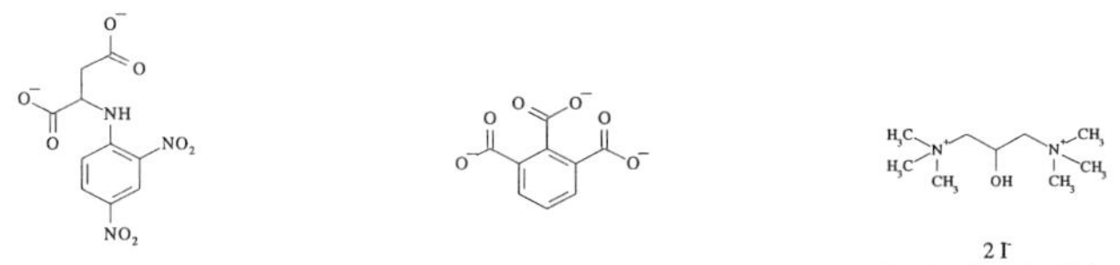

DNP-Asp Hemimellitic acid Proloniumiodide

Fig. 5. Examples for the structural variety among the IgE binding haptens.

Table 3 *Results from a CoMFA based on automated docking of molecules to the receptor as an alignment procedure, reference [40]*

Model	Probe atom (charge)	Grid spacing	r^2_{cv}	k[a]	r^2 [b]	Steric contribution[c]
A	C sp^3 (+)	2 Å	0.785	5	0.974	0.467
B	C sp^3 (+)	1 Å	0.773	5	0.972	0.498
C	C sp^3 (+)	2 Å	0.809	6	0.994	0.513
D	H (+)	2 Å	0.773	6	0.991	0.468

[a] Optimum number of components.
[b] Conventional r^2 determined with this number of components.
[c] The sum of steric and electrostatic contributions equals 1.0.

tance maps in an 'active analog approach') [42] or field-fitting approaches did not yield satisfactory QSAR analyses. Therefore, the results of docking experiments were used instead, a procedure that proved to be very successful. Remarkably, this alignment method yielded highly consistent QSAR models, as shown in Table 3.

In some cases, the docking program had delivered several docked orientations for a particular molecule. In these instances, the orientation yielding the best r^2_{cv} value was included in the model. Therefore, the question was raised whether the high consistency of the initial QSAR model generated was an artefact in the sense that the alignment of each compound was chosen with respect to a constant grid definition. In order to address this question, several analyses with altered grids were carried out (models A through C, Table 3), but all showed good internal consistency.

In addition to the grid variations, an analysis was carried out using a proton as probe atom. This was done in order to obtain an estimate of the importance of hydrogen bonding in the ligand–receptor interactions. The corresponding r^2_{cv} was of similar magnitude as the other models.

The best test for the general validity of a QSAR analysis is to predict the activity of molecules which were not members of the training set. Therefore, the activity of three additional compounds was predicted. Despite the fact that the new structures were unique compared to the training set, all CoMFA models were able to predict the activities of these molecules rather accurately, indicating a high predictive quality of the analyses. This was also confirmed by comparing root mean square errors of training and test sets.

In conclusion, the most important aspect of this study was the fact that conventional alignment had failed, but an automated docking procedure was able to provide a basis for a consistent and predictive CoMFA.

4.3.3. Incorporation of receptor flexibility

One of the basic ideas behind CoMFA is to align the structures in the way they are thought to bind to the receptor. However, normally a rigid alignment rule is applied which does not account for receptor flexibility. This implies that even identical parts in different molecules will not be aligned perfectly when these compounds bind to the receptor. In contrast, a reason why two in principle identical parts of different molecules might not overlap perfectly is that their superposition results from aligning pharmacophore elements 'at the other end of the molecules'.

Steric interaction energies in CoMFA are normally calculated using a Lennard-Jones 6-12 potential, characterized by a very steep increase in energy at short distances [43]. Therefore, slight deviations in the alignment of two molecules (as caused by receptor flexibility, or by the alignment rule) may give rise to significantly different energy values at grid-points close to the molecules. This is of particular importance, as these points have the highest variance in energy, consequently significantly influencing the statistics of the PLS analysis.

In order to investigate this alignment problem, an automated procedure was devised which systematically reorientates the compounds in a training set, with the aim to improve the predictive quality of the corresponding CoMFA [44]. As an example, the classical QSAR dataset of Hansch and co-workers was used [45]. From this ensemble of 256 dihydrofolate reductase inhibitors, two training sets consisting of 80 structures each and a test set of 70 compounds were randomly chosen. Initial alignment was performed by a standard procedure — i.e. by pairwise fitting of a common structural element of the molecules to a reference compound. The resulting CoMFAs were of mediocre inner consistency. The reorientation procedure which was applied subsequently is outlined in Fig. 6.

Each compound was excluded once and its activity was predicted by the CoMFA-model derived from the remaining ones. The residual is defined as:

$$\text{residual} = \text{activity}_{\text{real}} - \text{activity}_{\text{predicted}} \qquad (1)$$

Molecules with a positive residual were then systematically reoriented by translations and rotations in order to reduce their residual. The translation increments (T-INC) were set to 0.1 Å and those for the rotations (R-INC) to 1.0°, thus making up a maximum translation of 0.3 Å along one direction and a maximum rotation of 3.0° about one axis of a Cartesian coordinate system.

For the training sets, this procedure gave very good results. For set A, the r^2_{cv} was improved from 0.582 to 0.860. In the case of the second set (set B), r^2_{cv} rose from 0.328 to 0.796.

However, an important caveat should be made at this point. Clearly, the inner consistency of the CoMFA could be improved by the procedure but, at the same time, the original alignment rule was destroyed. Therefore, the question was which rule or pro-

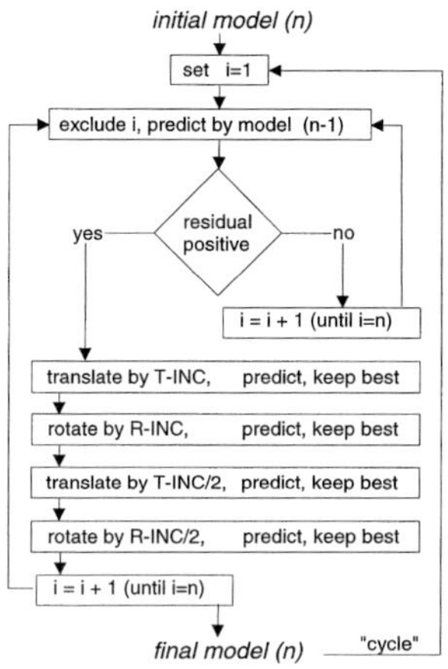

Fig. 6. Flowchart for improvement of the consistency of the training set. The dataset consists of n *compounds;* i *refers to the* id *of the molecule to exclude,* i ∊ 1 ... n. T-INC *and* R-INC *are the increments for translation and rotation, respectively.*

cedure to apply for the prediction of novel molecules; and this question will be addressed below.

4.4. Methods to improve predictive quality: improvement of statistics

There are also methods to enhance the quality of the CoMFA procedure by improving the underlying statistics. The aim is to determine and use only those variables which are relevant for a proper description of the molecules.

GOLPE is an advanced variable selection method developed by Clementi et al. [46]. Based on a number of reduced models, the variable selection is driven by a fractional factorial design strategy. For further details see the chapter by Cruciani et al. in this volume.

Clark and Cramer discussed the noise sensitivity of PLS analyses and its influence on CoMFA results [9]. It was suggested to use PLS-derived expressions like modelling power or discriminate power to preselect variables of importance. Another approach based on cross-validated sub-models is described by Tropsha et al. in this volume.

4.5. Prediction of novel compounds

In section 4.3.3, we have described a method to improve the internal consistency of a CoMFA by slight reorientations of the molecules in the training set. This leads us to the other part of the problem of predictive quality in CoMFA. In particular, we would like to address two points: (i) how far can we extrapolate, that is make reliable predictions for compounds which are dissimilar to the ones in the training set; and (ii) in the case of a method to generate a higher internal consistency, how can we also improve the prediction of test compounds?

4.5.1. *The general extrapolation problem*

The problem of extrapolation became quite obvious in the recent study on HIV-protease inhibitors [10]. After having generated an internally consistent CoMFA using a training set of 100 compounds, a test set of 75 inhibitors was predicted. The predictive r^2 value for the whole set of test compounds was rather low (0.094–0.258 for the three models established). However, removal of only eight compounds from this test set yielded q^2 values of comparable magnitude to the respective r^2_{cv}. Analysis of the 8 'outliers' revealed some unique features not present in the training set.

In general, two conclusions could be drawn from this problem. First, if test compounds contain certain features in a region not explored by the training set, the prediction becomes highly unreliable. This is directly related to the similarity problem. Therefore, similarity should be considered before making predictions of test compounds. In this context, one could envisage two methods for assessment of similarity. The first would be an assessment via a similarity index (cf. the chapter by Good in this volume). The other method could be investigation of the so-called sigma fields (i.e. fields indicating the variance at the grid-points for a particular dataset) in CoMFA. In the case that test compounds exhibit unique features in areas not represented well by the training set, one has to be careful with the predictions.

Another problem which is not related so much to the structural properties — i.e. the binding enthalpy of the compounds — was also highlighted in this study, namely he problem of entropy. CoMFA is a method which correlates enthalpies with target properties. In the case that novel compounds possess totally different degrees of internal freedom, or if there is a significant change in solvation/desolvation energy, then prediction becomes a difficult task.

4.5.2. *Flexible alignment for test sets*

In section 4.3.3, we have described a method to improve the internal predictive quality by slight but methodical reorientations of the molecules in the training set. However, this method created a problem for the prediction of novel compounds because the original alignment rule (pairwise fitting of common structural elements to a reference compound) had been destroyed.

Therefore, a procedure had to be introduced in order to improve the predictive quality for test molecules as well. This fully automated procedure consisted of several steps: first, for each test molecule, the most similar structures in the training set were identified by pairwise fitting of the test compound to all training set molecules. Two fitting methods, namely 'point fitting' (i.e. pairwise fitting of atoms) and 'field fitting' (i.e. maximizing the similarity of two SYBYL CoMFA fields), were applied. Those orientations of the test molecule corresponding to a fit to the most similar compounds were then used for predicting its activity, and the mean value was calculated from these values. In addition, the prediction was corrected by the residuals of the corresponding (most similar) structures in the training set. Thus, four different prediction methods were compared (Table 4).

The best method was the one which included field fitting and correction of the prediction. In fact, this method was able to improve the predictive r^2 as well. Nevertheless,

Table 4 *CoMFA results for an automated alignment procedure involving systematic realignment of the structures, reference [44]*

	A		B		A′		A″	
	r^2_{cv}	q^2	r^2_{cv}	q^2	r^2_{cv}	q^2	r^2_{cv}	q^2
Initial analysis	0.582	0.444	0.328	0.546	–0.099		–0.174	
Final analysis	0.860		0.796		0.624		0.596	
Prediction methods								
Point-fitted, uncorrected		0.484		0.604		–0.447		–0.405
Point-fitted, corrected		0.492		0.625		–0.456		–0.419
Field-fitted, uncorrected		0.569		0.626				
Field-fitted, corrected		0.598		0.645				

some caveats need to be pointed out and deserve further investigation: The biggest concern is certainly the fact that the reorientation procedure was able to create a pseudo-consistency for training sets with randomized activities (Table 4, A′ and A″) — i.e. the procedure is able to overfit the data significantly. However, in this case the corresponding q^2 value could not be improved, thus making it possible to distinguish between a real and a pseudo-improvement.

Another point might be the problem of very diverse datasets where fitting of the test molecule(s) could lead to unexpected orientations. Also the procedure for improvement of q^2 is rather complicated and computationally intensive, leaving room for further improvement.

5. Outlook

We are challenged today with larger and larger amounts of data originating from high-throughput chemistry and screening. This has severe implications on the quality of the data and also on the methods of analysis. We are confident that CoMFA will play its part in the processing of these data. However, there are a number of open questions/problems, which have an impact on the predictive value of the resulting models. One task will be to establish consistent alignment rules for large and diverse sets of compounds in an automated fashion. Another problem will be the fairly low accuracy of structural and biological data generated. Here one could envisage the use of inhibition threshold data rather than accurate activity values.

We will also face new challenges in the effective use of CoMFA results. Up to now, after the successful establishment of a CoMFA model, information about potentially active compounds was derived and the most promising candidates were subsequently synthesized. The advent of combinatorial chemistry allows us to determine all the potential products which can possibly be synthesized with a particular reaction type.

With this information, a virtual library of all potential products can be generated. Subsequently, CoMFA models could be used to select and predict compounds of the greatest interest which could be subsequently synthesized and tested. Nevertheless, such a strategy will put more emphasis not only on the automated prediction of compounds, but also on automatic procedures critically to access the reliability of the prediction. Therefore, it will be of great interest to monitor the progress in this area; and hopefully, first results will be presented soon.

Acknowledgement

The authors express their gratitude to Elisa Boccaletti for her invaluable help in the preparation of this manuscript.

References

1. Cramer III, R.D., Patterson, D.E. and Bunce, J.D., *Comparative molecular field analysis (CoMFA): 1. Effect of shape on binding of steroids to carrier proteins*, J. Am. Chem. Soc., 110 (1988) 5959–5967.
2. Wold, S., Albano, C., Dunn, W.J., Edlund, U., Esbenson, K., Geladi, P., Hellberg, S., Lindberg, W. and Sjöström, M., *Multivariate data analysis in chemistry*, In Kowalski, B. (Ed.) Chemometrics: Mathematics and statistics in chemistry, Reidel, Dordrecht, The Netherlands, 1984, p. 17–95.
3. Dunn, W.J., III, Wold, S., Edlund, U., Hellberg, S. and Gasteiger, J., *Multivariate structure–activity relationship between data from a battery of biological tests and an ensemble of structure descriptors: The PLS method*, Quant. Struct.-Act. Relat., 3 (1984) 131–137.
4. Geladi, P., *Notes on the history and nature of partial least squares (PLS) modeling*, J. Chemometrics, 2 (1988) 231–246.
5. Wold, S., *Cross-validatory estimation of the number of components in factor and principal component models*, Technometrics, 4 (1978) 397–405.
6. Diaconis, P. and Efron, B., *Computer-intensive methods for statistics*, Sci. Am., 116 (1984) 96–117.
7. Cramer III, R.D., Bunce, J.D. and Patterson, D.E., *Cross-validation, bootstrapping and partial least squares compared with multiple regression in conventional QSAR studies*, Quant. Struct.-Act. Relat., 7 (1988) 18–25.
8. Thibaut, U., *Applications of CoMFA and related 3D QSAR approaches*, In Kubinyi, H. (Ed.) 3D QSAR in drug design: Theory, methods and applications, ESCOM, Leiden, The Netherlands, 1993, pp. 661–696.
9. Clark, M. and Cramer III, R.D., *The probability of chance correlation using partial least-squares (PLS)*, Quant. Struct.-Act. Relat., 12 (1993) 137–145.
10. Kroemer, R.T., Ettmayer, P. and Hecht, P., *3D-quantitative structure–activity relationships of human immunodeficiency virus type-1 proteinase inhibitors: comparative molecular field analysis of 2-hetero-substituted statine derivatives — implications for the design of novel inhibitors*, J. Med. Chem., 38 (1995) 4917–4928.
11. Kellog, G.E., Semus, F.E. and Abraham, D.J., *HINT: A new method of empirical hydrophobic field calculation for CoMFA*, J. Comput.-Aided Mol. Design, 5 (1991) 545–552.
12. Kim, K.H. and Martin, Y.C., *Direct prediction of dissociation-constants (PKAS) of clonidin-like imidazolines, 2-substituted imidazoles, and 1-methy-2-substituted-imidazoles from 3D structures using a comparative molecular-field analysis (CoMFA) approach*, J. Med. Chem., 34 (1991) 2056–2060.
13. Greco, G., Novellino, E., Silipo, C. and Vittoria, A., *Comparative molecular-field analysis on a set of muscarinic agonists*, Quant. Struct.-Act. Relat., 10 (1991) 289–299.
14. Klebe, G. and Abraham, U., *On the prediction of binding-properties of drug molecules by comparative molecular-field analysis*, J. Med. Chem., 36 (1993) 70–80.

15. Floersheim, P., Nouzlak, J. and Weber, H.P., *Experience with comparative molecular-field analysis*, In Wermuth, C.G. (Ed.) Trends in QSAR and molecular modeling 92, ESCOM, Leiden, The Netherlands, 1993, pp. 227–232.
16. Marsili, M., Floersheim, P. and Dreiding, A.S., *Generation and comparison of space-filling molecular-models*, Comput. Chem., 7 (1983) 175–181.
17. Kroemer, R.T. and Hecht, P., *Replacement of steric 6-12 potential-derived interaction energies by atom-based indicator variables in CoMFA leads to models of higher consistency*, J. Comput.-Aided Mol. Design., 9 (1995) 205–212.
18. Cramer III, R.D., Patterson, D.E. and Bunce, J.D., *Cross-validation, bootstrapping, and partial least-squares compared with multiple-regression in conventional QSAR Studies*, Quant. Struct.-Act. Relat., 7 (1988) 18–25.
19. Cramer III, R.D., DePriest, S.A., Patterson, D.E. and Hecht, P., *The developing practice of comparative molecular-field analysis*, In Kubinyi, H., (Ed.) 3D QSAR in drug design, ESCOM, Leiden, The Netherlands, 1993, pp. 465–485.
20. Calder, J.A., Wyatt, J.A., Frenkel, D.A. and Casida, J.E., *CoMFA validation of the superposition of 6 classes of compounds which block GABA receptors noncompetitively*, J. Comput.-Aided Mol. Design, 7 (1993) 45–60.
21. Rault, S., Bureau, R., Pilo, J.C. and Robba, M., *Comparative molecular-field analysis of CCK-A antagonists using field-fit as an alignment technique — a convenient guide to design new CCK-A ligands*, J. Comput.-Aided Mol. Design, 6 (1992) 553–568.
22. Allen, M.S., Tan, Y.-C., Trudell, M.L., Narayanan, K., Schindler, L.R., Martin, M.J., Schultz, C., Hagen, T.J., Koehler, K.F., Codding, P.W., Skolnick, P. and Cook, J.M., *Synthetic and computer-assisted analyses of the pharmacophore for the benzodiazepine receptor inverse agonist site*, J. Med. Chem., 33 (1990) 2343–2357.
23. Allen, M.S., LaLoggia, A.J., Dorn, L.J., Martin, M.J., Costatino, G., Hagen, T.J., Koehler, K.F., Skolnick, P. and Cook, J.M., *Predictive Binding of beta-carboline inverse agonists and antagonists via the CoMFA GOLPE approach*, J. Med. Chem., 35 (1992) 4001–4010.
24. Kroemer, R.T., Liedl, K.R. and Hecht, P., *Different electrostatic descriptors in comparative molecular field analysis (CoMFA): A comparison of molecular electrostatic and coulomb potentials*, J. Comput. Chem., 17 (1996) 1296–1308.
25. Gasteiger, J. and Marsilli, M., *Iterative partial equalization of orbital electronegativity — a rapid access to atomic charges*, Tetrahedron, 36 (1980) 3219–3228.
26. Dewar, M.J.S. and Thiel, W., *Ground states of molecules: 38. The MNDO method — approximations and parameters*, J. Am. Chem. Soc., 99 (1977) 4899–4907.
27. Dewar, M.J.S., Zoebisch, E.G., Healy, E.F. and Stewart, J.J.P., *AM1: A new general purpose quantum chemical mechanical molecular model*, J. Am. Chem. Soc., 107 (1985) 3902–3909.
28. Stewart, J.J.P., Optimization of parameters for semiempirical methods: 1. Method, J. Comp. Chem., 10 (1989) 209–220.
29. Mulliken, R.S., *Electronic population analysis on LCAO–MO molecular wave functions*, I., J. Chem. Phys., 23 (1955) 1833–1840.
30. Singh, U.C. and Kollman, P.A., *An approach to computing electrostatic charges for molecules*, J. Comp. Chem., 5 (1984) 129–145.
31. Besler, B.H., Merz, K.M., Jr. and Kollman, P.A., *Atomic charges derived from semiempirical methods*, J. Comp. Chem., 11 (1990) 431–439.
32. Chirlian, L.E. and Francl, M.M., *Atomic charges derived from electrostatic potentials — a detailed study*, J. Comp. Chem., 8 (1987) 894–905.
33. Breneman, C.M. and Wiberg, K.B., *Determining atom-centred monopoles from molecular electrostatic potentials — the need for high sampling density in formamide conformational analysis*, J. Comp. Chem., 11 (1990) 361–373.
34. Debnath, A.K., Jiang, S., Strick, N., Lin, K., Haberfield, P. and Neurath, A.R., *Three-dimensional structure–activity analysis of a series of porphyrin derivatives with anti-HIV-1 activity targeted on the V3 loop of the gp120 envelope glycoprotein of the human immunodeficiency virus type 1*, J. Med. Chem., 37 (1994) 1099–1108.

35. Avery, M.A., Gao, F., Chong W.K.M., Mehrotra, S. and Milhous, W.K., *Structure–activity relationships of the antimalarial agent artemisinin: 1. Synthesis and comparative molecular field analysis of C-9 analogs of artemisinin and 10-deoxoartemisinin*, J. Med. Chem., 36 (1993) 4264–4275.
36. Carroll, F.I., Mascarella, S.W., Kuzemko, M.A., Gao, Y., Abraham, P., Lewin, A.H., Boja, J.W. and Kuhar, M.J., *Synthesis, ligand binding, and QSAR (CoMFA and classical study of 3β-3′*-substituted phenyl)-, 3β-(4′-substituted phenyl)-, and 3β-(3′,4′-disubstituted phenyl) tropane-2β-carboxylic acid methyl esters, J. Med. Chem. 37 (1994) 2865–2873.
37. Tong, W., Collantes, E.R., Chen, Y. and Welsch, W.J., *A comparative molecular-field analysis study of N-benzylpiperidines as acetylcholesterinesterase inhibitors*, J. Med. Chem., 39 (1996) 380–387.
38. Kroemer, R.T., Koutsilieri, E., Hecht, P., Liedl, K.R., Riederer, P. and Kornhuber, J., *Quantitative analysis of the structural requirements for blockade of the NMDA receptor at the PCP binding site*, J. Med. Chem., (in press).
39. Martin, Y.C., Bures, M.G., Dahaner, E.A., DeLazzer, J., Lico, I. and Pavlik, P., *A fast approach to pharmacophore mapping and its application to dopaminergic and benzodiazepine agonists*, J. Comput.-Aided Mol. Des., 7 (1993) 83–102.
40. Gamper, A.M., Winger, R.H., Liedl, K.R., Sotriffer, C.A., Varga, J.M., Kroemer, R.T. and Rode, B.M., *Comparative molecular field analysis (CoMFA) of haptens docked to the multispecific antibody IgE(Lb4)*, J. Med. Chem., 39 (1996) 3882–3888.
41. Goodsell, D.S. and Olson A.J., *Automated docking of substrates to proteins by simulated annealing*, Proteins: Struct. Funct. Genet., 8 (1990) 195–202.
42. Marshall, G.R., Barry, C.D., Bosshard, H.E., Dammkoehler, R.A. and Dunn, D.A., *The conformational parameters in drug design*, In Olson, E.C. and Christoffersen, R.E. (Eds.) Computer-assisted drug design, ACS Symp. Series, Vol. 112, American Chemical Society, Washington, DC, 1979, pp. 205–226.
43. Thibaut, U., Folkers, G., Klebe, G., Kubinyi, H., Merz, A. and Rognan, D., *Recommendations for CoMFA studies and 3D QSAR publications*, Quant. Struct.-Act. Relat., 13 (1994) 1–3.
44. Kroemer, R.T. and Hecht, P., *A new procedure for improving the predictiveness of CoMFA-models and its application to a set of dihydrofolate reductase inhibitors*, J. Comput.-Aided Mol. Des., 9 (1995) 396–406.
45. Silipo, C. and Hansch, C., *Correlation analysis: Its application to the structure–activity relationship of triazines inhibiting dihydrofolate reductase*, J. Am. Chem. Soc. (1975) 6849–6861.
46. Baroni, M., Constantino, G., Cruciani, G., Riganelli, D., Valigi, R. and Clementi, S., *Generating optimal linear PLS estimations (GOLPE): An advanced chemometric tool for handling 3D-QSAR problems*, Quant. Struct.-Act. Relat., 12 (1993) 9–20.

Cross-Validated R^2 Guided Region Selection for CoMFA Studies

Alexander Tropsha and Sung Jin Cho
Laboratory for Molecular Modelling, Division of Medicinal Chemistry and Natural Products, School of Pharmacy, University of North Carolina, Chapel Hill, North Carolina 27599, U.S.A.

1. Introduction

The Comparative Molecular Field Analysis (CoMFA) [1] approach was introduced in 1988. Since then, it has rapidly become one of the most widely used tools for three-dimensional quantitative structure–activity relationship (3D QSAR) studies. Over the years, this approach has been applied to a wide variety of receptor and enzyme ligands (recently reviewed by Cramer et al. [2] and Thibaut [3]). Undoubtedly, the further development of this method is of great importance and interest to many scientists working in the area of rational drug design.

CoMFA methodology is based on the assumption that since, in most cases, the drug–receptor interactions are noncovalent, the changes in the biological activities or binding affinities of sample compounds correlate with changes in the steric and electrostatic fields of these molecules. In a standard CoMFA procedure, all molecules under investigation are structurally aligned first, and the steric and electrostatic fields around them are then sampled with probe atoms, usually sp^3 carbon with +1 charge, on a rectangular grid that encompasses aligned molecules. The results of the field evaluation in every grid-point for every molecule in the dataset are placed in the CoMFA QSAR table which, therefore, contains thousands of columns. The analysis of this table by the means of standard multiple regression is practically impossible; however, the application of special multivariate statistical analysis routines, such as partial least squares (PLS) analysis and cross-validation ensures the statistical significance of the final CoMFA equation [1]. A cross-validated R^2 (q^2) which is obtained as a result of this analysis serves as a quantitative measure of the predictability of the final CoMFA model. The statistical meaning of the q^2 is different from that of the conventional R^2: the q^2 value greater than 0.3 is considered significant [4].

Despite obviously successful and growing application of CoMFA in molecular design, several problems intrinsic to this methodology have persisted. Studies done by us [5] and others [1,6–9] revealed that CoMFA results can be extremely sensitive to a number of factors such as alignment rules, overall orientation of aligned compounds, lattice shifting, step size and the probe atom type. The problem of three-dimensional alignment has been the most notorious among others. Even with the development of automated and semiautomated alignment protocols, such as Active Analog Approach [10,11] and DISCO [12], and the opportunity to use, in some cases, the structural information about the target receptor [6,13], there is generally no standard recipe to align all molecules under consideration in a unique and unambiguous fashion. Our recent QSAR analysis of 60 acetylcholinesterase inhibitors is particularly illustrative with respect to

H. Kubinyi et al. (eds.), 3D QSAR in Drug Design, Volume 3. 57–69.

this point [13]. In that paper, we employed the combination of structure-based alignment and CoMFA to obtain three-dimensional QSAR for 60 chemically diverse inhibitors of acetylcholinesterase (AChE). The great structural diversity of the AChE inhibitors, ranging from choline to decamethonium, makes it practically impossible structurally to align all the inhibitors in any unbiased way and generate a unique three-dimensional pharmacophore. As a result, earlier SAR studies were limited to series of structurally congeneric ligands [14,15–18]. Recent X-ray crystallographic analysis of AChE from *Torpedo californica* (EC 3.1.1.7) [19], followed by X-ray determination of the complexes of the enzyme with three structurally diverse inhibitors, tacrine, edrophonium and decamethonium [20], provided crucial information with respect to the orientation of these inhibitors in the active site of the enzyme (Fig. 1). The crystallographic data indicated that each of the three inhibitors had a unique binding orientation in the active site of the enzyme (Fig. 1). Their natural structural alignment would probably never have been predicted by any of the existing automated algorithms for ligand alignment, or even by the researcher's imagination based on the ligand chemical structure alone.

The 3D alignment problem will most likely remain as a source of ambiguity in CoMFA, especially in the case of structurally diverse compounds. However, as we recently discovered [5], even if the structural alignment is fixed, the resulting q^2 value could also be sensitive to the orientation of the whole set of superimposed molecules on the computer screen. The circumstances preceding this discovery were somewhat anecdotal. We first noticed this phenomenon during the laboratory sessions of the introductory molecular modelling class taught by the first author of this paper at the University of North Carolina. All students were given the same series of compounds, 20 5-HT_{1A} receptor ligands [4] — i.e. we conducted, as we later called it, the most statistically significant 'student test' of CoMFA. However, the final q^2 values differed by up

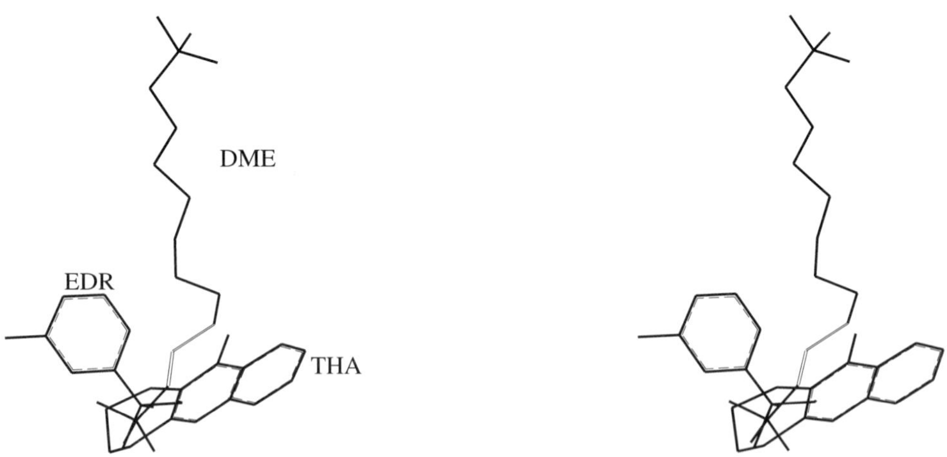

Fig. 1. Superposition of tacrine (THA), edrophonium (EDR) and decamethonium (DME) in the active site of acetylcholinesterase.

to 0.5 q^2 units, even when all students were finally given the same molecular database with rigidly aligned receptor ligands (the database was kindly sent to us electronically by Professor E.W. Taylor). Puzzled by this result, we examined closely each student's report and found that the only difference among the analyses was the orientation of superimposed molecules on the student's monitor.

In this chapter, we first briefly discuss the possible origin of this phenomenon. We then concentrate on the development and application of q^2-Guided Region Selection method (q^2-GRS) that was designed in this laboratory. We emphasize the ability of this algorithm to deal effectively with the problems related to overall orientation, lattice placement and step size. Finally, we discuss future application of this methodology and related methods of QSAR.

2. Orientation Dependence of q^2

In the initial publication [5], we have analyzed three datasets of model compounds of different sizes: 7 cephalotaxine esters, [21] 20 5-HT_{1A} receptor ligands [22], and 59 inhibitors of human immunodeficiency virus (HIV) protease [6]. The alignment rules for the first dataset were described elsewhere [5]. The files with prealigned inhibitors of HIV protease and 5-HT_{1A} receptor ligands, as used in the original publications [6,22], were kindly provided by Drs. Waller and Taylor, respectively.

Conventional CoMFA was performed with the QSAR option of SYBYL [23]. The steric and electrostatic field energies were calculated using sp^3 carbon probe atoms with +1 charge. The CoMFA grid spacing was 2.0 Å in all three dimensions within the defined region, which extended beyond the van der Waals envelopes of all molecules by

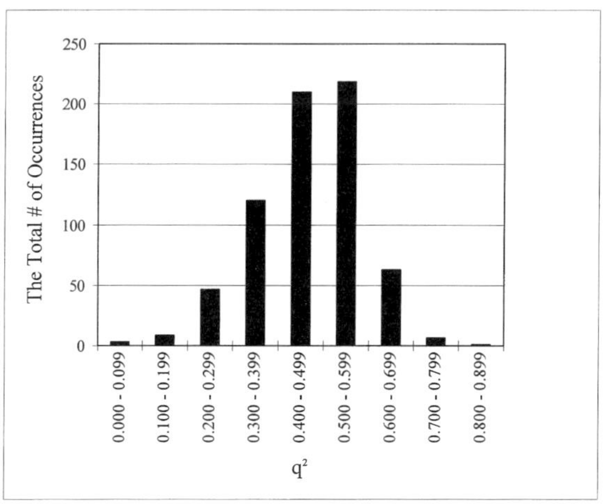

Fig. 2. The q^2 *values observed for different orientations of 7 superimposed cephalotaxine esters. The set was systematically rotated by 30° at a time around x, y and z axes.*

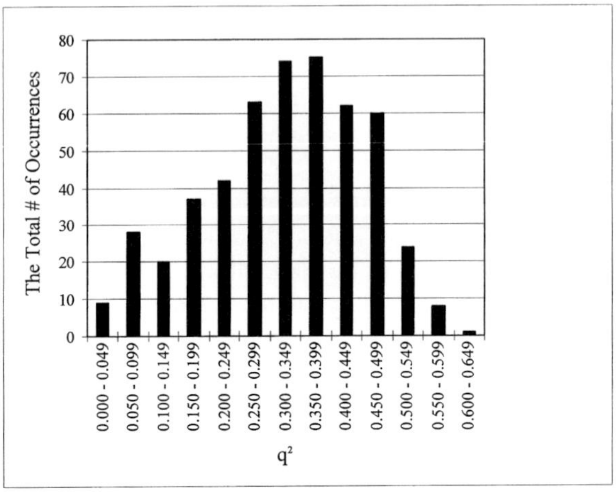

Fig. 3. The q^2 *values observed for different orientations of 20 superimposed 5-HT*$_{1A}$ *receptor ligands. The set was systematically rotated by 30° at a time around x, y and z axes.*

at least 4.0 Å. The CoMFA QSAR equations were calculated with the PLS algorithm. The optimal number of components (ONC) in the final PLS model was determined by the q^2 value, obtained from the leave-one-out cross-validation technique. For small datasets, in order to maximize the q^2 value and minimize the standard error of prediction, the number of components was increased only when adding a component raised the q^2 value by 5% or more [24]. For HIV protease inhibitors, the number of components with the lowest standard error of prediction (SDEP) was selected as the ONC.

The overall orientation of superimposed molecules was varied as follows. Starting from an arbitrary orientation, the whole set of molecules was rotated by 30° at a time around x, y and z axes using SYBYL STATIC command . For each orientation, the conventional CoMFA was performed with 10 components, using 7 cross-validation groups for cephalotaxine esters, 20 cross-validation groups for 5-HT$_{1A}$ receptor ligands and 59 cross-validation groups for HIV protease inhibitors. The region files were generated automatically. After each CoMFA analysis, the q^2 value and the ONC were recorded.

The frequency distribution of q^2 values observed for different datasets as a result of rotations are given in Figs. 2–4 (due to the large number of CoMFA runs, the number of components with the highest q^2 is selected as the ONC rather than employing 5% increase rule). For cephalotaxine esters, the highest (0.819) and lowest (0.050) q^2's were obtained with the ONC of 6 (Fig. 2). For 5-HT1A receptor ligands, the highest (0.607) and lowest (–0.015) q^2's were obtained with the ONC of 10 and 1, respectively (Fig. 3). For HIV protease inhibitors, the range of q^2 value was much more narrow (Fig. 4). The highest (0.802) and lowest (0.586) q^2's were obtained with the ONC of 10. It is obvious from these results, that a single orientation gives an arbitrary value of q^2 which most probably would fall into the region with the highest frequency of occurrences of the q^2 values. For instance, the reported q^2 values for 5-HT1A receptor ligands and HIV protease inhibitors were 0.481 [4] and 0.778 [6], respectively. In both cases, these values

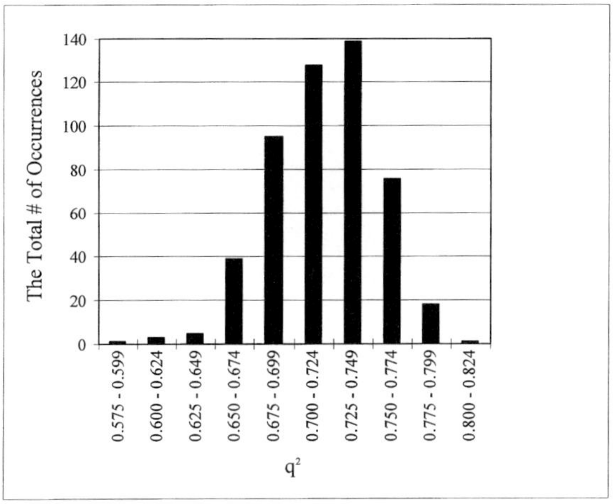

Fig. 4. The q^2 *values observed for different orientations of 59 superimposed HIV-protease inhibitors. The set was systematically rotated by 30° at a time around x, y and z axes.*

lay within the highest frequency regions of the q^2 distribution (cf. Figs. 3 and 4, respectively).

It was suggested [2] that increasing the grid resolution may improve the CoMFA results. Table 1 shows q^2's obtained as a result of CoMFA with the grid spacing of 1.0 Å versus. 2.0 Å for 5-q^2's. HT_{1A} receptor ligands (the results for other datasets follow the same trend [5]). For comparison, we have included the results obtained with the different number of components. Indeed, lowering the step size from 2.0 Å to 1.0 Å narrowed the distribution of q^2s (cf. the differences between the lowest and the highest values of q^2 for 2.0 Å CoMFA runs versus. 1.0 Å CoMFA runs in Table 1). However, for each dataset, the highest q^2 obtained with 1.0 Å grid resolution was consistently lower than the highest q^2 obtained with the 2.0 Å step size.

3. CoMFA/q^2-GRS method

This method was originally proposed in 1995 [5] and was modified later [25] to incorporate different types of probe atoms (Fig. 5). The current version of the q^2-GRS routine consists of the following steps: (1) a conventional CoMFA is performed initially using an automatically generated region file; (2) the rectangular grid encompassing aligned molecules is then broken into 125 small boxes of equal size (this number can vary), and the Cartesian coordinates of the upper right and lower left corners of each box are calculated; (3) the coordinates calculated from step 2 are used to create region files with different probe atoms; for instance, we used C (sp^3, +1), C (sp^3, 0), H (+1) and O (sp^3, −1) (see reference [25]); (4) for each of these newly generated region files, a separate CoMFA is performed using each probe atom independently with the step size of 1.0 Å to improve sampling; (5) the resulting q^2 values are compared to select the best probe atom for each sub-region; (6) the best q^2 values for each sub-region are compared to a specified threshold, and only those regions with the q^2 greater than the threshold are

Table 1 *The comparison of conventional CoMFA and the* q^2-*GRS method as applied to 20 5-HT*$_{1A}$ *receptor ligands; the numbers in bold type represent the* q^2 *values for the optimal number of components*

Orientation[a]	CoMFA method	Step size (Å)	q^2					
			1[b]	2[b]	3[b]	4[b]	5[b]	6
Random	Conventional	2.0	0.127	0.274	0.358	0.418	**0.460**	0.477
Worst	Conventional	2.0	**–0.015**	–0.135	–0.383	–0.645	–0.674	–0.660
Best	Conventional	2.0	0.204	0.347	0.436	0.487	0.557	**0.589**
Random	Conventional	1.0	0.129	0.231	0.304	0.334	**0.383**	0.396
Worst	Conventional	1.0	0.119	0.215	0.266	0.284	**0.329**	0.345
Best	Conventional	1.0	0.137	0.219	0.275	0.292	**0.341**	0.357
Random	q^2-GRS[c]	1.0	0.209	0.405	0.512	0.567	**0.603**	0.612
Worst	q^2-GRS[c]	1.0	0.225	0.358	0.405	0.459	**0.511**	0.525
Best	q^2-GRS[c]	1.0	0.237	0.394	0.506	0.560	**0.590**	0.592

[a] The 'best' and 'worst' orientations are defined as those that generate the highest and lowest q^2's, respectively, as a result of systematic rotation using 2.0 Å step size. No orientation dependence of q^2 was investigated using 1.0 Å step size; the later calculations were done on three fixed orientations ('random', 'worst' and 'best') obtained from conventional CoMFA studies using 2.0 Å step size. Thus, the 'best' orientation obtained with 2.0 Å grid may not be the best that would be obtained if we performed the systematic rotation using a 1.0 Å grid. This consideration explains why the q^2 value for 'random' orientation using 1.0 Å grid appears higher than the value for the 'best' orientation.
[b] The number of components.
[c] q^2 cutoff = 0.

selected for further analysis; (7) the selected regions are combined to generate a master region file; and (8) the final PLS is performed.

This method has been successfully applied in our laboratory to a number of different datasets, including 7 cephalotaxine esters [5], 20 5-HT$_{1A}$ receptor ligands [5], 59 inhibitors of HIV protease [5], 21 steroids [5], 101 Topoisomerase II inhibitors [25], 60 acetylcholinesterase inhibitors [13] and several other unpublished series of compounds. Other groups also applied this method to the inhibitors of cytochrome P4502C9 [26] and PLA inhibitors [9]. In all reported cases, the q^2-GRS generated an orientation independent, high q^2, exceeding the one obtained with the conventional CoMFA. This is illustrated by the data presented in Table 1 for 5-HT1A receptor ligands. We have applied the q^2-GRS routine to three different orientations of these ligands obtained in the course of the systematic rotation of superimposed molecules (see previous section): 'random' (i.e. some arbitrary initial orientation; in this case, an orientation used in the original publication [4]), 'best' (i.e. the one with the highest value of the q^2); and 'worst' (i.e. the one with the lowest value of q^2). The results presented in Table 1 were obtained with the q^2 threshold value of zero. Apparently, the application of the q^2-GRS led to very consistent values of q^2, regardless of the orientation of superimposed molecules. With the q^2 cutoff of zero, the resulting q^2 values were fairly close to the best q^2 values obtained with the 2.0 Å step size (cf. Table 1).

The effect of various cutoff values on the resulting q^2 can be best illustrated by our analysis of acetylcholinesterase inhibitors [13] which also allows us to discuss here some important aspects of the method. The predictability of the QSAR model was initially assessed by conventional CoMFA (Table 2). The q^2-GRS routine was then applied

Step 1:	Perform CoMFA using an automatically created region file	Probe atom: C.3 Charge: +1 Step size: 2.0	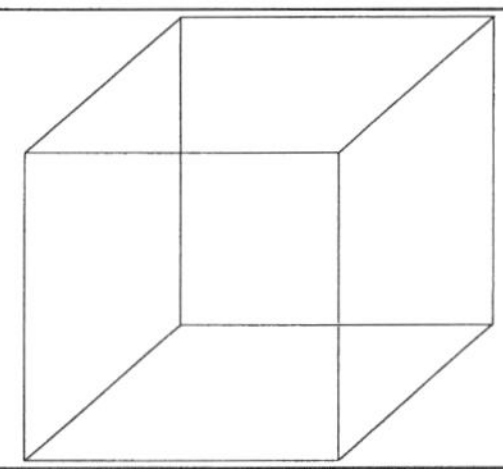
Step 2:	Use the region file generated from step 1 to calculate the Cartesian coordinates of the upper-right and lower-left corners of 125 small region files		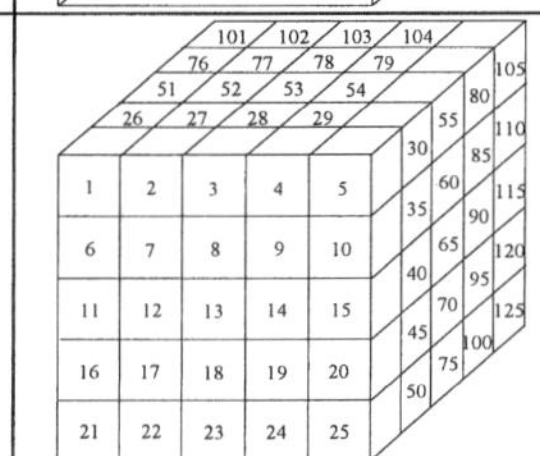
Step 3:	Use the coordinates calculated from step 2 to create region files with different probe atoms and charges	Set 1: C.3, +1, 1.0 Set 2: C.3, 0, 1.0 Set 3: H, +1, 1.0 Set 4: O.3, −1, 1.0	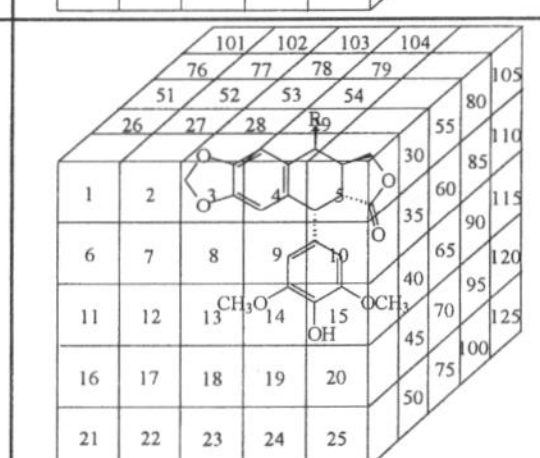
Step 4:	Perform CoMFA using region files generated from step 3	Set 1: C.3, +1, 1.0 Set 2: C.3, 0, 1.0 Set 3: H, +1, 1.0 Set 4: O.3, −1, 1.0	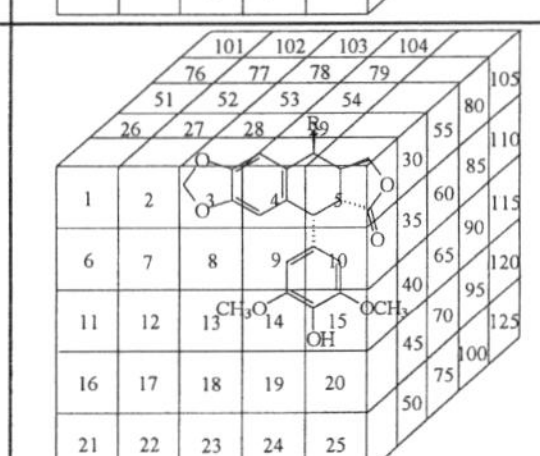
Step 5:	Select the best probe atom and charge for each region using the q^2 value as a guide	Set 1: Regions 1–20 Set 2: Regions 21–39 Set 3: Regions 40–70 Set 4: Regions 71–125	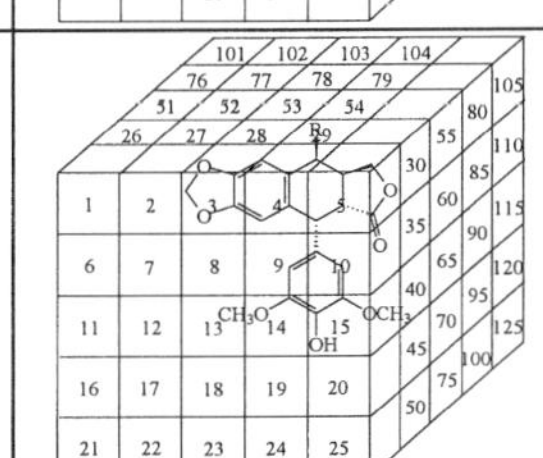

Fig. 5. The illustration of the q^2-GRS *procedure that can incorporate different probe atoms and charges. Regions depicted in this figure are shown as an example.* *(continued on next page)*

Step 6:	Select those regions where the q^2 value is greater than a specified cutoff value	Set 1: 5, 10, 15–17, 20 Set 2: 22–24, 26–29, 30, 35 Set 3: 40, 45, 50, 53–55, 65 Set 4: 75, 77, 79, 85, 95	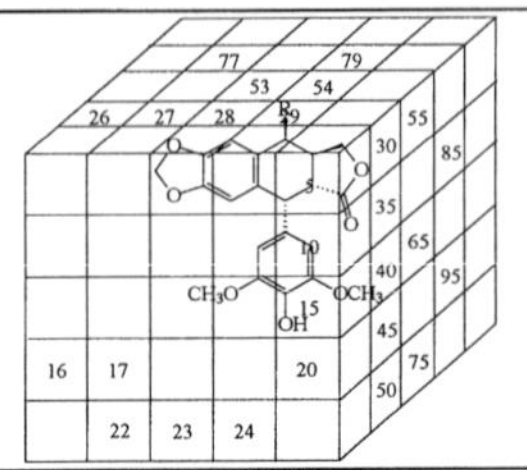
Step 7:	Create a master region file	Set 1: 5, 10, 15–17, 20 Set 2: 22–24, 26–29, 30, 35 Set 3: 40, 45, 50, 53–55, 65 Set 4: 75, 77, 79, 85, 95	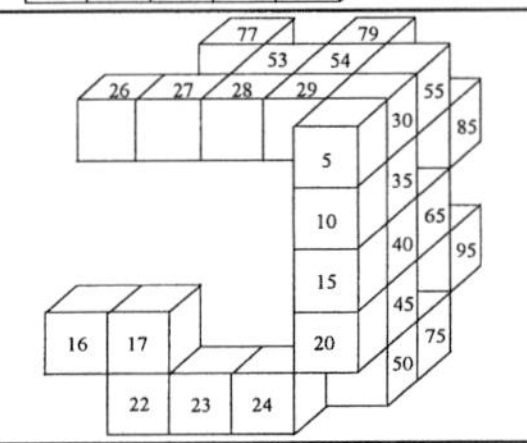
Step 8:	Perform PLS using the master region file created from Step 7	Set 1: 5, 10, 15–17, 20 Set 2: 22–24, 26–29, 30, 35 Set 3: 40, 45, 50, 53–55, 65 Set 4: 75, 77, 79, 85, 95	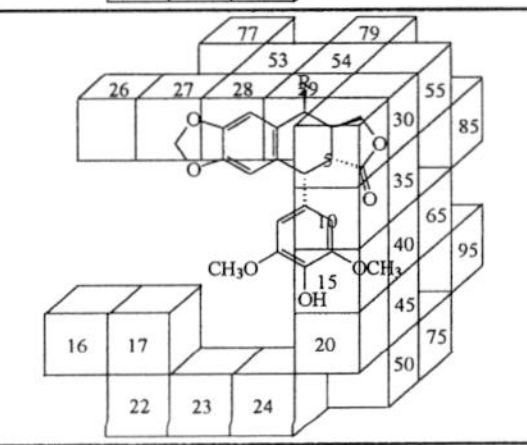

Fig. 5. (continued)

Table 2 *The* q^2*'s obtained after performing CoMFA/*q^2*-GRS with different* q^2 *threshold values for 60 acetylcholinesterase inhibitors; numbers in bold type represent the* q^2 *values at the optimal number of components*

q^2 threshold	Number of components								
	1	2	3	4	5	6	7	8	9
None[a]	0.475	0.551	**0.621**	0.624	0.624	0.636	0.630	0.638	0.653
0.1	0.472	0.566	0.650	0.658	0.666	0.688	**0.701**	0.698	0.677
0.2	0.472	0.565	0.648	0.657	0.663	0.686	**0.699**	0.697	0.682
0.3	0.497	0.572	0.644	0.662	0.658	0.680	**0.694**	0.686	0.680
0.4	0.498	0.583	0.644	0.683	0.670	0.714	**0.735**	0.718	0.715
0.5	0.518	0.598	0.683	0.698	0.701	0.718	**0.734**	0.737	0.720
0.6	0.430	0.588	0.676	0.706	0.707	0.703	0.707	0.724	**0.743**

[a] The results of conventional CoMFA for an arbitrary orientation of superimposed molecules.

to optimize the initial CoMFA model. Various q^2 thresholds (0.1–0.6) were used to isolate the regions of the lattice surrounding the aligned molecules where the change in the field values correlated strongly with biological activity. This procedure can be interpreted as elimination of the irrelevant variables in the PLS analysis. As the q^2 threshold

Table 3 *Summary of CoMFA/q²-GRS results*

	Conventional CoMFA	CoMFA/q^2-GRS	
q^2 threshold	None	0.4	0.5
Minimum sigma	0	0	0
Number of small boxes	None	18	11
Number of lattice points	2652	3150	1925
Optimal number of components	3	7	7
q^2	0.621	0.735	0.734
Standard error of prediction	1.029	0.893	0.894
Standard error of estimate	0.739	0.478	0.451
r^2	0.804	0.924	0.932
F values	76.805[a]	90.259[b]	102.337[b]
Prob. of $R^2 = 0$	0[a]	0[b]	0[b]
Relative contributions			
Steric	0.772	0.683	0.670
Electrostatic	0.228	0.317	0.330

[a] n1 = 3, n2 = 56.
[b] n1 = 7, n2 = 52.

increases from 0.1 to 0.6, the q^2 values for the ONC increase, reaching a maximum at 0.4 and 0.5 q^2 threshold, and then decrease again (cf. Table 2).

Since the values of both q^2 and SDEP for both 0.4 and 0.5 q^2 thresholds were very close to each other, we have examined both models. The results obtained from CoMFA/q^2-GRS at 0.4 and 0.5 q^2 thresholds are summarized in Table 3. Non-cross-validated CoMFA calculations showed that the 0.5 q^2 threshold exhibits slightly better overall statistics compared to that with the 0.4 q^2 threshold. Table 3 also presents the number of lattice points for the two different CoMFA runs; obviously, a significant number of lattice points are excluded from the analysis as the q^2 threshold value increases (3150 versus. 1925 lattice points at 0.4 and 0.5 q^2 thresholds, respectively). This suggests that 1225 additional lattice points (i.e. 2450 variables) present in 0.4 q^2 threshold model most likely do not contribute to the predictability of the CoMFA model. Based on the above considerations, we have finally selected a 0.5 q^2 threshold at 7 principal components as the final CoMFA model. This example emphasizes that the careful choice of the threshold is an important component of every q^2-GRS study.

4. Why the Conventional CoMFA Results May Be Orientation Dependent?

In the conventional CoMFA implementation, the steric and electrostatic fields, which theoretically form a continuum, are sampled on a fairly coarse grid. As a result, these fields are represented inadequately, and the results are not strictly reproducible [2]. Intuitively, decreasing the grid spacing may increase the adequacy of sampling, as was suggested by Cramer et al. [2]. Indeed, we report in this paper that decreasing the grid spacing from 2.0 Å to 1.0 Å minimizes the fluctuation in the observed q^2 values. Most probably, the reason for this phenomenon is that the decrease in grid spacing increases the number of probe atoms which, in turn, should raise the probability of placing the probe atoms in a region where the steric and electrostatic field changes can be best correlated with biological activity. However, as was noticed by Cramer et al. [2], the increase in the number of probe atoms also increases the noise in PLS analysis and leads to a less statistically significant q^2 [27]. Furthermore, as mentioned above, decreasing the grid spacing from 2.0 Å to 1.0 Å decreased the highest q^2 value obtained for each dataset.

The grid orientation in CoMFA is fixed in the coordinate system of the computer; thus, every time when the orientation of superimposed molecules is changed, the size of the grid may change, but not its orientation. The orientation of the assembled molecules, therefore, affects the placement of probe atoms which, in turn, influences the results of the field sampling process. This leads to the variability of the q^2 values, mostly due to the reasons outlined above. We also noticed that the variability of q^2 as a function of the orientation of superimposed molecules is more pronounced in the case of structurally diverse compounds, such as cephalotaxine esters and 5-HT1A receptor ligands, than in the case of much less structurally diverse molecules, such as HIV protease inhibitors [5]. This effect may be due to the fact that the pattern of probe atom placement with respect to the aligned molecules changes more dramatically when one changes the orientation of more structurally diverse molecules than it does when the dataset is comprised of structurally similar molecules.

5. Why q^2-GRS is Effective?

An important feature of conventional CoMFA routine is that it assumes equal sampling and *a priori* equal importance of all lattice points for PLS analysis, whereas the final CoMFA result actually emphasizes the limited areas of three-dimensional space as important for biological activity. We have realized that the deficiencies of conventional CoMFA routine mentioned above may be effectively dealt with by eliminating from the analysis those areas of three-dimensional space where changes in steric and electrostatic fields do not correlate with changes in biological activity. Thus, we devised the q^2-GRS routine which eliminates those areas based on the (low) value of the q^2 obtained for such regions individually. The major feature of this routine is that it optimizes the region selection for the final PLS analysis. In this regard, it is intellectually analogous to the recently proposed GOLPE [27] approach (see also the chapter by Cruciani et al. in this volume) and PLS region focusing [23]. The relative efficiency of all these algorithms shall be compared using the same datasets as was done recently for comparing q^2-GRS and GOLPE [9]. One advantage of the q^2-GRS method is that it is very straightforward, and it is implemented entirely within the SYBYL working environment. The latter feature makes the application of this routine transparent for SYBYL users: the scripts to run q^2-GRS routine are written in SYBYL Programming Language and are available from our QSAR WWW server (http://mmlin1.pha.unc.edu/~jin/QSAR/).

6. Conclusions and Prospective

The successful development and application of the q^2-GRS method to several datasets illustrates several important aspects of the present and future applications of CoMFA in drug design. Our discovery that the results of conventional CoMFA are sensitive to the overall orientation of superimposed molecules on computer terminal shows that, for a given alignment, the single q^2 value obtained from standard CoMFA will most likely fall within the region of the highest frequency of q^2 (cf. Fig. 2–4). On the other hand, the low q^2 value obtained from conventional CoMFA (which, in many cases, will not be reported in the literature) may not necessarily be a result of a poor alignment, but may be caused merely by the poor orientation of superimposed molecules on the computer screen. Thus, simple reorientation of the set may significantly improve the results. For instance, Agarwal et al. [4] have reported the q^2 value of 0.481 which, as we have shown (Table 1), is lower by 0.3 q^2 units than the best value possible for their alignment.

Another important aspect of our work is that reporting the single value of q^2 and associated CoMFA fields as a result of standard CoMFA method appears inadequate. In general, scientists who use standard CoMFA routines should present the range of possible q^2 values (similar to our Fig. 2–4) instead of one number. Furthermore, the presentation of associated CoMFA fields becomes ambiguous because the shape of CoMFA fields varies with the q^2.

The successful development and implementation of the q^2-GRS [5,13,25], and related procedures [23,27], emphasizes one of the deficiencies of the standard CoMFA procedure — i.e. orientation dependence of the CoMFA results. Nevertheless, the 3D alignment rules in preparation for CoMFA remain one of the major sources of ambiguity. This problem can be circumvented by the development of alignment-free 3D structure-based descriptors that can be used in existing or novel QSAR protocols. New methods based on such descriptors are emerging [28,29], and this trend, in our opinion, should continue. The development of fast and fully automated procedures for descriptor generation and QSAR analysis is especially important today when the drug development process is characterized by the rapid accumulation of structural and bioactivity data through the means of combinatorial chemistry and high-throughput screening.

In summary, the new q^2-GRS routine developed in our laboratory, generates an orientation-independent, high q^2, generally exceeding the one obtained with the conventional CoMFA. We conclude that this novel routine that eliminates the major deficiency of the conventional CoMFA method shall be applied both to the future analyses and, perhaps, even to previously reported CoMFA studies in order to ensure the reproducibility of CoMFA results.

References

1. Cramer R.D., III, Patterson, D.E. and Bunce, J.D., *Comparative molecular field analysis (CoMFA): 1. Effect of shape on binding of steroids to carrier proteins*, J. Am. Chem. Soc., 110 (1988) 5959–5967.
2. Cramer, R.D., III, DePriest, S.A., Patterson, D.E. and Hecht., P., *The developing practice of comparative molecular field analysis*, In Kubinyi, H. (Ed.) 3D QSAR in drug design: Theory, methods and applications, ESCOM, Leiden, The Netherlands, 1993, pp. 443–485.
3. Thibaut, U., *Applications of CoMFA and related 3D QSAR approaches*, In Kubinyi, H. (Ed.) 3D QSAR in drug design: Theory, methods and applications, ESCOM, Leiden, The Netherlands, 1993, pp. 661–696.
4. Agarwal, A., Pearson, P.P., Taylor, E.W., Li, H.B., Dahlgren, T., Herslof, M., Yang, Y., Lambert, G., Nelson, D.L., Regan, J.W. and Martin, A.R., *Three-dimensional quantitative structure–activity relation-*

ships of 5-HT receptor binding data for tetrahydropyridinylindole derivatives: A comparison of the Hansch and CoMFA Methods, J. Med. Chem., 36 (1993) 4006–4014.

5. Cho, S.J. and Tropsha, A., *Cross-validated R^2-guided region selection for comparative molecular field analysis (CoMFA): A simple method to achieve consistent results*, J. Med. Chem., 38 (1995) 1060–1066.
6. Waller, C.L., Oprea, T.I., Giolitti, A. and Marshall, G.R., *Three-dimensional QSAR of human immunodeficiency virus (I) protease inhibitors: 1. A CoMFA study employing experimentally-determined alignment rules*, J. Med. Chem., 36 (1993) 4152–4160.
7. Debnath, A.K., Hansch, C., Kim, K.H. and Martin, Y.C., *Mechanistic interpretation of the genotoxicity of nitrofurans (antibacterial agents) using quantitative structure–activity relationships and comparative molecular field analysis*, J. Med. Chem., 36 (1993) 1007–1016.
8. Brusniak, M.Y., Pearlman, R.S., Neve, K.A. and Wilcox, R.E., *Comparative molecular field analysis-based prediction of drug affinities at recombinant D1A dopamine receptors*, J. Med. Chem., 39 (1996) 850–859.
9. Ortiz, A.R., Pastor, M., Palomer, A., Cruciani, G., Gago, F. and Wade, R.C., *Reliability of comparative molecular field analysis models: Effects of data scaling and variable selection using a set of human synovial fluid phospholipase A2 inhibitors*, J. Med. Chem., 40 (1997) 1136–1148.
10. Marshall, G.R., Barry, C.D., Bosshard, H.E., Dammkoehler, R.A. and Dunn, D.A., *The conformational parameter in drug design: The active analog approach*, In Olsen, E.C. and Christoffersen, R.E. (Eds.), *Computer-assisted drug design*, ACS Symp. Series, Vol. 112, American Chemical Society, Washington, DC, 1979, pp. 205–226.
11. Martin, Y.C., *Overview of concepts and methods in computer-assisted rational drug design*, Methods Enzymol., 203 (1991) 587–613.
12. Martin, Y.C., Bures, M.G., Danaher, E.A., DeLazzer, J., Lico, I. and Pavlik, P.A., *A fast new approach to pharmacophore mapping and its application to dopaminergic and benzodiazepine agonists*, J. Comput. Aided Mol. Des., 7 (1993) 83- 102.
13. Cho, S.J., Serrano, M.G., Bier, J. and Tropsha, A., *Structure based alignment and comparative molecular analysis of acetylcholinesterase inhibitors*, J. Med. Chem., 39 (1996) 5064–5071.
14. Villalobos, A., Blake, J.F., Biggers, C.K., Butler, T.W., Chapin, D.S., Chen, Y.L., Ives, J.L., Jones, S.B., Liston, D.R. and Nagel, A.A., *Novel benzoisoxazole derivatives as potent and selective inhibitors of acetylcholinesterase*, J. Med. Chem., 37 (1994) 2721–2734.
15. Ishihara, Y., Hirai, K., Miyamoto, M. and Goto, G., *Central cholinergic agents: 6. Synthesis and evaluation of 3-[1-(phenylmethyl)-4-piperidinyl]-1-(2,3,4,5-tetrahydro-1H-1-benzazepin-8-yl)-1-propanones* and their *analogs as central selective acetylcholinesterase inhibitors*, J. Med. Chem., 37 (1994) 2292–2299.
16. Chen, Y.L., Liston, D., Nielsen, J., Chapin, D., Dunaiskis, A., Hedberg, K., Ives, J., Johnson, J. Jr. and Jones, S., *Syntheses and anticholinesterase activity of tetrahydrobenzazepine carbamates*, J. Med. Chem., 37 (1994) 1996–2000.
17. Vidaluc, J.L., Calmel, F., Bigg, D., Carilla, E., Stenger, A., Chopin, P. and Briley, M., *Novel [2-(4-piperidinyl) ethy](thio)ureas: Synthesis and antiacetylcholinesterase activity*, J. Med. Chem., 37 (1994) 689–695.
18. Sasho, S., Obase, H., Ichikawa, S., Kitazawa, T., Nonaka, H., Yoshizaki, R., Ishii, A. and Shuto, K., *Synthesis of 2-imidazolidinylidenepropanedinitrile derivatives as stimulators of gastrointestinal motility*, J. Med. Chem., 36 (1993) 572–579.
19. Sussman, J.L., Harel, M., Frolow, F., Oefner, C., Goldman, A., Toker, L. and Silman, I., *Atomic structure of acetylcholinesterase from* Torpedo californica: *A prototypic acetylcholine-binding protein*, Science, 253 (1991) 8872–8879.
20. Harel, M., Schalk, I., Ehret-Sabatier, L., Bouet, F., Goeldner, M., Hirth, C., Axelsen, P.H., Silman, I. and Sussman, J.L., *Quaternary ligand binding to aromatic residues in the active-site gorge of acetylcholinesterase*, Proc. Natl. Acad. Sci. USA, 90 (1993) 9031–9035.
21. Huang, M.T., *Harringtonine, an inhibitor of initiation of protein biosynthesis*, Molecular Pharmacol. 11 (1975) 511–519.

22. Taylor, E.W. and Agarwal, A., *3-D QSAR for intrinsic activity of 5-HT_{1A} receptor ligands by the method of comparative molecular field analysis*, J. Comp. Chem., 14 (1993) 237–245.
23. The program SYBYL 6.3 is available from Tripos Associates, 1699 South Hanley Road, St Louis, MO 63144, U.S.A.
24. David E. Patterson (Tripos Associates), personal communications.
25. Cho, S.J., Tropsha, A., Suffness, M., Cheng, Y.C. and Lee, K.H., *Antitumor agents: 163. Three-dimensional QSAR study of 4′-O-demethylepipodophyllotoxin analogs using the modified CoMFA/q^2-GRS approach*, J. Med. Chem., 39 (1996) 1383–1395.
26. Jones, J.P., He, M., Trager, W.F. and Rettie, A.E., *Three-dimensional quantitative structure–activity relationship for inhibitors of cytochrome P4502C9*, Drug Metabol. Dispos., 24 (1996) 1–6.
27. Baroni, M., Costantino, G., Cruciani, G., Riganelli, D., Valigi, R. and Clementi, S., *Generating optimal linear PLS estimations (GOLPE): An advanced chemometric tool for handling 3D-QSAR problems*, Quant. Struct.-Act. Relat., 12 (1993) 9–20.
28. Silverman, B.D. and Platt, D.E., *Comparative molecular moment analysis (CoMMA): 3D-QSAR without molecular superposition*, J. Med. Chem., 39 (1996) 2129–2140.
29. Ginn, C.M.R., Turner, D.B. and Willett, P., *Similarity searching in files of three-dimensional chemical structures: Evaluation of the EVA descriptor and combination of rankings using data fusion*, J. Chem. Inf. Comput. Sci., 37 (1997) 23–37.

GOLPE-Guided Region Selection

Gabriele Cruciani[a], Sergio Clementi[a] and Manuel Pastor[b]

[a] *Laboratory for Chemometrics, Chemistry Department, University of Perugia, Via Elce di Sotto 10, I-06123 Perugia, Italy*

[b] *Department of Physiology and Pharmacology, University of Alcala, Campus Universitario, E-28871, Alcala de Henares, Spain*

1. Introduction

One of the most important tasks of computer chemistry in drug design is the graphical representation of molecular properties. Nowadays, molecules can be precisely represented in the computer and ligand–receptor interactions can be simulated in a sophisticated way. Force fields and docking procedures can be of help to highlight the regions around the receptors where the ligand–receptor interactions are more favorable, thus leading to a discrete partitioning of the surrounding space. Therefore, computer simulations provide a numerical description of the phenomena under investigation which can be used by the medicinal chemist in order to design better ligands or more selective compounds [1].

An important drawback of computer chemistry is that the interpretation of the data and graphics given by such an exhaustive description can be overwhelming. Moreover, accompanying the increased number of descriptors, there is usually a decrease in the overall signal–noise ratio, with the result that important information may be hidden in the middle of the data. Appropriate chemometric tools can be applied to extract from the noise all the useful information.

However, although chemometrics have been used for a long time in drug design, no method can handle the information contained at explicit spatial regions as a whole, and this information has to be coded into isolated grid-point *variables*. 3D QSAR methods such as CoMFA [2], CoMPA [3], CoMSIA [4] and others describe molecules by means of variables which represent steric and electrostatic interaction energies with probes at single, definite positions. This description has two deficiencies: first, it lacks the *continuity constraints* that arise because neighboring grid-point variables contain similar chemical information. Second, the information is often spread out in several contiguous yet isolated independent variables.

New procedures are emerging that use the information given by the positions of the variables around the molecules. However, so far these procedures use only geometric criteria to build the regions around the molecules. This gives rise to inhomogeneity in terms of the amount of information embedded in these regions. In fact, some regions often do not contain information at all, or alternatively, a single piece of chemical information is spread out in many different regions. The problem is that, while it is simple to define regions containing homogeneous chemical information for a single molecule, it is very difficult to do so for a series of compounds, as in a 3D QSAR study.

The aim of this chapter is to present a novel 3D QSAR approach that aims to define homogeneous regions around the molecules of the series under study. This allows

H. Kubinyi et al. (eds.), 3D QSAR in Drug Design, Volume 3. 71–86.

correlating the information given by these regions with the biological activity of the compounds by selecting only these regions strongly related to the property under investigation.

2. The Meaning of a 3D Region

Chemically speaking, a three-dimensional region can be defined as an assembly of positions, close to one another in Euclidean space, where the structural, energetic or chemical properties of a molecule are similar and defined. For instance, the hydrophobic region reproduced by the side chain of a tryptophan amino acid or the negative electrostatic potential region induced hereby an aspartate are good examples of three-dimensional regions with a precise chemical meaning. Similarly, in docking procedures a binding-site region is defined as a place where the structural and energetic properties of the macromolecule favor the interaction with an explicit chemical group or ligand molecule. However, it should be noted that the first two examples of regions express the properties of actual molecules while the binding-site regions represents the potential interaction of the receptor with different ligands.

Problems arise when a number of molecules are studied at the same time as in the case of 3D QSAR strategies. In 3D QSAR the molecules are superimposed and occupy equivalent positions in the space. As shown in Fig. 1, three different molecules might induce different effects in space, for instance, different molecular electrostatic potential (MEP), which define different regions in different positions in the space. The effect of the three molecules can be seen from the point of view of a hypothetical receptor, which is feeling the interaction that comes out from the different compounds. Since the receptors are chemical entities as well, the ligands induce effects on whole regions of the space and not in a few single points. For each of the molecules, it is relatively simple to identify regions with appropriate chemical meaning. However, when all three molecules are considered simultaneously, the regions so clearly identified around each of the isolated molecules lose their chemical meaning and are no longer useful.

Conversely, it often happens that in other zones the molecules exhibit the same chemical characteristics (see Fig. 1). In this case, the aforementioned region around the isolated molecules maintains its chemical meaning, even in the global model in which the molecules are considered together.

In conclusion, in the context of 3D QSAR, a 3D region may be defined as a portion of the space surrounding the compounds which is affected in the same way by the structural variation in the series of the molecules. As a consequence of the definition, all such regions contain homogeneous chemical information and may represent, ideally, putative residues of the receptors, which interact in a similar way with all the compounds of the series.

3. How to Define 3D Regions in 3D QSAR

In 3D QSAR methodologies the compounds are described by a large number of isolated grid-field variables. Depending on the force field and on the computational procedure

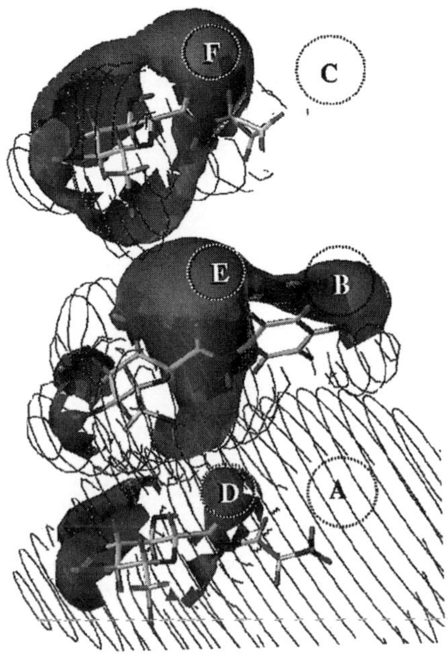

Fig. 1. Molecular Electrostatic Potential (MEP) plots of three glucose analog inhibitors experimentally superimposed in the active site of the GPb enzyme. Lines represent positive MEP values, while surfaces represent negative MEP values and A, B and C label the same 3D region. In A the MEP is highly positive, in B it is highly negative and in C it is neutral. Conversely, D, E and F represent a homogeneous region where the MEP is negative, although with different magnitude.

used, these grid-field values may represent total interaction energies [5], steric and electrostatic interactions [2], molecular electrostatic potential [3], hydrophobic interactions [6] or a mixture of some of them. In this context, defining 3D regions of homogeneous variables means finding a criterion on which one could extract, from a matrix of descriptors, groups of neighboring variables bearing the same information.

This is not a trivial point: it is clear that variables belonging to the same region should be close in 3D space; however, the Euclidean distance is a necessary but not sufficient criterion to discriminate between regions. Indeed, variables that are very close in the 3D space often carry opposite chemical information. This is particularly common at the molecular surface where the interaction energy of adjacent grid-points (variables) changes sharply from attractive to repulsive. In other words, not only the distances in Euclidean space, but also the amount and type of information contained in the variables, should be taken into account in defining a region.

The region definition (RD) procedure [7–8] described here works by extracting a subset of highly informative X-descriptors and then partitioning the space around the molecules among them.

Our computational algorithm involves three major steps: (1) selecting the most informative variables (seeds) from an initial PCA or PLS model; (2) building polyhedra around the seeds containing variables which are close in 3D space; and (3) merging together polyhedra that contain similar information.

It should be noticed that step 1 is performed on the chemometric space of PCA loadings or PLS weights of the descriptor matrix, while steps 2 and 3 are performed in the real Euclidean space around the molecules. These two steps are repeated separately for each probe or field (steric, electrostatic, hydrophobic) used to describe the compounds.

1. *Seed selection*: Fig. 2 illustrates steps 1 and 2. An initial PLS or PCA model is made on the X-matrix and a given umber of variables are extracted following a D-optimal design criterion from the chemometric space of PLS weights or PCA loadings. These selected variables are called *seeds*. Variables selected in such a way are guaranteed as being of high statistical importance. Moreover, the D-optimal criterion assures that most of them contain independent information.
2. *Voronoi polyhedra*: the seeds selected in the previous step are placed back in the real 3D space around the molecules, in the field to which they belong (see Fig. 2). Then each X-variable in the dataset is assigned to the nearest seed in 3D space, thus producing a number of Voronoi polyhedra (VPs). The Voronoi polyhedra are the first attempt to produce 3D region. They have a shape and size which depends upon the amount of information they contain. For instance, those placed near to the molecules in areas rich in information tend to be smaller, while those far away grow larger. Usually these regions around the molecules where no interaction is possible, or positions where the compounds in the series exhibit no chemical variation. In this case, the variables belonging to these areas are assigned to a special group called *group 0*. Therefore this group 0 contains variables that are far away from any seed and that are impossible to group in steps 1 and 2.
3. *Collapsing of polyhdedra*: the Voronoi polyhedra can be used directly as 3D regions, but if neighboring regions contain the same information, they can be profitably combined together to produce larger regions. In order to check if the neighboring regions actually contain the same information or not, the algorithm computes the correlation of the information contained in the regions. Only the regions for which this information is strongly correlated are merged into a single new common 3D region. The operation is called *collapsing*: it first computes, for each polyhedron, three more vectors that describe the numerical content of the polyhedron. The algorithm then looks for the two nearest polyhedra and makes pair-wise comparisons of the vector sign patterns. If the patterns are different, no collapsing is performing. However if the patters are similar, the algorithm computes the correlation coefficient between the vectors. The polyhedra are merged into a new region only if the correlation coefficient is greater than a certain cutoff value. The procedure is explained in detail in reference [9].

Such procedure ensures obtaining single, independent pieces of information. Regions rich in information contain many informative seeds, which compete for the space, thus producing many small polyhedra in step 2 of the algorithm. Conversely, areas poor in

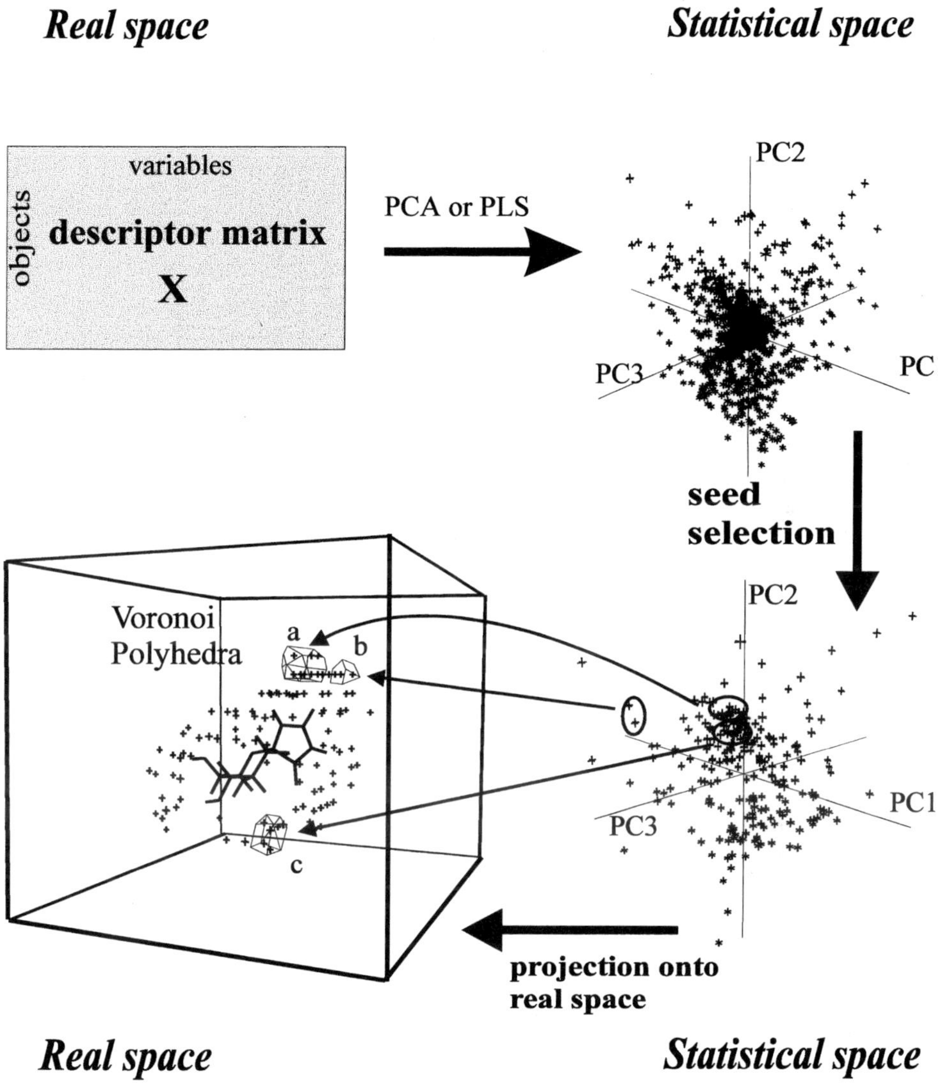

Fig. 2. Scheme of the first and second steps of the RD algorithm. The information flow starts from the descriptor matrix, goes along the statistical space and finally ends in the real 3D space of molecules. Seeds close in the statistical space can be variables far away in the real space. Polyhedra a and b can be merged in the third step of the RD algorithm only if they contain the same chemical information.

information will contain few seeds, thus generating a few larger polyhedra. It is important to point out that the regions formed are strictly dependent upon the probe used; different probes describe different interactions and generate different regions, as is the case in the real world and not only in the simulations phase.

4. How to Check the Correlation between the 3D Regions

Any empirical model is highly dependent on the information contained in the structural data. Often, the information given by different 3D regions is correlated, just as the substitution pattern of a poorly designed QSAR series can be correlated. This is a consequence of the fact that two or more 3D regions contain the same information for the statistical model and their effect on the response cannot be separated, nor independently quantified (see Fig. 3). Moreover, if the number of the correlated 3D regions increases, the chance of finding misleading models increases accordingly. From a different point of·view, the knowledge of the correlation between the 3D regions is a valuable source of information of the amount of chemical variability contained in the data and very illustrative of the structural characteristics of the molecules that can be further investigated.

The third step of the RD algorithm checks the correlation between the 3D regions. When the collapsing Euclidean distance value is increased, groups far way from one another (even in opposite corners of the grid cage, if the cutoff distance is enough) are merged together (see Fig. 3). There is nothing wrong in this phenomenon, which highlights the presence of at least two areas, say, A and B, that contain correlated information in the actual series. It means that a change in the structure of area A is always accompanied by a similar change in the B area structure. In this case, it will not be possible to know if an increase of the interactions in the area A or area B, or in both areas, will result in a corresponding modification of the biological response. In this case, it is advisable to de-correlate such A and B areas by adding appropriate molecules to the dataset.

5. Advantages of Working with 3D Regions

Although defining homogeneous regions is not simple, working with regions, instead of isolated variables, can be advantageous for several reasons:

1. In a typical PLS analysis, the three-dimensional matrices of energies are unfolded into vectors to build the matrix of descriptors X. The result is that the variables are considered individually and neighboring variables are spread out in different (often distant) positions of the **X** matrix. Thus, the spatial relationships of the variables are lost and the spatial continuity constraints are ignored. In contrast, with the use of the 3D regions, the spatial correlation and the continuity constraints are implicitly incorporated into the chemometric analysis. This adds stability to the models.
2. Regions do exist, and any attempt to predict their effects must take into account this simple fact. Even the smallest structural change in a compound will be reflected not in a single variable only, but rather in a group of spatially contiguous variables. These contiguous groups of variables represent portions of the space surrounding the compounds that are affected in the same way by the structural variations in the series. As a consequence, all variables inside the group bear the same information [8] and, hence, the use of groups can clarify the chemical interpretation [8] of the models.

3. New 3D QSAR approaches [10] are being developed explicitly to address the effect of water molecules in receptor–ligand interactions and to quantify their importance in the activity. The effect of the water molecules is not sectionable and there are advantages in describing them by homogeneous regions of joined variables, instead of by a set of isolated variables.
4. Finally, as reported above, by considering the correlation between distant 3D regions, one can identify a poor design of the series and suggest exploring new structural characteristics of the molecules.

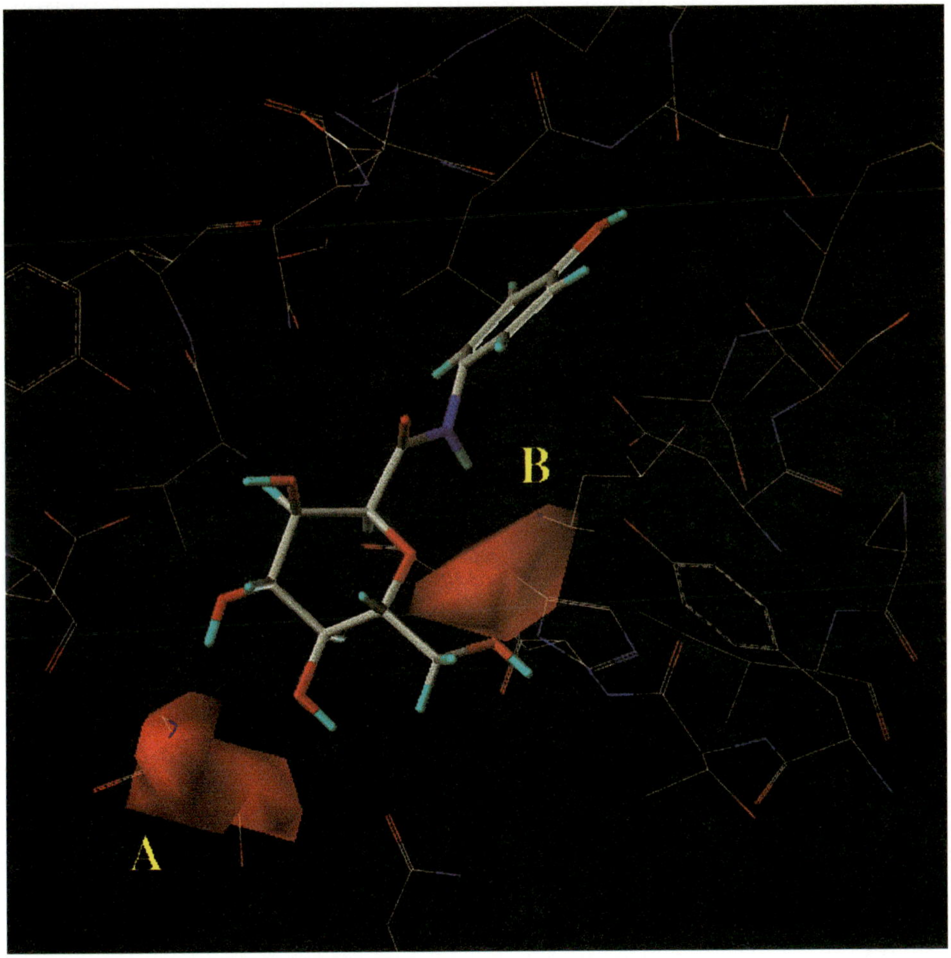

Fig. 3. A glucose analog inhibitor in the GPb pocket. The RD algorithm reveals two areas, say, A and B, containing correlated information. Area A describes a homogeneous region where all the 47 compounds of Table 2 interact in a similar way with Ser674. Area B represents another region where the 47 compounds interact with His337 backbone. A and B regions are more than 9 Å apart, but since they show internal correlation, their effect in the biological activity cannot be properly dissected.

6. How to Relate the 3D Regions to the Biological Response

The 3D regions are groups of neighboring variables in real 3D space bearing the same information. These regions can be correlated with the biological properties of the compounds using an adapted partial least squares (PLS) or other chemometric models.

When a 3D region contains a large number of variables, the dimensionality of the model can benefit from the data reduction obtained from the replacement of all these variables with their weighted average. A more sophisticated data reduction can be made performing a Principal Component Analysis (PCA) of the variables within each 3D region and substituting the variable values in the 3D region with the principal component scores. These approaches, especially the second one, are very promising, although the procedure is still under development and not so far sufficiently tested.

It should be borne in mind that the region definition RD algorithm does not render a model, nor introduce new information; indeed, it only uses the information present in the series to group the isolated variables into regions. For this reason, the models obtained from isolated variables do not present large differences with respect to those obtained from 3D regions. However, the interpretation of models obtained from 3D regions is straightforward and the variable selection performed on regions is more robust than the classical variable selection procedures, as is shown in the next section.

7. How to Select the Most Important 3D Regions

The 3D regions generated by the RD algorithm can be used directly to replace the individual variables in the GOLPE [11] variable-selection method. Once the 3D regions are defined, a modified GOLPE procedure [7,8] evaluates the effect of these regions of joined variables on the predictive ability of the PLS model. The procedure is able in the end to retain the 3D regions that increase the predictive ability of the model, and to remove those 3D regions that do not improve the model.

Different procedures for region selection have been suggested [12,13]. However, they use non-homogeneous regions, and the validation and selection criteria deserve further discussion. The GOLPE-guided region selection strategy, on the other hand, is based on use of reduced models made with combinations of 3D regions according to a FFD where each of the two levels (plus and minus) corresponds to the presence and absence of the regions (see Fig. 4). The flowchart of the procedure is reported in Table 1.

The first step of the procedure is to build the design matrix. The design matrix proposed to test the prediction ability of these reduced models involves combinations of 3D regions. In the combination matrix, each column represents a 3D region; for each combination (i.e. for each row of the combination matrix), regions are included in the model if the *plus* is present and excluded if the *minus* sign is present in the row according to a fractional factorial design.

In the second step, some dummy regions can be inserted in the combination matrix to better evaluate the effect of the real 3D region. Then, in the third step, for each such combination, the prediction ability of the corresponding PLS model can be evaluated by cross-validation using the leave-many-more-out method implemented in the GOLPE

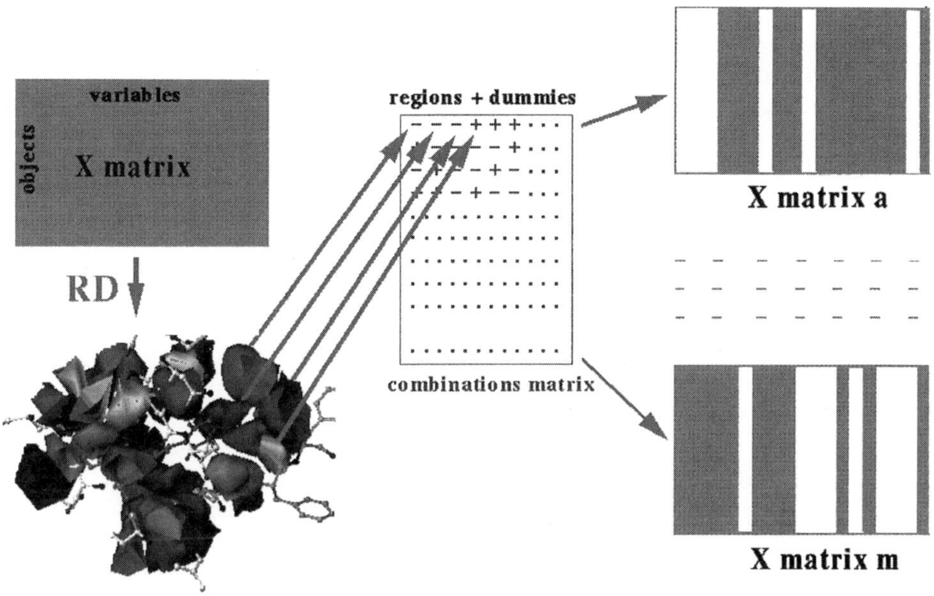

Fig. 4. Once the 3D regions are generated by the RD algorithm, these regions are temporarily kept out (−) or used (+). This means that for each row of the combination matrix a different perturbation is made in the original **X** *matrix.*

Table 1 *Flow chart of the GOLPE region selection procedure*

Step 0:	The RD algorithm defines the 3D regions
Step 1:	Build a design matrix as in Figure 4.
Step 2:	Include in the design matrix a number of dummy regions
Step 3:	Compute the predictive ability for each region combination
Step 4:	Evaluate the effect of each region on predictive ability
Step 5:	Divide regions into positive, detrimental and uncertain for predictive ability.
Step 6:	Compute the model with the reduced matrix using the positive and uncertain regions

procedure. It should be pointed out that for each row of the combination matrix step 3 produces a standard deviation of error of prediction (SDEP). SDEP is exactly reproducible only for leave-one-out or leave-two-out cross-validation, while for leave-more-out it is not exactly reproducible, even if it converges to an asymptotic value. The fourth step is used to compute, by means of the Yates algorithm, the effects of the 3D regions and those of the dummy regions on the predictive ability of the models. Once the effects of 3D regions computed, the fifth step is used to classify the 3D regions into three main categories (helpful, detrimental for the model or with an uncertain effect). The final step selects the helpful and the uncertain 3D regions and discards the detrimental regions. The reduced matrix produced by the algorithm can be used for statistical modelling, or for another region selection procedure that starts from this point.

The advantage of using 3D regions in variable selection is two-fold: first, the analysis takes into account the information about their 3D position, thus introducing a new constraint (the spatial continuity constraint) which minimizes the risk of chance effects and leads to more predictive models [7]. Second, the selected variables are grouped in space, and so are the results of the PLS analysis, thus greatly increasing their interpretability. Moreover, the method represents a compromise between the requirement to simplify models and plots and to minimize undesirable oversimplifications. In addition, since the number of regions is significantly smaller than the number of variables, the combined RD/GOLPE method does not require variable pre-selection. From a computational point of view, the algorithm is completed in a fraction of the time required for the regular FFD variable selection.

8. Alternative Methods that Generate 3D Regions

There are other ways in which the X-variables (grid nodes) can be grouped. The first attempt to group isolated variables [12] used squared boxes of fixed size following only a geometrical criterion. The regions formed following such a scheme have a fixed shape and a size that does not depend upon the amount of information given by the variables. This does not guarantee that each box contains a single different piece of information, expressing that effect of a structural modification; some boxes will contain little or no information, while others will express the effect of diverse structural changes in the series. Even worse, some pieces of information can be split in two or more contiguous boxes [7,8].

Consequently, it is doubtful that the boxes generated by this method can be successfully used in a box-selection procedure because, as mentioned above, they do not contain unique information. Moreover, this method can be further criticized because the effect of the variables included in each box on the predictive ability is evaluated individually (one box at a time) without using any design criteria for selecting a representative number of box combinations.

Other authors [13] have used the same approach to define the boxes around the molecules, although using a design criterion in a GOLPE-like fashion, reporting only marginal improvements on the predictive ability.

9. Case Study

In this contribution, we wish to show some results obtained in a GRID/GOLPE CoMFA-like study on a set of recently synthesized glucose analog inhibitors [7] of the glycogen phosphorylase *b* (GP*b*) enzyme, reported in Table 2.

This set is especially suitable for 3D QSAR methodological research, because high-resolution crystallographic structures of the enzyme–ligand complexes are available for every compound in the series. Therefore, the conformation and the superposition of the compounds have been experimentally determined and it is possible to investigate the effect of different parameters on the quality of the models.

Table 2 *Series of glucose analog inhibitors of glycogen phosphorylase b (reference [7])*

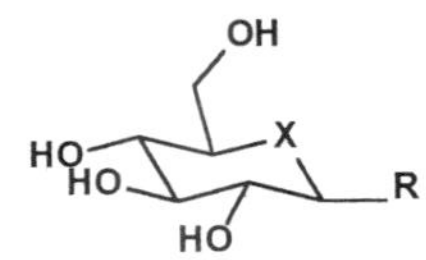

	X	Alpha	Beta	K_i (mM)		X	Alpha	Beta	K_i (mM)
1	O	OH	H	1.7	**23**	O	H	$NHC(=O)NH_2$	0.14
2	O	$C(=O)NH_2$	H	0.37	**24**	O	H	$NHC(=O)CH_3$	0.032
3	O	$C(=O)NHNH_2$	H	3.0	**25**	O	H	$NHC(=O)CH_2CH_3$	0.039
4	O	$COOCH_3$	H	24.2	**26**	O	H	$NHC(=O)CH_2CH_2CH_3$	0.094
5	O	$CH_2NH_3^+$	H	34.5	**27**	O	H	$NHC(=O)CH_2Cl$	0.045
6	O	CH_2N_3	H	22.4	**28**	O	H	$NHC(=O)CH_2Br$	0.044
7	O	CH_2OH	H	1.5	**29**	O	H	$NHC(=O)CH_2NH_2$	0.37
8	O	H	O-(1-6)-D-Glucose	16.3	**30**	O	H	NHC(=O)-Ph	0.081
9	O	H	$C(=O)NH_2$	0.44	**31**	O	H	$NHC(=O)CH_2NHCOCH_3$	0.99
10	O	H	$C(=O)NHCH_3$	0.16	**32**	O	H	$NHCOOCH_2$-Ph	0.35
11	O	H	$C(=O)NHCH_2CH_2OH$	2.6	**33**	O	H	$CH_2OSO_2CH_3$	4.8
12	O	H	C(=O)NH-Ph	5.4	**34**	O	H	1H-Indol-2-yloxy	2.6
13	O	H	C(=O)NH-4-OH-Ph	4.4	**35**	O	$C(=O)NH_2$	$NHCOOCH_3$	0.016
14	O	H	$C(=O)NHNH_2$	0.4	**36**	O	OH	CH_2OH	15.8
15	O	H	$C(=O)NHNHCH_3$	1.8	**37**	O	OH	CH_2N_3	7.4
16	O	H	$C(=O)NHCH_2CF_3$	8.1	**38**	O	OH	CH_2CN	7.6
17	O	H	C(=O)NH-Ciclopropyl	1.3	**39**	O	OH	$CH_2OSO_2CH_3$	3.7
18	O	H	$COOCH_3$	2.8	**40**	O	H	SH	1.0

Table 2 *(Continued)*

	X	Alpha	Beta	K_i (mM)		X	Alpha	Beta	K_i (mM)
19	O	H	$CH_2NH_3^+$	16.8	**41**	O	H	$SCH_2C(=O)NH_2$	21.1
20	O	H	$CH_2CH_2NH_3^+$	4.5	**42**	O	H	$SCH_2C(=O)NH$-Ph	3.6
21	O	H	CH_2N_3	15.2	**43**	O	H	$SCH_2C(=O)NH$-2,4-$(F)_2$-Ph	18.9
22	O	H	CH_2CN	9.0	**44**	S	OH	H	2.0

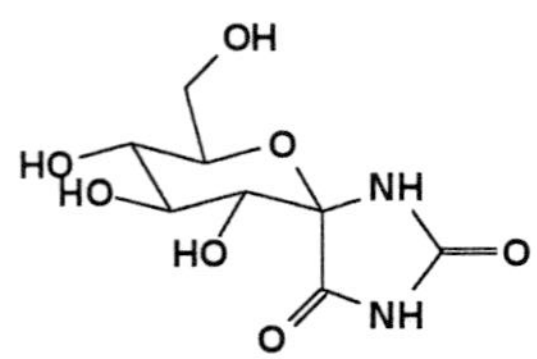

45 K_i = 0.003 mM

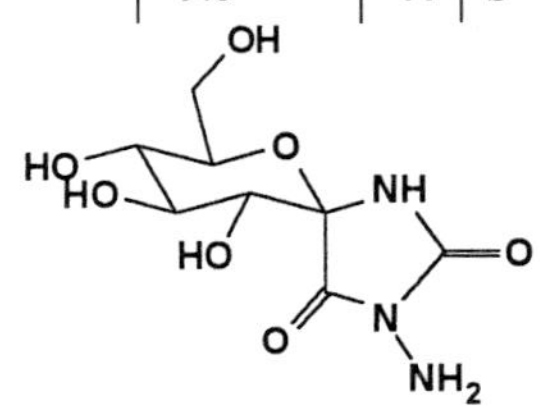

46 K_i = 0.146 mM

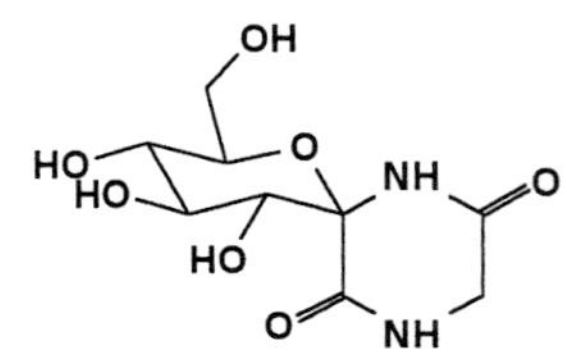

47 K_i = 0.0597 mM

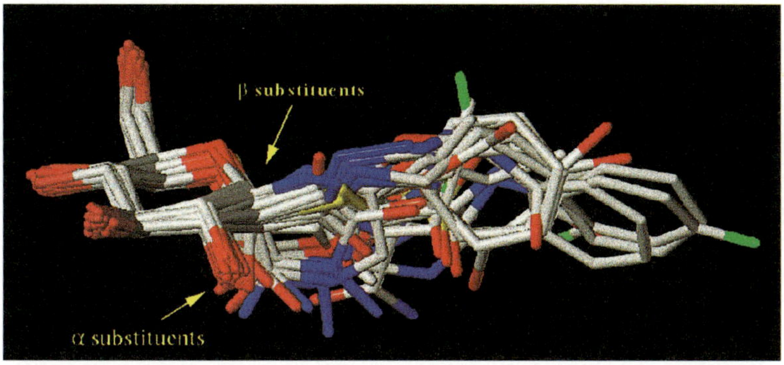

Fig. 5. All the glucose analog inhibitors of Table 2, superimposed in the conformations found in the X-ray crystallographic structures. Arrows indicate substituents at C1 in alpha and beta configurations.

The inhibitors were considered in the conformation and position found in the crystal, and no further superposition operation was applied. All inhibitors superimposed in the GP*b* active site are reported in Fig. 5; further details are given in references [14–18]. The energy calculations were carried out using the GRID [5] program and the phenolic hydroxyl group probe (OH). The size of the box was defined in such a way that it extends about 4 Å from the structure of the inhibitors. GRID calculations were carried out using 1 Å grid spacing, thus giving 7920 probe–target interactions for each compound, which were unfolded to produce a one-dimensional vector of variables. A cutoff of +20.9 kJ/mol (5 kcal/mol) was applied to produce a more symmetrical distribution of the **X** matrix. The matrix was imported into GOLPE 3.0.3 and further pre-treated zeroing values having absolute values smaller than 0.42 kJ/mol (0.1 kcal/mol), deleting variables with standard deviation below 0.1 and removing variables with skewed distribution (two- and three-level variables).

On this matrix, we applied the RD algorithm, described above, with the following parameters: 450 seeds selected on the PLS weights space, critical distance cutoff of 1.0 Å and collapsing distance cutoff of 2.0 Å. These regions were used in a later step in

Table 3 *Results of PLS modeling with different variable/region selection procedures*

Var. sel.	Vars.	Dimen.[b]	R^{2c}	Q^{2d}	SDEP[e]
None	2087	4	0.92	0.50	0.69
FFD	379	4	0.93	0.73	0.51
RD/FFD	457	4	0.95	0.79	0.45
Tropsha	336	4	0.82	0.59	0.63

[a] Number of variables used in the PLS model.
[b] Dimensionality of the model.
[c] Squared correlation coefficient.
[d] Cross-validated squared correlation coefficient.
[e] Standard deviation of error of predictions.

an FFD-selection procedure. PLS analysis was carried out without variable selection, with regular GOLPE variable selection and with SRD/GOLPE region selection (a single FFD selection performed on regions).

The model produced by RD/GOLPE is the best from the point of view of its interpretability. Figure 6 shows the coefficients grid plot for plain PLS model and for RD/GOLPE variable selection. Active site residues are superimposed for reference. From Fig. 6a, it can be seen that the model contains so many small coefficients that this model is not useful for interpretation; conversely, Fig. 6b is simpler to interpret. Although the RD/GOLPE retains only 20% of the original variables (see Table 3), such variables highlight all the major effects and are clustered in space. The numerical results, listed in Table 3, indicate that PLS models obtained with both variable and region selection are better than the simple PLS model. It is noteworthy that the RD/GOLPE method produces a slightly better model than GOLPE itself, although without variable pre-selection and in a single run.

The same dataset was used to evaluate the predictive ability of the models obtained using the Tropsha method. In this approach, the grid cage was split into 125 ($5 \times 5 \times 5$) boxes and singular PLS models were derived using only the variables inside of each box, one at a time. In order to be able to compare the results, the predictive ability of such models was assessed using the leave-more-out cross-validation method, as

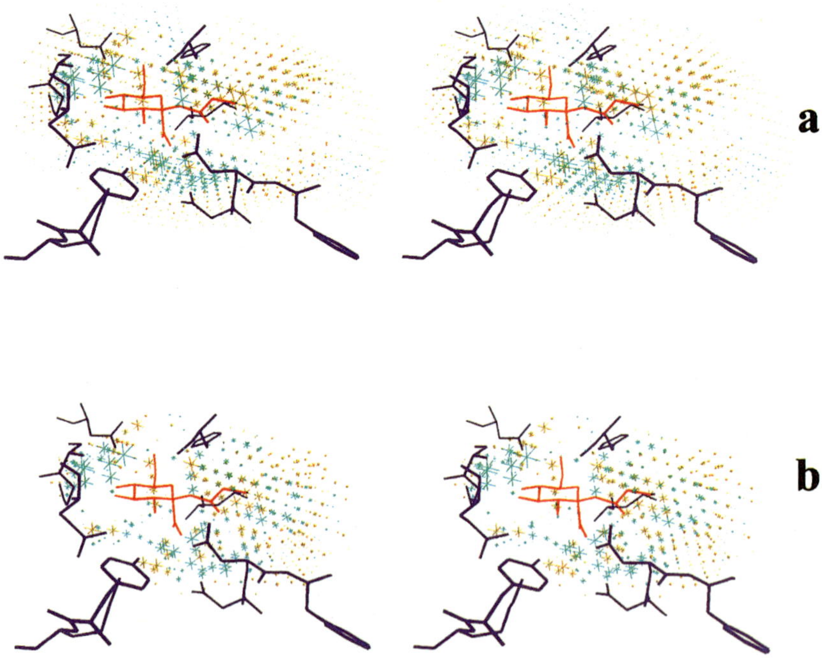

Fig. 6. Stereoview of the coefficient grid plot for (a) a plain PLS model, and (b) SRD/GOLPE model. Residues of the GPb active site have been included for reference.

opposed to the LOO procedure described in the original method. Only the 12 boxes with a Q^2 higher than 0.2 were used in the final model. The overall model has a slightly better predictive ability than the original PLS model, but the prediction error (SDEP) is about 40% larger than that obtained with our FFD/RD procedure. Moreover, a graphical analysis reveals that the Tropsha procedure removes all the variables in one of the pockets of the active site, hence excluding any possible interpretation of the effects of the substituents in these positions.

In order to compare the methods of variable and region selection, it is of critical importance to make sure that the cross-validation procedures actually reflect the real predictive quality of the models. Therefore, external validation was carried out using six newly synthesized GP*b*-inhibitor compounds. The results are presented in Table 4.

It should be noted that the models obtained using both GOLPE FFD procedures produce better external predictions (smaller SDEP). The best results were obtained with the GOLPE procedure applied to regions, whereas the Tropsha [12] method, in this dataset, fails to improve the external prediction, compared with the plain CoMFA model.

In conclusion, the numerical results listed in Tables 3 and 4 indicate that PLS models obtained with the region-selection procedure RD/GOLPE are better than the simple PLS model, both in internal and external validation. The RD/GOLPE method, in this dataset, produces models that are more stable and simpler to interpret. In our opinion, the power of the procedure is a consequence of the chemical and statistical homogeneity of the regions selected by the RD algorithm, together with the design criteria method used to select the regions in the validation phase.

Table 4 *External predictions obtained using models reported in Table 3*

Compound	Experimental	None	FFD	RD/FFD	Tropsha
bnac2	4.06	4.04	4.09	4.10	3.54
cmpd	3.50	3.30	3.32	3.21	1.97
hyda	3.26	1.39	1.57	1.83	2.04
hydo	4.40	3.38	3.31	3.24	2.62
old13	1.70	2.44	2.32	2.26	1.86
old29	2.00	1.16	1.55	1.75	2.67
SDEP[a]	–	0.98	0.88	0.80	1.15

[a] External standard deviation of error of predictions.

Acknowledgements

We thank our colleagues L.N. Johnson, K.A. Watson, M. Gregoriou, G.W.J. Fleet and N.G. Oikonomakos for sending data regarding some of the compounds in the training set and compounds in Table 4 prior to their publication. We thank the EC for providing financial support (project BIO2-CT943025), including a grant for one of us (M.P.). The Italian funding agencies of MURST and CNR are also thanked for financial support.

References

1. Kunz, I.D., Meng, E.C. and Shoichet, B.K., *Structure-based molecular design*, Acc. Chem. Res., 27 (1994) 117–123.
2. Cramer, R.D. III, Patterson, D.E. and Bunce, J.D., *Comparative molecular field analysis (CoMFA): 1. Effect of shape on binding of steroids to carrier proteins*. J. Am. Chem. Soc., 110 (1988) 5959–5967.
3. Floersheim, P., Nozulak, J. and Weber, H.P., *Experience with comparative molecular fields analysis*, In Wermuth, C.G. (Ed.) Trends in QSAR and molecular modeling 92, ESCOM, Leiden, The Netherlands, 1993, pp. 227–232.
4. Klebe, G., Abraham, U. and Mietzner, T., *Molecular similarity indices in a comparative analysis (CoMSIA) of drug molecules to correlate and predict their biological activity*, J. Med. Chem., 37 (1994) 4130–4146.
5. Boobbyer, D.N.A., Goodford, P.J. and McWhinnie, P.M., *New hydrogen-bond potentials for use in determining energetically favorable binding sites of molecules of known structure*, J. Med. Chem., 32 (1989) 1083–1094.
6. Kellogg, G.E., Semus, S.F. and Abraham, D.J., *HINT: A new method of empirical field calculation for CoMFA*, J. Comput.-Aided Mol. Design, 5 (1991) 545–552.
7. Pastor, M., Cruciani, G. and Clementi, S., *Smart region definition (SRD): A new way to improve the predictive ability and interpretability of 3D-QSAR models*, J. Med. Chem. 40 (1997) 1455–1464.
8. Cruciani, G., Pastor, M. and Clementi, S., *Region selection in 3D QSAR*, In Computer-assisted lead finding and optimization, VCH Weinheim 1997 p. 379–395, 1996 (in press).
9. GOLPE Version 3.0.3., *Multivariate infometric analysis*, Perugia, Italy, 1996.
10. Pastor, M. and Cruciani, G., *The role of water in receptor–ligand interactions: A 3D-QSAR approach*, In Computer-assisted lead finding and optimization, VCH Weinheim 1997 p. 473–484.
11. Baroni, M., Costantino, G., Cruciani, G., Riganelli, D., Valigi, R. and Clementi, S., *Generating optimal linear PLS estimations (GOLPE): An advanced chemometric tool for handling 3D-QSAR problems*, Quant. Struct.-Act. Relat. 12 (1993) 9–20.
12. Cho, S.J. and Tropsha, A., *Cross-validated R^2-guided region selection for comparative molecular field analysis: A simple method to achieve consistent results*, J. Med. Chem., 38 (1995) 1060–1066.
13. Norinder, U., *Single and domain mode variable selection in 3D QSAR applications*, J. Chemom., 10 (1996) 95–105.
14. Watson, K.A., Mitchell, E.P., Johnson, L.N., Son, J.C., Bichard, C.J.F., Orchard, M.G., Fleet, G.W.J., Oikonomakos, N.G., Leonidas, D.D., Kontou, M. and Papageorgioui, A., *Design of inhibitors of glycogen phosphorylase: A study of* α- and β-C-glucosides and 1-thio-β-D-glucose compounds, Biochemistry 33 (1994) 5745–5758.
15. Cruciani, G. and Watson, K.A., *Comparative molecular field analysis using GRID force-field and GOLPE variable selection methods in a study of inhibitors of glycogen phosphorylase b*, J. Med. Chem., 37 (1994) 2589–2601.
16. Bichard, C.J.F., Mitchell, E.P., Wormald, M.R., Watson, K.A., Johnson, L.N., Zographos, S.E., Koutra, D.D., Oikonomakos, N.G. and Fleet, G.W.J., *Potent inhibition of glycogen phosphorylase by a spirohydantoin of glucopyranose: First pyranose analogues of hydantocidin*, Tetrahedron Lett., 36 (1995) 2145–2148.
17. Krülle, T.M., Watson, K.A., Gregoriou, M., Johnson, L.N., Crook, S., Watkin, D.J., Griffiths, R.C., Nash, R.J., Tsitsanou, K.E., Zographos, S.E., Oikonomakos, N.G. and Fleet, G.W.J., *Specific inhibition of glycogen phosphorylase by a spirodiketopiperazine at the anomeric position of glucopyranose*, Tetrahedron Lett., 36 (1995) 8281–8294.
18. Watson, K.A., Mitchell, E.P., Johnson, L.N., Cruciani, G., Son, J.C., Bichard, C.J.F., Fleet, G.W.J., Oikonomakos, N.G., Kontou, M. and Zographos, S.E., *Glucose analogue inhibitors of glycogen phosphorylase: From crystallographic analysis to drug prediction using GRID force-field and GOLPE variable selection*, Acta Cryst., D51 (1995) 458–472.

Comparative Molecular Similarity Indices Analysis: CoMSIA

Gerhard Klebe
Institute of Pharmaceutical Chemistry, University of Marburg, Marbacher Weg 6, D 35032 Marburg, Germany

1. The Prerequisites: Structural Alignment and Binding Affinity

Previously, in this volume, we have drawn our focus on the alignment of drug molecules in order to compare, correlate and predict their biological properties [1]. As dependent property variable, the binding affinity of the drug molecules toward a common receptor has been selected. It has been pointed out that a structural alignment is mainly required because information about the 3D structure of the target protein is not available (Fig. 1). In such a case, no direct estimate on the binding affinity of a particular ligand toward a given receptor is possible. Affinities are based on structural features of both, the ligands and the proteins. As a consequence, in the absence of the protein structure, only variations of binding affinity can be related with relative differences between the ligands. These differences are expressed in terms of some appropriate descriptors, in particular those describing gradual changes in structural and energetic features. However, in order to compute and compare them, we do require a mutual alignment or superposition of the drug molecules involved. This alignment determines to what extent the descriptors differ from one molecule to the next. Hence, it influences substantially the results of the evaluation. Accordingly, we can expect only significant and relevant results from such an analysis if the selected superposition approximates best the experimentally given alignment in the protein-binding pocket of an (unfortunately) structurally unknown receptor.

2. Structural Alignments to Reproduce Experimentally Observed Binding Modes

In the literature, a remarkable number of crystallographically determined protein–ligand complexes has been published over the last years [2], including many examples where a particular protein has been co-crystallized with a series of different ligands [3,4]. In several of these complexes, ligands with related bonding skeletons also occupy similar regions in the binding pocket. They suggest that molecules with common or related skeletons also show similar binding modes [4]. However, also a substantial number of examples is available that indicates a more complex and less clear-cut relationship. For example, different amino acid residues are involved in the binding or distinct functional groups of the ligands are engaged in the protein–ligand interface. These cases are usually addressed as 'alternative' binding modes. Even minor modifications with respect to the topology of the underlying bonding skeleton can substantially modify the molecular properties, so that alternative binding modes result [3,4].

Nevertheless, molecular comparisons require in the absence of a detailed structure of the receptor a structural alignment. In such a case, is it possible to describe and predict

H. Kubinyi et al. (eds.), 3D QSAR in Drug Design, Volume 3. 87–104.

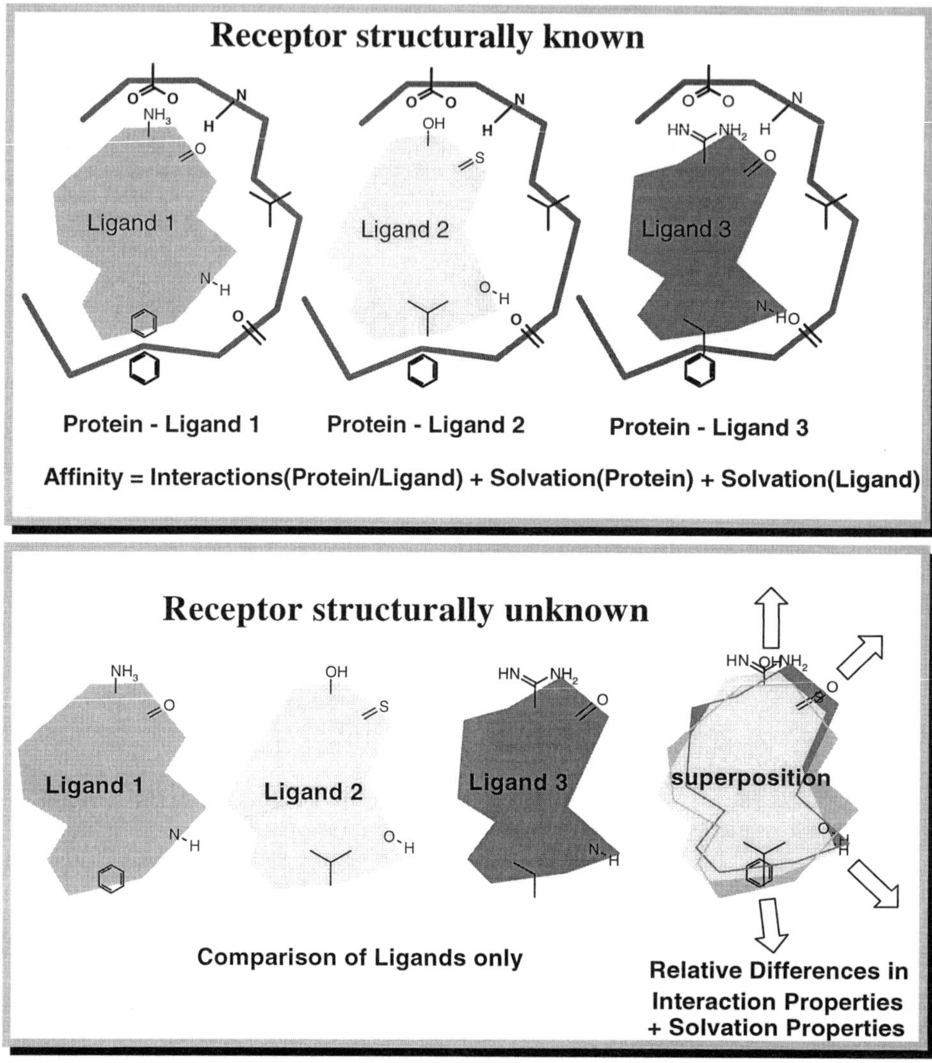

Fig. 1. If the structure of the target protein is known, the binding modes of various ligands can either be determined experimentally by X-ray crystallography or approximated by computer docking. Subsequently, affinity predictions can be performed by computing for each complex separately individual contributions from protein–ligand interactions and solvation effects. In the absence of a receptor structure, ligands can only be compared relative to each other. In order to evaluate these relative differences, molecular descriptors are required that map differences in interaction and solvation properties. A prerequisite for such relative comparisons is a mutual superposition of the ligands that approximates the experimentally given binding modes.

binding modes by comparing ligand properties only? These properties have to be features that determine molecular recognition of ligands at protein binding sites.

As ultimate goal, computational approaches handling this problem have to generate a spatial superposition of the ligands that reproduces experimentally given binding modes. Several approaches have been described in the literature to compute such alignments; however, only very rarely is a rigorous validation using experimental results performed.

We have extended the procedure SEAL, originally described by Kearsley and Smith [5], to consider simultaneously steric, electrostatic, hydrophobic and hydrogen-bonding properties [6–8]. To quantify the similarity of two molecules, their shape is approximated by a set of spatial Gaussian-type functions centered at the atomic positions. For each molecule, these functions are associated with a vector of physico-chemical properties derived from atom-based descriptors. To compute the similarity of two molecules in space, the scalar product of these vectors corresponding to the two molecules is determined and weighted by the overlap of the associated Gaussian functions. The obtained quantity is used to maximize spatial similarity. Starting from random orientations, it is subsequently optimized by minimizing the mutual distances between molecular portions having similar physico-chemical properties. This method does not require predefined pairs of matching centers associated with the molecular framework — e.g. in terms of a 'pharmacophore pattern'. Accordingly, also strongly deviating bonding skeletons can be compared and aligned.

To validate the achieved results, the above-described alignment function has been applied to a dataset of 184 ligand pairs binding to the same protein [7]. Their actual binding modes and accordingly their relative structural alignments are known from protein crystallography. Across this reference set, the observed alignments could be reproduced in one-third of the cases with an rms deviation below 0.7 Å, 51% below 1 Å and nearly 90% below 2 Å. Considering the inherent accuracy limits of about 0.7 Å for such a superposition of two experimentally determined protein–ligand complexes, the obtained residuals appear rather satisfactory. The alignment function exhibits several minima. Thus, the approach suggests not only the global minimum, but additional solutions, with a lower similarity scoring, however. In two-thirds of the test cases, the best solution also approximates the experimentally observed alignment. For 91%, the experimental situation is found among the best and second-best solution. These different solutions can propose alternative binding modes, especially if their relative similarity scorings do not differ by more than 5% from the best solution.

The alignment procedure described so far does not consider molecular flexibility. In order to reflect some 'local' flexibility in the superposition process, the alignment function mentioned above has been introduced as an additional term into the potential function used in the optimization step of the heuristical conformation search program MIMUMBA [6,9]. Since no predefined fit centers directly associated with the molecular framework are required, strongly deviating bonding skeletons also can be successfully compared and aligned. Nevertheless, this local optimization method needs an initial orientation and starting conformation. This can be a guess, based upon a putative pharmacophoric pattern, or, more objectively, the result of a previous rigid alignment with SEAL [6].

To allow for a global search, simultaneously including molecular flexibility, the described alignment function has been combined with the conformational searching technique applied in MIMUMBA [9]. In this combined approach, sets of up to 150 conformers, well spread in conformational space, are subjected to mutual comparisons with a reference structure. Subsequently, those of the conformers receiving the highest similarity scorings in this rigid superposition are subjected to a local conformational relaxation, together with a similarity maximization. Again, similarity scoring is extensively used in this comparison. Taking sets of structurally deviating ligands, binding to common proteins, the approach has been validated. In a convincing number of cases, the experimentally observed alignment could be reproduced in conformation and relative orientation with an rms deviation below 1.5 Å.

3. Binding Affinity: A Summation over Several Energetic and Entropic Contributions?

Binding affinity, the predominant dependent property variable to be correlated and predicted in 3D QSAR studies, can be calculated from the experimentally observed binding constants. It is related to Gibbs free enthalpy of binding ΔG which itself is composed by an enthalpic and entropic contribution:

$$\Delta G = -RT\ln K_{\mathrm{i}} = \Delta H - T\Delta S$$

How does the binding constant relate to structural properties of a complex, and what are the important properties that allow a protein to bind a ligand tightly and selectively? The binding process is governed by various effects determining the binding affinity [4]. The ligand and the protein binding site are fully solvated before binding. Polar groups form hydrogen bonds with the solvent. The ligand is usually flexible with several rotatable bonds and can, in principle, adopt a potentially large number of low-energy conformations. The protein is also flexible and its conformation in the unbound state can be significantly different from that in the protein–ligand complex. Upon binding to the protein, the ligand looses part of its solvation shell and replaces the water molecules occupying the binding site. This process involves the breaking of several hydrogen bonds with water molecules. The ligand is then able to form favorable direct interactions with the protein. As a consequence of binding, the ligand and also the protein may change their conformation and also lose some internal flexibility. Due to steric restrictions of the binding site, certain parts of conformation space of the ligand are no longer accessible.

For the understanding and prediction of ligand-binding affinity, a partition of the free energy of binding into individual, physically interpretable terms is desirable. However, these attempts are not without problems [4]. Especially, the relative calibration of the individual contributions against each other is difficult. The additivity of non-bonded protein–ligand interactions is usually assumed; however, it is only a non-proven postulate. Nevertheless, several studies have been described in the literature where a simple function composed by different additive contributions to ΔG achieves a reasonable correlation of structural features with binding affinities. In these approaches, most important are hydrogen bonds, ionic and lipophilic interactions. The latter are assumed to be proportional to the lipophilic contact surface between the protein and the ligand. Furthermore, contributions arising from the conformational immobilization at the binding site and the release of bound water molecules also contribute substantially. With respect to comparative 3D QSAR studies, it can be assumed — at least as a first approximation — that binding affinities as free energy values can be reasonably well described by an additive summation over several molecular descriptors.

4. Molecular Fields as Descriptors to Quantify Binding Affinities of Aligned Molecules

As mentioned above, the target property to be correlated and predicted in a comparative analysis of ligands is a free energy value. It can be imagined that enthalpic contributions to the binding constant are covered by molecular descriptors that explore the capabilities of molecules to perform intermolecular interactions such as hydrogen bonds or ionic interactions with a putative receptor (Fig. 1). In the CoMFA method [10], gradual changes of the interaction properties are mapped by evaluating the potential energy at regularly spaced grid-points surrounding the mutually aligned molecules. The forces involved between molecules are frequently described by Lennard-Jones and Coulomb-type potentials.

Entropic contributions to the binding affinity are more difficult to describe. A major factor arises from the solvent-to-protein transfer. As shown in several studies, this portion of the entropic contributions appears to consider changes of the water structure around ligands and in the active site. The first part approximately correlates with the size of the hydrophobic surface area of the drug molecules [1,4]. Accordingly, descriptors should be useful that appropriately quantify relative differences of the hydrophobic surface area of ligands. The second aspect, the release of water molecules from the active site, is more difficult to handle. In the absence of the protein structure, we can only suppose, assuming ligands of comparable size, that an equivalent number of water molecules is replaced. Furthermore, in a dataset covering molecules with distinct conformational flexibility, differences in the degree of conformational freedom have to be considered since the immobilization at the binding site involves important entropy changes.

The CoMFA approach uses in its standard implementation only Lennard-Jones and Coulomb potentials [10]. Evidence has been collected that these potentials solely describe the energetic contributions to the binding constants [11]. Entropic influences seem to be neglected or insufficiently covered. In order to include entropic contributions, some kind of field considering the differences in hydrophobic surface contributions is required. Hydrophobic fields have been described by Kellog and Abraham [12,13] and are implemented into the program HINT. Furthermore, using a water probe in Goodford's GRID program [14] allows one to map hydrophobic surface regions in terms of a field. These fields and other potential fields with various functional forms have been applied in CoMFA analyses [15].

5. Shortcomings and Problems with the Usually Applied Interaction Fields

The fields presently used in CoMFA [16] imply some problems. For example, the Lennard-Jones potential is very steep close to the van der Waals surface (Fig. 2). As a

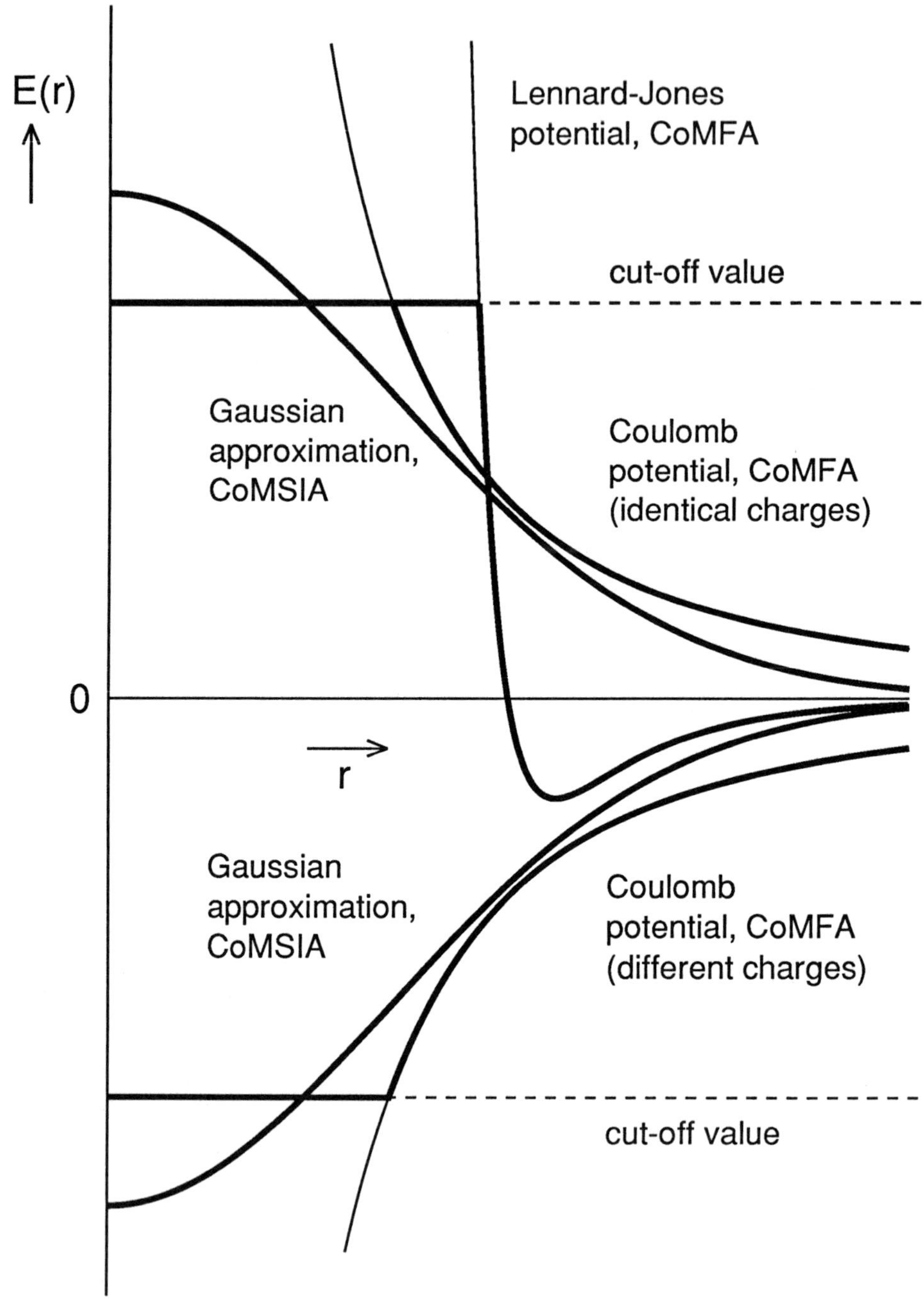

Fig. 2. The Lennard-Jones potential is very steep close to the molecular surface. The Coulomb potential shows a smoother slope. Both potentials show singularities at the atomic positions (here at the origin). To avoid unacceptably large values, the potential evaluation is restricted to an arbitrarily fixed cutoff value. Using a Gaussian-type functional form reveals a smoother distance dependence and any singularities are avoided.

consequence, the potential energy expressed at grid-points in the proximity of the molecular surface changes dramatically. Nevertheless, it is likely that especially values from this region display significant descriptors in a QSAR [17,18]. Accordingly, just some small mutual shifts of the molecules or minor conformational changes can result in strong variations of these descriptors. Nevertheless, these shifts can be so small that they are easily accepted as 'nearly identical' by visual inspection.

Furthermore, the Lennard-Jones and Coulomb potentials show singularities at the atomic positions (Fig. 2). To avoid unacceptably large values, the potential evaluations are normally restricted to the regions outside the molecules, and some arbitrarily fixed cutoff values are defined. Due to differences in the slope of the potentials (e.g. Lennard-Jones and Coulomb), these cutoff values are exceeded for the different terms at different distances from the molecules [18]. This requires further arbitrary settings to adjust the two fields in a simultaneous evaluation and can involve the loss of information about one of the fields. For the interpretation of CoMFA results, in particular with respect to the design of novel compounds, contour maps of the relative spatial contributions of the different fields are extremely useful tools [17]. However, due to the described cutoff settings and the steepness of the potentials close to the molecular surfaces, these maps are often not contiguously connected and accordingly difficult to interpret.

6. Similarity Indices Fields to Describe Similarities and Differences between Aligned Molecules

To overcome the outlined problems, we have developed an alternative approach to derive molecular descriptors for a comparative analysis [19]. Based on that what we learned from the alignment function used in SEAL, which reveals convincing results for a spatial comparison of molecules, similarity indices are calculated in space. Using a common probe, these similarity indices are enumerated for each of the aligned molecules in the dataset at regularly spaced grid-points (Fig. 3). They do not exhibit a direct measure of similarity determined between all mutual pairs of molecules. Instead, they are indirectly evaluated via the similarity of each molecule in the dataset with a common probe atom that is placed at the intersections of a surrounding lattice. In determining this similarity, the mutual distance between the probe atom and the atoms of the molecules of the dataset is considered. As functional form Gaussian-type functions with no singularities have been selected to describe this distance dependence (Fig. 2), no arbitrary definition of cutoff limits is any longer required. Indices can be calculated at all grid-points. In principle, any relevant physico-chemical property can be considered in this approach to calculate a 'field' of similarity indices. We have tested steric, electrostatic, hydrophobic and hydrogen-bond donor and acceptor properties. According to the considerations above, it is supposed that the most important contributions responsible for binding affinity are covered by these properties. The distance dependence of the different properties is equivalently handled in all cases. The applied Gaussian-type functional form defines a significantly smoother distance dependence compared to, for example, the Lennard-Jones potential. The obtained indices are evaluated in a PLS analysis [20] according to the usual CoMFA protocol [16]. This *C*omparative *M*olecular

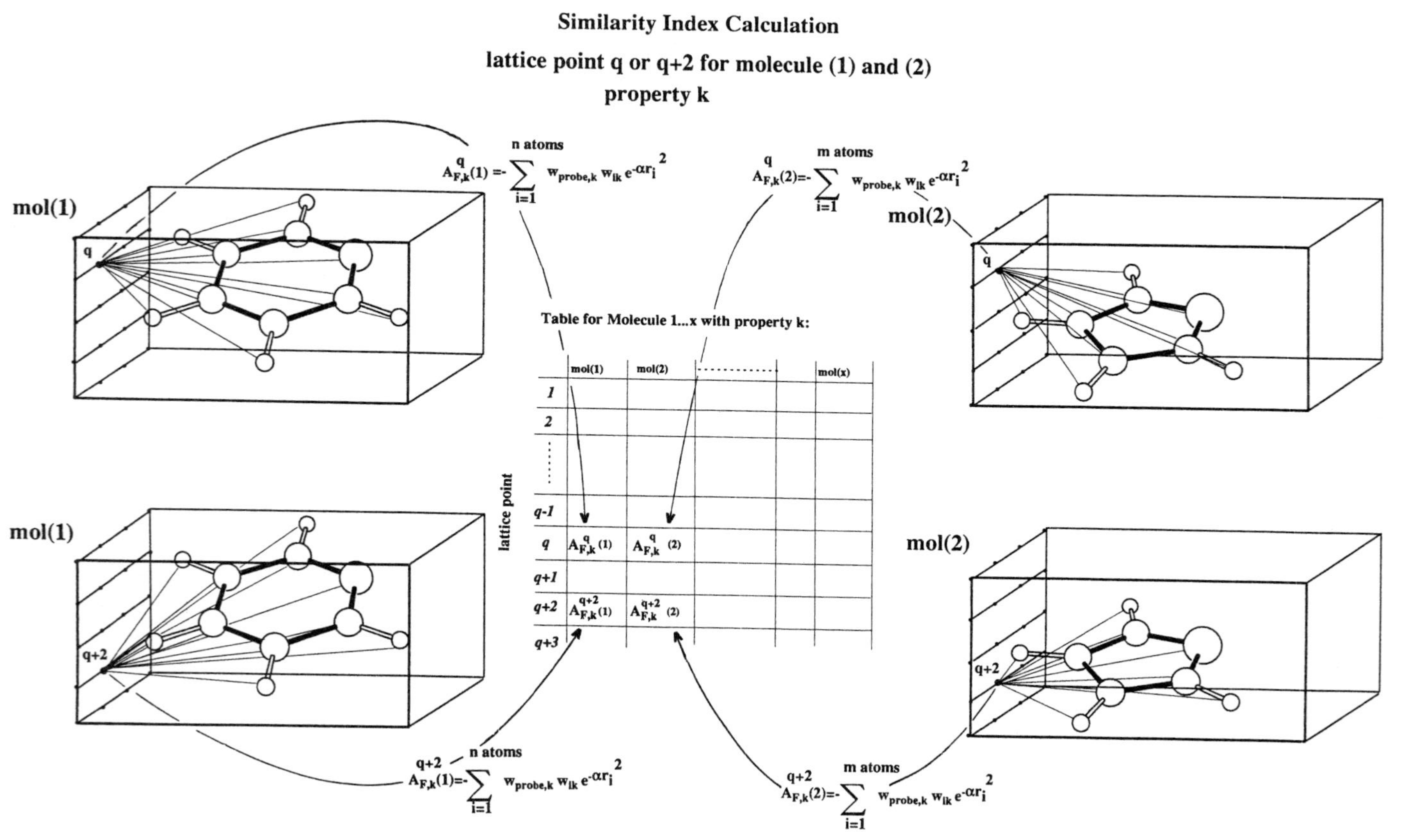

Fig. 3. In the CoMSIA approach, in relation to the CoMFA method, the similarity of molecules is evaluated via the similarity of each molecule in the dataset with a common probe. The first molecule (mol(1)) is placed into a regularly spaced lattice. From each of the lattice intersections (here: at q *and* q + 2*), a similarity index is calculated for the different probes* ($w_{probe,k}$)*. Each obtained value is written up on a large data table. Subsequently, all molecules (mol(2) ... mol(x)) in the dataset are proceeded. The obtained data table is accomplished by the affinity data, and examined by a PLS procedure.*

*S*imilarity *I*ndices *A*nalysis (CoMSIA) has been applied to several datasets [19,21]. Applying CoMFA and CoMSIA to the same datasets, in our experience, results in similar statistical significance being obtained. This alone would not justify the introduction of a new method; however, the major improvement is achieved with respect to the contour maps derived from the results. The relative spatial contributions of the different fields are much easier (and more intuitive) to interpret.

The CoMSIA approach implies moving from field descriptors based on well-established and generally accepted potentials (Lennard-Jones and Coulomb) to some arbitrary descriptors considering the spatial similarity or dissimilarity of molecules. Perhaps, on first sight, this could be seen as a step backwards. However, we have to remember that a statistical approach such as a 3D QSAR analysis seeks to correlate *relative* differences of discriminating molecular descriptors with a dependent property — e.g. the binding affinity. In that respect, 3D QSAR is a method to map and pin down similarities or dissimilarities of molecules. The descriptors used in 3D QSAR need not necessarily display partitions of interaction energy terms. They have only to correlate in a uniform manner with contributions determining binding affinity. Good et al. [22] reported on the successful evaluation of similarity indices in correlating and predicting the activity of aligned molecules. Since the authors used only integral similarity indices of entire molecules in the analysis, limited information about spatial features and characteristics is available, responsible for the variation of the activity with the 3D structure. Keeping the design of novel molecules in mind, this spatial interpretation of 3D QSAR results is of utmost importance; it allows us to understand what really matters in terms of structural features. With CoMSIA, substantially improved contour maps are obtained. They can easily be interpreted and used as a visualization tool in designing novel

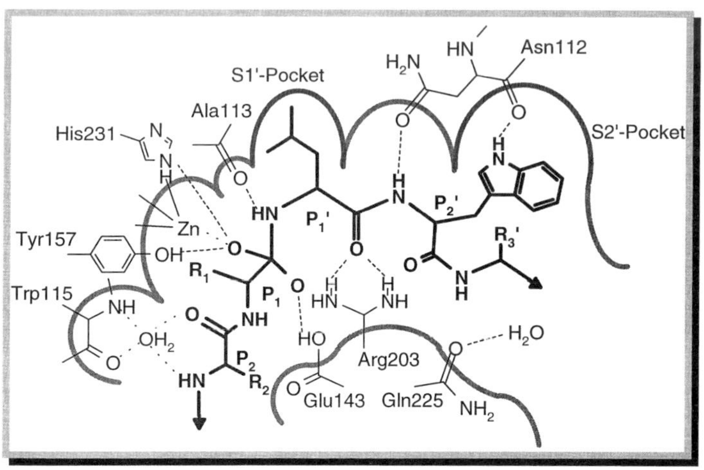

Fig. 4. Schematic diagram of the binding site of thermolysin with a substrate-like ligand. Orientation of the binding site corresponds to that in Fig. 8.5–9. The peptide cleavage proceeds via a tetrahedral transition state coordinating to the zinc. Typical interactions between the ligand and key active site residues or structurally important water molecules are indicated.

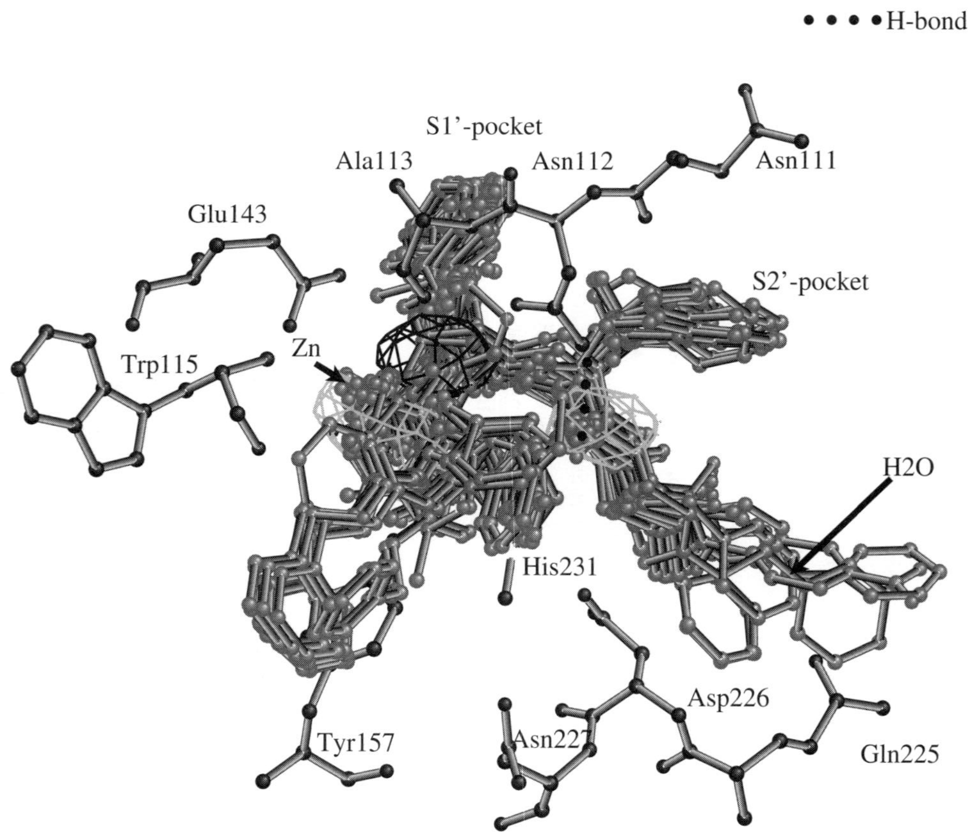

Fig. 5. Isocontour plot of field contributions of the electrostatic properties based on CoMSIA. Superimposed are the aligned inhibitors and some key residues in the active site of thermolysin according to the crystal structure determination. Areas contoured in gray correspond to regions where negatively charged residues enhance binding affinity. In areas surrounded by black isopleths increasingly positive charge will enhance affinity.

compounds. Whereas the level-dependent contouring of usual CoMFA-field contributions highlights those regions in space where the aligned molecules would favorably or unfavorably interact with a possible environment, the CoMSIA-field contributions denote those areas within the region occupied by the ligands that 'favor' or 'dislike' the presence of a group with a particular physico-chemical property. This association of required properties with a possible ligand shape is a more intuitive guide to check whether all features important for activity are present in the structures under consideration.

7. CoMSIA Applied to Thermolysin Inhibitors: A Case Study

To demonstrate the advantages of a CoMSIA study, especially with respect to the interpretation of field contributions, a dataset of thermolysin inhibitors already studied by DePriest et al. [23] will be used. The crystal structure of this metalloprotease is known [24]. Accordingly, for some of the inhibitors, crystallographically determined binding geometries are available. They have been used as a starting point to reveal an alignment of all 61 ligands in the training set [19]. In parallel, CoMFA and CoMSIA have been applied to this dataset. In all cases, q^2 values of 0.59–0.64 have been obtained. In CoMSIA, five different fields have been considered [25].

Usually, 3D QSAR methods are not applied if the 3D structure of the target protein is known. In such cases, more powerful design tools are available. However, for the present test example, the knowledge of the receptor protein provides the opportunity to interpret and understand features indicated in the contour maps with respect to a protein environment.

In the following, the isocontour plots of the steric, electrostatic, hydrophobic and H-bonding properties will be discussed. Since reference is taken to the protein environment of thermolysin, the binding geometry of a representative substrate-like ligand is sketched in Fig. 4. In Figs. 5–9 the aligned ligands are shown, together with some key residues in the active site and gray or black isopleths contouring the different field contributions.

Figure 5 shows the electrostatic properties. In the gray contoured areas, negatively charged groups enhance affinity, whereas groups with increasing positive charge improve affinity in regions enclosed by black isopleths. A gray contour is found close to the zinc-binding site. This indicates that negatively charged functional groups of the ligands serve as potent coordinating groups for the metal ion. A second gray contour matches with the position of the substrate's amide bond adjacent to the P2′ position (Fig. 4). Some of the potent ligands show a charged carboxy terminus at this location, apparantly the presence of this group improves affinity.

The steric contour map highlights the S1′ and S2′ pocket for preferred steric occupancy (black isopleths in Fig. 6). As in the natural substrate, filling of the specificity pockets is important for ligand binding. An additional extended region requiring steric bulk falls close to the protein–solvent interface close to the P2 position. Ligands with bulky groups occupying this area show enhanced binding affinity. Three regions unfavorable for steric occupancy are indicated, above zinc (P1 position), at the rim of the S2′ pocket and where the binding site opens to the solvent. Ligands with extended

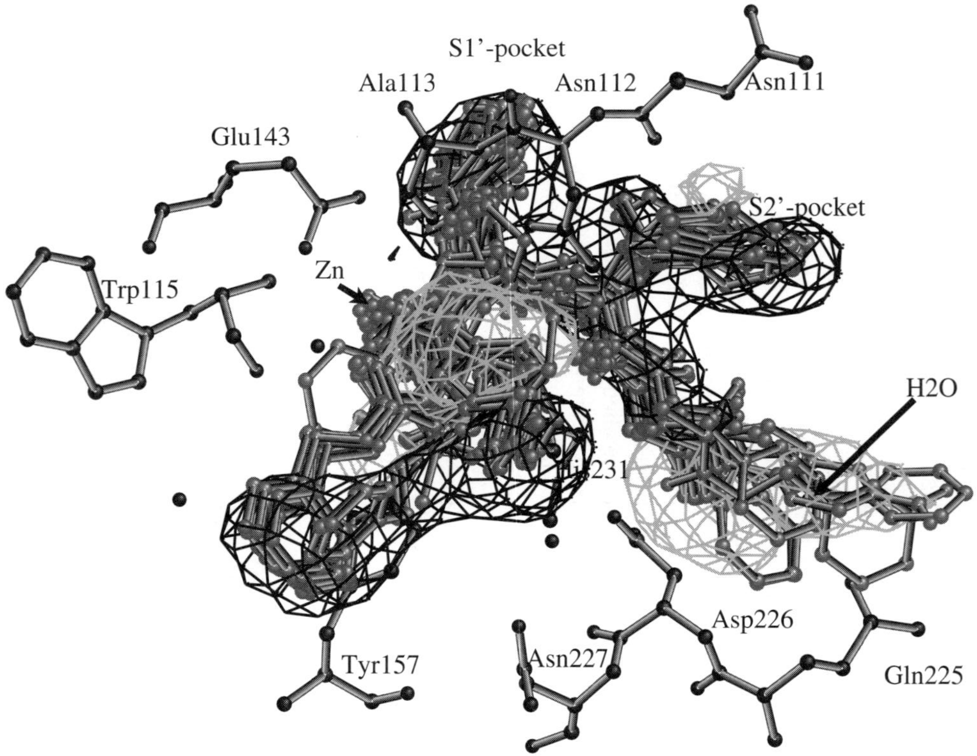

Fig. 6. Isocontour plot of field contributions of the steric properties based on CoMSIA. Areas contoured in black indicate regions favorable for steric occupany. Areas where steric bulk reduces binding affinity are contoured in gray.

substituents occupied this latter area (beyond the P2′). The crystal structure of thermolysin with the potent inhibitor phosphoramidon shows a water molecule, bound to Gln 225, in this sterically unfavorable region. Phosphoramidon does not extend into this area beyond P2′; however, larger ligands requiring this space would have to replace this water molecule. It could well be that this replacement is energetically very unfavorable; therefore, the extended ligands lose part of their affinity.

This effect is also traced by the hydrophobic field (Fig. 7), where gray isopleths point toward the requirement for hydrophilic groups. Close to the binding site of the above-mentioned water, a gray contour points to the necessity for the presence of polar groups.

The field contributions of the hydrogen-bond acceptor properties are summarized in Fig. 8. A gray contour in this map indicates that the occurrence of an acceptor group will be favorable for binding, whereas a black contour highlights that this property should be absent. A gray isopleth surrounds the carbonyl oxygen in the side chain of Asn112. Obviously, this area is favorable for a hydrogen-bond acceptor. In fact, the carbonyl oxygen of the Asn112 side chain is frequently involved as acceptor in hydrogen bonds toward potent inhibitors. The black contour encompassing the amide group of the side chain indicates that this area should lack hydrogen-bond acceptor capabilities.

In the donor field (Fig. 9), black isopleths indicate areas unlikely for hydrogen-bond donor properties. One encloses the backbone carbonyl oxygen of Ala113. This group accepts a hydrogen bond from many of the potent inhibitors. Regions of the donor map, highlighted in gray, are favorable for hydrogen-bond donor groups in the protein. One area surrounds an adjacent water molecule. In the case of this water, the position of a protein residue is not suggested as bonding partner, but a structurally important water molecule mediating a hydrogen bond between a ligand and Trp115.

8. Conclusion and Outlook

The present example has shown that the CoMSIA field contributions can be interpreted very easily. Taking the protein environment of thermolysin as a reference, the various contributions can even be attributed to some physical meaning. Steric, electrostatic and hydrophobic features are highlighted in the maps where ligands require or should miss these properties. Characteristics for H-bonding are contoured beyond the molecules in areas where in the receptor a donor or acceptor group should be located. The obtained map can be used as a first step toward the development of a pseudoreceptor model. Since the CoMSIA approach can also be extended to various kinds of similarity fields, other intermolecular interaction properties can be mapped in order to obtain a more detailed receptor model. With respect to *de novo* design and lead optimization, the obtained contour plots mark the areas where to alter and improve particular molecular properties.

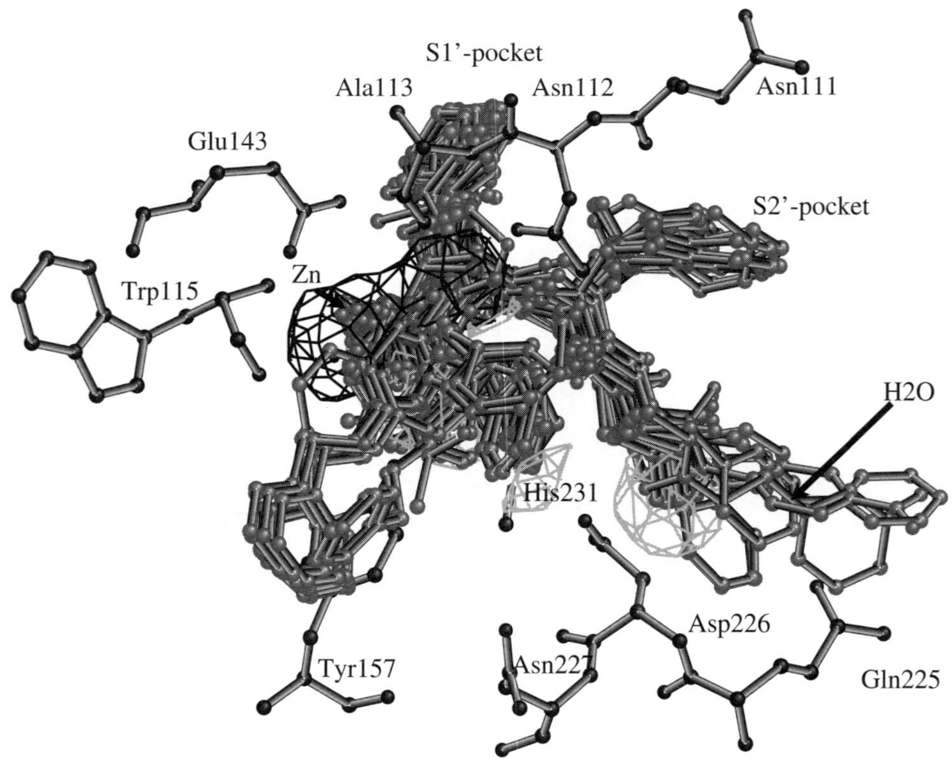

Fig. 7. Isocontour plot of field contributions of hydrophobic properties based on CoMSIA. Gray isopleths encompass regions where hydrophilic group improve binding; in the black-contoured areas hydrophobic substituents are more favorable.

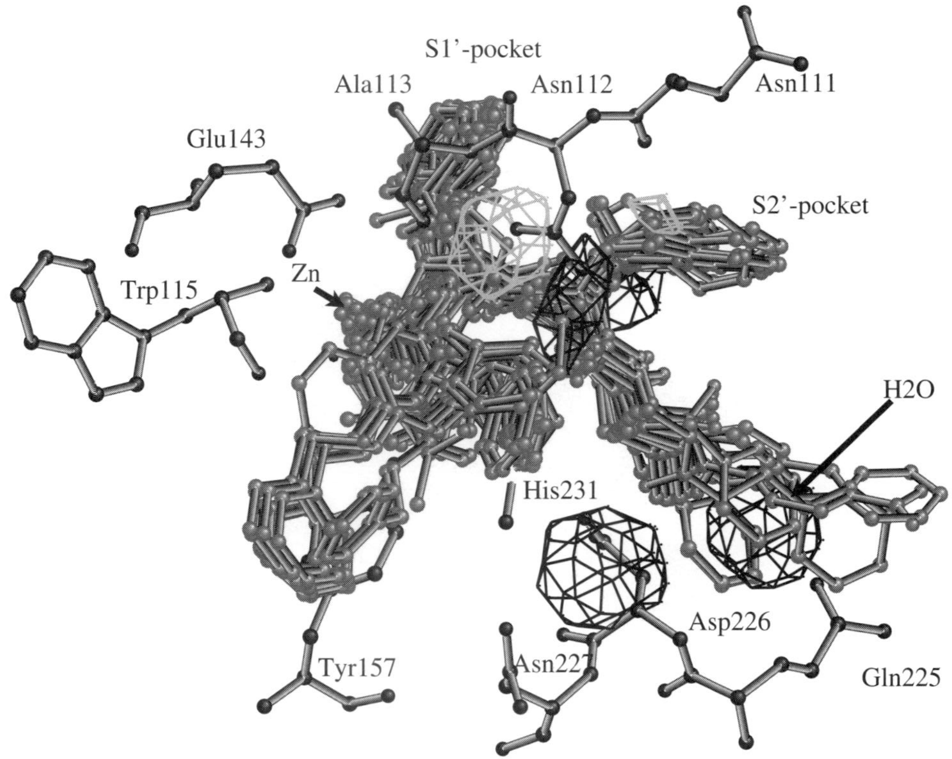

Fig. 8. Isocontour plot of field contributions of hydrogen-bond acceptor properties based on CoMSIA. Gray contours map areas beyond the ligands where an acceptor group in the receptor will be favorable for binding; regions surrounded by black isopleths show that hydrogen-bond acceptor capabilities should be absent in this particular part of the receptor.

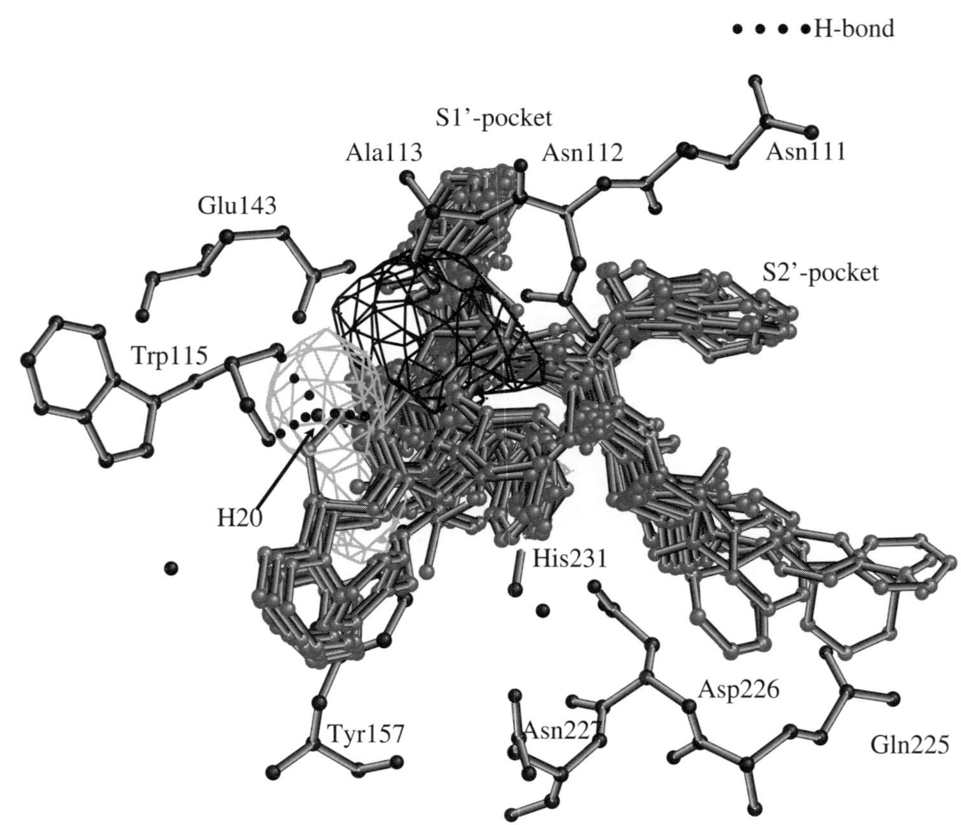

Fig. 9. *Isocontour plot of field contributions of hydrogen-bond donor properties based on CoMSIA. Black isopleths indicate areas for which the presence of a donor group in the receptor is unlikely. Gray-contoured regions point toward the required presence of a donor group in the protein.*

Acknowledgement

The author is grateful to Ute Abraham (BASF AG) for a very productive and creative collaboration on various developments and applications of 3D QSAR methods over several years. Furthermore, the many stimulating discussions with Hugo Kubinyi (BASF AG) are gratefully acknowledged. They helped to pave the ground for the development of the present method. The author also thanks Hugo Kubinyi for making available a copy of Fig. 2.

References

1. Klebe, G., *Structural alignment of molecules*, In Kubinyi, H. (Ed.) 3D QSAR in drug design, ESCOM, Leiden, The Netherlands, 1993, pp. 173–199.
2. Bernstein, F.C., Koetzle, T.F., Williams, G.J.B., Meyer, E.F., Jr., Brice, M.D., Rodgers, J.R., Kennard, O., Shimanouchi, T. and Tasumi, T., *The protein data bank: a computer-based archival file for Macromolecular structures*, J. Mol. Biol., 112 (1977) 535–542.
3. Meyer, E.F., Botos, I., Scapozza, L. and Zhang, D., *Backward binding and other structural surprises*, Persp. Drug Discov. Design, 3 (1996) 168–195.
4. Böhm, H.J. and Klebe, G., *What can we learn from molecular recognition in protein–ligand complexes for the design of new drugs?*, Angew. Chem. Int. Ed. Engl., 35 (1996) 2588–2614.
5. Kearsley, S.K. and Smith, G.M., *An alternative method for the alignment of molecular structures: Maximizing electrostatic and steric overlap*, Tetrahed. Comput. Meth., 3 (1990) 615–633.
6. Klebe, G., Mietzner, T. and Weber, F., *Different approaches toward an automatic alignment of drug molecules: Applications to sterol mimics, thrombin and thermolysin inhibitors*, J. Comput.-Aided Mol. Design, 8 (1994) 751–778.
7. Klebe, G., *Toward a more efficient handling of conformational flexibility in computer-assisted modeling of drug molecules*, Persp. Drug Discov. Design, 3 (1995) 85–105.
8. Klebe, G., Mietzner, W. and Weber, F., *Methodological developments and strategies for a fast flexible superposition of drug-size molecules* (in preparation).
9. Klebe, G. and Mietzner, T., *A fast and efficient method to generate biologically relevant conformations*, J. Comput.-Aided Mol. Design, 8 (1994) 583–606.
10. Cramer III, R.D., Patterson, D.E. and Bunce, J.D., *Comparative molecular field analysis (CoMFA): 1. Effect of shape on binding of steroids to carrier proteins*, J. Am. Chem. Soc., 110 (1988) 5959–5967.
11. Klebe, G. and Abraham, U., *On the prediction of binding properties of drug molecules by comparative molecular field analysis*, J. Med. Chem., 36 (1993) 70–80.
12. Kellogg, G.E. and Abraham, D.J., *KEY, LOCK, and LOCKSMITH: Complementary hydrophathic map predictions of drug structure from a known receptor–receptor structure from known drugs*, J. Mol. Graph., 10 (1992) 212–217.
13. Kellog, G.E., Joshi, G.S. and Abraham, D.J., *New tools for modeling and understanding hydrophobicity and hydrophobic interactions*, Med. Chem. Res., 1 (1992) 444–453.
14. Goodford, P.J., *A computational procedure for determining energetically favorable binding sites on biologically important macromolecules*, J. Med. Chem., 28 (1985) 849–857.
15. Thibaut, U., *Applications of CoMFA and related 3D QSAR approaches*, In Kubinyi, H. (Ed.), 3D QSAR in drug design, ESCOM, Leiden, The Netherlands, 1993, pp. 661–696.
16. SYBYL Molecular Modeling System (Version 5.40), Tripos Ass., 1699 Hanley Road, St. Louis, MO 63144, U.S.A.
17. Cramer, R.D. III, DePriest, S.A., Patterson, D.E. and Hecht, P., *The developing practice of comparative molecular field analysis*, In Kubinyi, H. (Ed.), 3D QSAR in drug design, ESCOM, Leiden, The Netherlands, 1993, pp. 443–485.
18. Folkers, G., Merz, A. and Rognan, D., *CoMFA: Scope and limitations*, In Kubinyi, H. (Ed.) 3D QSAR in drug design, ESCOM, Leiden, The Netherlands, 1993, pp. 583–618.

19. Klebe, G., Abraham, U. and Mietzner, T., *Molecular similarity indices in a comparative analysis (CoMSIA) of drug molecules to correlate and predict their biological activity*, J. Med. Chem., 37 (1994) 4130–4146.
20. Stahle, L. and Wold, S., *Multivariate data analysis and experimental design in biomedical research*, Prog. Med. Chem., 25 (1988) 292–334.
21. Klebe, G. and Abraham, U., results obtained with proprietory datasets.
22. Good, A.C., So, S.-S. and Richards, W.G., *Structure–activity relationships from molecular similarity matrices*, J. Med. Chem., 36 (1993) 433–438.
23. DePriest, S.A., Mayer, D., Naylor, C.B., Marshall, G.R., *3D QSAR of angiotensin-converting enzyme and thermolysin inhibitors: A comparison of CoMFA models based on deduced and experimentally determined active site geometries*, J. Am. Chem. Soc., 115 (1993) 5372–5384.
24. Matthews, B.W., *Structural basis of the action of thermolysin and related zinc peptidases*, Acc. Chem. Res., 21 (1988) 33–340.
25. Klebe, G. and Abraham, A. *Comparative Molecular Similarity Index Analysis (CoMSIA) to study hydrogen bonding properties and to score combinatorial libraries* (submitted).

Alternative Partial Least-Squares (PLS) Algorithms

Fredrik Lindgren[a] and Stefan Rännar[b]
[a] *Department of Medicinal Chemistry, Astra Draco AB, P.O. Box 34, S-22100 Lund, Sweden*
[b] *Umetri AB, P.O. Box 7960, S-907 19 Umeå, Sweden*

1. Introduction

Mathematical treatments and modelling of large data structures have always created problems. From the infancy of computers to the late 1980s, the limiting factor when modelling large data structures was often the size of the computer memory. Due to the strong evolution in the field of computer technology, this problem is steadily decreasing. Consequently, when hardware restrictions are becoming less significant, one allows for the development of new, interesting but also calculation-intensive techniques. Typical examples within the area of drug design are techniques like 3D QSAR and molecular library characterization and modelling. However, improved hardware puts the focus on other limiting factors such as speed and efficiency of the mathematical operations performed when processing data. Algorithms and programs must be refined and optimized to meet the demands of today. The desired 'interactiveness' in data processing and molecular modelling serves as a good example of the needs of a modern drug design chemist.

A group of data-analytical tools which steadily increase their applicability are the latent variable based ones, such as Principal Components analysis (PCA) [1,2]; Principal Components Regression (PCR) [3]; and Partial Least-squares Regression (PLS) [4–18]. Especially in the disciplines of natural science, their impact has been large during the past few decades, even if statistical methods based on diagonalization of covariance matrices have been used earlier. The usefulness and advantages of projection methods have been discussed by several authors, and for their introduction and applicability we refer to the vast literature [1–22]. However, these methods are frequently studied and their algorithms have been subjects for refinement and optimization.

In this chapter, we will focus on the further developments of the PLS algorithm, using the classical algorithm as a reference for comparison. During the past years, several authors have published modified PLS algorithms with the main aim of increasing the computational speed. Often the code is optimized for a certain type of computational job or a special shape of data matrix. One common step which ties all new developments together is the calculation of some useful variance/covariance and association matrices. Our aim is to point out some commonalities and differences between the individual PLS algorithms in a simple and transparent way. No deep-penetrating computational evaluation was carried out. Instead, the paper will provide a detailed reference list of original articles.

2. Background

Many users of PLS are familiar with its Non-linear Iterative Partial Least-squares (NIPALS) algorithm [5], often referred to as the 'classical' algorithm (Fig. 1). The

H. Kubinyi et al. (eds.), 3D QSAR in Drug Design, Volume 3. 105–113.

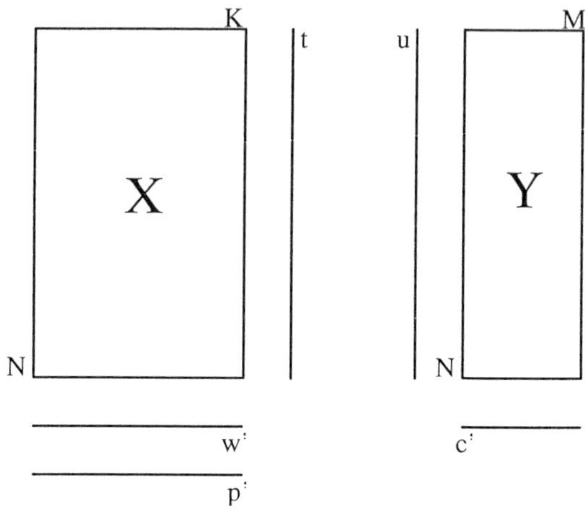

Fig. 1. A graphical illustration of the classical PLS algorithm. The predictor variable matrix **X** *(size* N × K*) and the response variable matrix* **Y** *(size* N × M*), together with the score vectors* **t** *and* **u**, *the weight vectors* **w** *and* **c**, *and the loading vector* **p**.

development was initiated by H. Wold [4–6] and later extended by S. Wold [7, 9]. Several authors have since then shown their interest in the method and many investigations and comparative studies have been performed. The most common topic for comparison is how the predictive properties of PLS relate to other regression methods, but this is not further discussed in this chapter.

Höskuldsson [14] was the first in reformulating PLS as an eigenvalue/eigenvector problem. He showed that the PLS score and weight vectors (**t, u, w, c**) can be determined as eigenvectors to a set of square variance/covariance matrices;

$$\mathbf{w}\, a_1 = (\mathbf{X'YY'X})\mathbf{w} \tag{1}$$

$$\mathbf{c}\, a_2 = (\mathbf{Y'XX'Y})\mathbf{c} \tag{2}$$

$$\mathbf{t}\, a_3 = (\mathbf{XX'YY'})\mathbf{t} \tag{3}$$

$$\mathbf{u}\, a_4 = (\mathbf{YY'XX'})\mathbf{u} \tag{4}$$

where a_1, a_2, a_3 and a_4 are all eigenvalues and the vectors **w, c, t** and **u**, all considered to have their norm equal to one. This evidence is the platform for all new developments.

The advantage of these matrices (Equations 1–4) is their sizes. The two matrices in Equations 1 and 2, (**X′YY′X**) and (**Y′XX′Y**), have the size of K × K (K is the number of **X**-variables) and M × M (M is the number of **Y**-variables), respectively. Hence, no matter how many observations (objects) there are in the original **X** and **Y** matrices, the size of the these matrices will only be dependent upon the number of **X** and **Y** variables (Fig. 2). The contrary situation holds for the matrices (**XX′YY′**) and (**YY′XX′**) (Equations 3 and 4). Their size is N × N (N is the number of observations), so therefore, the number of **X** and **Y** variables will be of no influence. Consequently, matrices with

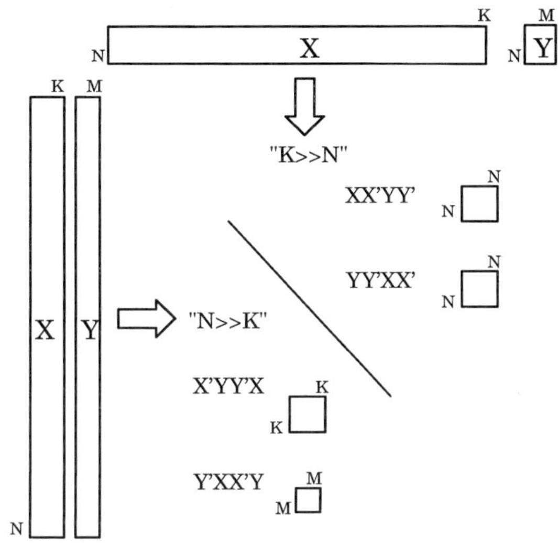

Fig. 2. The 'kernel' matrices of large data structures. For original matrices **X** *and* **Y** *with a large number of objects (case* 'N >> K' *data), the kernel matrices* **X'YY'X** *and* **Y'XX'Y** *are of size* K × K *and* M × M, *respectively. For matrices with a large number of* **X** *and/or* **Y** *variables (case* 'K >> N' *data), the kernel matrices* **XX'YY'** *and* **YY'XX'** *are both of size* N × N.

either a large number of objects or a large number of variables can be condensed into small matrices, containing all information necessary for developing a PLS model.

PLS builds up its model from sequentially calculated dimensions. Before estimating a new dimension, the variance explained by the last component must be removed in a so-called updating procedure. Normally, both **X** and **Y** are updated ($\mathbf{E}_0$ becomes $\mathbf{E}_1$, $\mathbf{E}_1$ becomes $\mathbf{E}_2$, etc., up to $\mathbf{E}_A$), but it has been shown that as long as either of the two is updated, the PLS vectors maintain their orthogonality [14, 23]. The updating procedure is one computation-intensive step and the new algorithms solve this in some alternative ways, either by using small updating matrices or through an orthogonalization procedure.

3. The Algorithms

The choice of algorithm depends strongly on the shape of the data matrices to be studied. In Multivariate Image Analysis [21,22], the number of observations is much larger than the number of variables. This leads to algorithms which utilize the variance/covariance matrices in Equations 1 and 2, since they are independent of the number of observations. An opposite situation occurs in 3D QSAR studies [24,25], where the number of variables usually widely exceeds the number of samples. In this case, one chooses an algorithm based on the association matrices in Equations 3 and 4, since their sizes are independent of the number of variables. In the following sections, we will present some alternative PLS algorithms which all have the advantage of being faster than the classical one for special cases of datasets. For a more thorough comparison of some of the algorithms, we refer to de Jong [26].

3.1. The UNIPALS algorithm

In 1989, Glen et al. [27,28] presented one of the first algorithms to utilize the smaller variance–covariance matrices for PLS computations. This algorithm is called UNIPALS (*UN*iversal *PA*rtial *L*east Squares) and is based on the matrix **Y′XX′Y** of size M × M. the eigenvector of **Y′XX′Y** with the largest eigenvalue is the first weight vector **c** for the **Y** block. From this weight vector and the original **X** and **Y** matrices, all other PLS vectors can be calculated without iteration. However, updating between dimensions is performed on the original **X** and **Y** matrices, equivalent to classical PLS. This implies that the **Y′XX′Y** matrix must be regenerated from the deflated **X** and **Y** for every new dimension. Since the original data matrices are deflated in the same way as in the classical algorithm, the results are identical.

The UNIPALS algorithm has been used in several QSAR studies [29–33] and is, according to the authors, implemented in at least two commercial softwares: the QSAR package from Molecular Simulations Inc. and in Molecular Analysis Pro. (For more detailed information please contact the authors directly.)

3.2. The kernel algorithms

The first kernel algorithm [34,35] developed by Lindgren et al. was an alternative to the classical algorithm for handling datasets where N >> K. Instead of working with **Y′XX′Y** (as in UNIPALS), one calculates the weight vector **w** (the eigenvector with the largest eigenvalue) for the **X** block from the K × K matrix **X′YY′X**. From the weight vector (**w**) and the sub-matrices **X′Y** and **X′X**, all other PLS vectors can be calculated in a straightforward manner. The novelty introduced by the first kernel algorithm was how to update the variance/covariance matrices directly, without interfering with the original **X** and **Y** matrices. By multiplication of an updating matrix (**I–wp′**) of size K × K, explained variance is removed from the variance/covariance matrices:

$$\mathbf{E'YY'E} = (\mathbf{I} - \mathbf{wp'})' \, \mathbf{X'YY'X} \, (\mathbf{I} - \mathbf{wp'}) \tag{5}$$

This simplification of the algorithm leads to major improvements in computational speed since the time-consuming step of creating the variance/covariance matrices has to be performed only once. One should note that only the **X** matrix is deflated. This will, however, not influence the results since deflation of **Y** is optional [14,23].

The second kernel algorithm [36,37] presented by Rännar et al. in 1994 is very much like the first kernel algorithm, but with the important difference that is optimized for datasets which K >> N. These types of matrices often occur in 3D QSAR and also in data from industrial processes. The association matrix **XX′YY′** is independent on the number of predictor variables and services, therefore, as a good start for this version of the kernel algorithm. The algorithm starts with the eigenvector analysis of **XX′YY′**, which gives the score vector **t** for the **X** matrix. From this vector and the small association matrix **YY′**, the score vector **u** for the **Y** block is calculated before proceeding to the next PLS dimension. Also in this kernel algorithm, the deflating is directly performed on the small variance/covariance matrices, now using the updating matrix (**I** – **tt′**). The last step is the calculation of all of the PLS weights (**w** and **c**) and loading (**p**) vectors using the original **X** and **Y** matrices. These vectors are needed to generate the regression coefficient matrix **B**:

$$\mathbf{B} = \mathbf{W}(\mathbf{P}'\mathbf{W})^{-1}\mathbf{C}' \tag{6}$$

One important point is that both kernel algorithms work well with multiple responses and give identical results as those from the classical PLS algorithm.

The kernel algorithms have lately been modified by de Jong et al. [26,38], resulting in faster and simplified kernel algorithms. Further modifications have been purposed by Dayal et al. [23,39]. They utilize the fact that only one of the matrices **X** or **Y** needs to be deflated. Since the **Y** variables often are few, deflating **Y** instead of **X** saves time.

Neither the original nor the modified kernel algorithms have been implemented in any commercial software, but the MATLAB [40] codes are available from the authors of the different versions.

3.3. The SAMPLS algorithm

SAMple-distance Partial Least Squares, or SAMPLS was presented by Bush et al. in 1993 [41,42] and is also focused on the special case of many descriptor variables and few objects (K >> N). However, the algorithm handles only one **Y** response variable, which is a limiting factor compared to other algorithms. Concerning computational time, the SAMPLS algorithm performs superior to both the classical algorithm and the kernel algorithms; however, the magnitude of the improvement will be noted in a later section. In the field of QSAR, and especially in CoMFA analysis where one only has one response variable, this algorithm is very fast and easy to use. The SAMPLS algorithm is available from QCPE [43] and this code, or a code that is supposed to be identical to the SAMPLS algorithm, is used by Tripos in the QSAR module (for further information we suggest contacting the original author).

The SAMPLS algorithm works with the association matrix **XX′** and the response vector **y** to calculate the score vector **t**, using ordinary matrix–vector multiplication without iteration. This algorithm does not give all the weight and loading vectors that come from other algorithms, but it can still be used for predictions. Not having weights and loadings can be a serious disadvantage since the inter-variable correlation information is lost. In the algorithm, Bush et al. also take advantage of the fact that one can choose to deflate either **X** or **Y** [23]; and in this case, where only one response variable exists, it is very fast to deflate only this vector. This construction makes the updating procedure performed essentially in the same way as in the classical PLS algorithm and, therefore, their results will be identical. However, in order to maintain the orthogonal PLS structure, new score vector (**t**'s) must be othogonalized to the previous ones without the algorithm.

3.4. The SIMPLS algorithm

The last algorithm to be mentioned in this chapter is the Straightforward Implementation of a Statistically Inspired Modification of the PLS method, or SIMPLS algorithm by de Jong [44]. This algorithm was first published in 1993 and the main difference between the above-mentioned algorithms and the SIMPLS algorithm is in the way the orthogonalization of the PLS components is performed. The SIMPLS algorithm aims at describing the scores as direct combinations of the original **X** matrix by a constrained optimization instead of using a deflated **X** matrix. This approach does not always give the same model as classical PLS, but the difference is very small and for most cases not significant. The results from SIMPLS are always identical to classical PLS in the first PLS component, but only in the case of one **Y** response are all components identical. The reason for this small difference is that the matrix **X′Y** is not deflated in the same sense as in the classical algorithm or the kernel algorithm. Instead, the eigenvector analysis is performed on the original **X′Y** matrix projected on the loading vectors from earlier components. This version of deflating will cause the small difference between the SIMPLS and the other PLS algorithm. The SIMPLS algorithm is, however, a very fast PLS algorithm for all kinds of shapes of data matrice (the MATLAB code is available from Dr. de Jong upon request).

4. Discussion and Concluding Remarks

The new PLS algorithms are often presented as revolutionary when comparing their speed to the classical algorithm [41]. This holds true in many cases, but sometimes the improvements are poor or even absent. Why is that? In principle the described algorithms contain one initial and rather time-consuming step, namely the computation of the variance/covariance or association matrices. In a comparative study with the classical algorithm, the time spent on calculating these condensed matrices must also be included. This is sometimes forgotten, which inevitably generates misleading results [41].

The classical PLS algorithm is always described as an iterative procedure. However, when only one **Y**-variable is modelled (most common case), the algorithm is non-iterative. This implies that only a fixed number of vector–matrix multiplications must be performed to generate the PLS model of a certain dimensionality.

Adding these two facts together (time-consuming matrix calculation and non-iterative PLS1 modelling), one quickly realizes that the classical PLS algorithm will outperform other algorithms in some cases. A typical situation is the calculation of a low-dimensional (1–3 dimension) PLS1 model without cross-validation [45,46]. In such a case, the calculation of the variance/covariance or association matrices will be more tedious than using the classical algorithm directly.

On the contrary, the new algorithms will prove advantageous in cases of repetitive modelling, as in cross-validation [45,46], bootstrapping [47] and in some variable selection techniques [48]. The great advantage of both variance/covariance and association matrices is that both objects and variables can be either added or removed, without recalculation of the condensed matrices. Other treatments, like mean-centering and scaling can also be performed directly on the condensed matrix form. These key features lead to considerable speed-up in the computation of repetitive modelling. A typical example is the cross-validation (CV) step, and the use of CV, or some other validation procedure, is strongly recommended in all types of PLS modelling. The only features which alter between consecutive runs in a CV loop are the division between training and test set objects, and some possible rescaling. Hence, CV can easily be performed on these condensed matrices directly. A 'leave-one-out' CV procedure for a typical 3D QSAR dataset would only take a limited number of seconds. The presented SAMPLS algorithm is now commonly used in CoMFA cross-validation runs and gives results identical to those from the classical algorithm, provided that no rescaling is performed within the CV procedure.

Other dataset-related features which favor the new algorithms are PLS2 modelling (more than one Y-variable) and the extraction of a large number of PLS components. Still, one has to remember that the major improvements are gained for datasets with either 'N >> K' or 'N << K'. When N and K are of similar size, no significant improvement is made.

A common problem among all alternative PLS algorithms is how to deal with missing values in the data. One cannot create the appropriate variance/covariance or association matrices without adding some type of an approximate value to fill the data gaps. One approach which deals with this problem was presented by Rännar et al. [37] and involved using the EM algorithm [49]. The classical PLS algorithm has no similar problem since it can deal with missing data in a straightforward way, without addition of approximate values.

Finally, one can conclude that there exist several alternative PLS algorithms, all optimized for different assignments. The choice of algorithm is very much related to questions like, 'What is my application area?' and 'What am I going to do?'. The answers to these questions will define if PLS1 or PLS2 modelling is needed, if K >> N or N >> K, if extensive cross-validation is foreseen, and so forth. These features will outline the computational task and one selects an algorithm which fulfils the defined requirements. The more specific the definition becomes, the more optimized algorithm can be chosen — e.g. the SAMPLS for 3D QSAR. For more general PLS modelling, the two complementary kernel algorithms and the classical algorithm are a sound choice.

References

1. Jackson, J.E., *A user's guide to principal components*, Wiley, New York, 1991.
2. Jolliffe, I.T., *Principle components analysis*, Springer-Verlag, New York, 1986.
3. Martens, H. and Naes, T., *Multivariate calibration*, Wiley, Chichester, U.K., 1989.
4. Wold, H., In David, F. (Ed.) *Research papers in statistics*, Wiley, New York, 1966, pp. 411–444.
5. Wold, H., *Path models with latent variables: The NIPALS approach*, In Blalock, H.M., Aganbegian, A., Borodkin, F.M., Boudon, R. and Capecchi, V. (Eds.) Quantitative sociology, Academic Press, New York, 1975, pp. 307–357.
6. Jöreskog, K.-G. and Wold, H. (Eds.) *System under indirect observation*, Vols 1 and 2, North-Holland, Amsterdam, The Netherlands, 1982.

7. Wold, S., Martens, M. and Wold, H., *The multivariate calibration problem in chemistry solved by the PLS method*, In Ruhe, A. and Kågström, B. (Eds.) Matrix Pencils, Springer-Verlag, Heidelberg, Germany, 1983, pp. 286–293.
8. Martens, H. and Jensen, S.-A., *Partial least squares regression: A new two-stage NIR calibration method*, In Holas, J. and Kratochvil, J. (Eds.) Progress in cereal chemistry and technology, Elsevier, Amsterdam, The Netherlands, 1983, pp. 607–647.
9. Wold, S., Ruhe, A., Wold, H. and Dunn III, W.J., *The collinearity problem in linear regression: The partial least squares approach to generalized inverses*, Siam J. Sci. Stat. Comput., 5 (1984), 735–743.
10. Geladi, P. and Kowalski, B.R., *Partial least squares regression (PLS): A tutorial*, Analyt. Chim. Acta, 185 (1986), 1–17.
11. Lorber, A., Wangen, L. and Kowalski, B., *The theoretical foundation for the PLS algorithm*, J. Chemometrics, 1 (1987) 19–31.
12. Manne, R., *Analysis of two partial squares algorithms for multivariate calibration*, Chemometrics Intell. Lab. Syst., 2 (1987) 187–197.
13. Helland, I.S., *The structure of partial least squares regression*, Commun. Stat. Simul. Comput., 17 (1988) 581–607.
14. Höskuldsson, A., *PLS regression methods*, J. Chemometrics, 2 (1988) 211–228.
15. Geladi, P., *Notes on the history and nature of partial least squares (PLS) modeling*, J. Chemometrics, 2 (1988) 231–246.
16. Phatak, A., *Evaluation of some multivariate methods and their applications in chemical engineering*, Ph.D. thesis, University of Waterloo, Ontario, Canada, 1993.
17. Garthwaite, P.H., *An interpretation of partial least squares*, J. Am. Stat. Assoc., 89 (1994) 122–127.
18. Wold, S., Albano, C., Dunn III, W.J., Edlund, U., Esbensen, K., Geladi, P., Hellberg, S., Johansson, E., Lindberg, W. and Sjöström, M., *Multivariate data analysis in chemistry*, In Kowalski, B.R. (Ed.) Chemometrics: Mathematics and statistics in chemistry, Reidel, Dordrecht, The Netherlands, 1984, pp. 17–95.
19. McGregor, J.F. and Nomikos, P., *Monitoring batch processes*, NATO Advanced Study Institute for Batch Processing Systems Engineering, Antalya, Turkey, Springer-Verlag, Heidelberg, Germany, 1992.
20. Forina, M., Armanino, C., Castino, M. and Ubigli, M., *Multivariate data analysis as a discriminating method of the origin of wines*, Vities, 25 (1986) 189–201.
21. Esbensen, K. and Geladi, P., *Strategy of multivariate image analysis (MIA)*, Chemometrics Intell. Lab. Syst., 7 (1989) 67–86.
22. Geladi, P. and Esbensen, K., *Regression on multivariate images: Principal component regression for modeling, prediction and visual diagnostic tools*, J. Chemometrics, 5 (1991) 97–111.
23. Dayal, B.S. and MacGregor, J.F., *Improved PLS algorithms*, J. Chemometrics, 11 (1997) 73–85.
24. Cramer III, R.D., Bunce, J.D., Patterson, D.E. and Frank, I.E., *Crossvalidation bootstrapping and partial least squares compared with multiple regression in conventional QSAR studies*, Quant. Struct..-Act. Relat., 7 (1988) 18–25.
25. Kubinyi, H., (Ed.), *3D-QSAR in drug design: Theory, methods and applications*, ESCOM, Leiden, The Netherlands, 1993.
26. De Jong, S., *A comparison algorithms for partial least squares regression*, J. Chemometrics, (1997) (submitted).
27. Glen, W.G., Dunn III, W.J. and Scott, D.R., *Principal components analysis and partial least squares regression*, Tetrahedron Comput. Methodol., 2 (1989) 349–376.
28. Glen, W.G., Dunn III, W.J., Sarker, M. and Scott, D.R., *UNIPALS: Software for principal components analysis and partial least squares regression*, Tetrahedron Comput. Methodol., 2 (1989) 377–396.
29. Hopfinger, A.J., Burke, B.J. and Dunn III, W.J., *A generalized formalism of three-dimensional quantitative structure–activity relationship analysis for flexible molecules using tensor representation*, J. Med. Chem., 37 (1994) 3768–3774.
30. Burke, B.J., Dunn III, W.J. and Hopfinger, A.J., *Construction of a molecular shape analysis: Three-dimensional quantitative structure–analysis relationship for an analog series of pyridobenzodiazepinone inhibitors of muscarinic 2 and 3 receptors*, J. Med. Chem., 37 (1994) 3775–3788.

31. Collantes, E.R. and Dunn III, W.J., *Amino acid side chain descriptors for quantitative structure–activity relationship studies of peptide analogues*, J. Med. Chem., 38 (1995) 2705–2713.
32. Dunn III, W.J., Hopfinger, A.J., Catana, C. and Duraiswami, C., *Solution of the conformation and alignment tensors for the binding of triethoprim and its analogs to dihydrofolate reductase: 3D-quantitative structure–activity relationship study using molecular shape analysis, 3-way partial least-squares regression, and 3-way factor analysis*, J. Med. Chem. 39 (1996) 4825–4832.
33. Dunn III, W.J. and Rogers, D., *Genetic partial least squares in QSAR*, In Devillers, J. (Ed.) Genetic algorithms in molecular modeling, Academic Press, London, 1996, pp. 109–130.
34. Lindgren, F., Geladi, P. and Wold, S., *The kernel algorithm for PLS.*, J. Chemometrics, 7 (1993) 45–59.
35. Lindgren, F., Geladi, P. and Wold, S., *Kernel-based PLS regression: Cross validation and applications to spectral data*, J. Chemometrics, 8 (1994) 377–389.
36. Rännar, S., Lindgren, F., Geladi, P. and Wold, S., *A PLS kernel algorithm for PLS, for data sets with many variables and less objects: Part 1. Theory and Algorithm.*, J. Chemometrics, 8 (1994) 111–125.
37. Rännar, S., Lindgren, F., Geladi, P. and Wold, S., *A PLS kernel algorithm for data sets with many variables and less objects: part 2. Cross-validation, missing data and examples*, J. Chemometrics, 9 (1995) 459–470.
38. De Jong, S. and Ter Braak, C.J.F., *Comments on the PLS kernel algorithm*, J. Chemometrics, 8 (1994) 169–174.
39. Dayal, B.S. and MacGregor, J.F., *Recursive exponentially weighted PLS and its applications to adaptive control and prediction*, J. Process Contr. (1997) (submitted).
40. MATLB™ Reference Guide, The Math Works Inc., Natick, U.S.A. (1992).
41. Bush, B.L. and Nachbar Jr., R.B., *Sample-distance partial least squares: PLS optimized for many variables, with application to CoMFA*, J. Comput.-Aided Mol. Design , 7 (1993) 587–619.
42. Sheridan, R.P., Nachbar Jr., R.B. and Bush, B.L., *Extending the trend vector: The trend matrix and sample based partial least squares*, J. Comput.-Aided Mol. Design, 8 (1994) 323–340.
43. QCPE 650: Ver. 1.3, 1994, Quantum Chemistry Program Exchange, Indiana University; Bloomington, IN 47404, U.S.A.: qcpe@indiana.edu.
44. De Jong, S., *SIMPLS: An alternative approach to partial least squares regression*, Chemometrics Intell. Lab. Syst., 18 (1993) 25–263.
45. Stone, M., *Cross-validatory choice and assessment of statistical predictions*, J. Royal Stat. Soc., B, 36 (1974) 111–133.
46. Geisser, S., *A Predictive approach to the random effect model*, Biometrika, 61 (1974) 101–107.
47. Leger, C., Politis, D.N. and Romano, J.P., *Bootstrap technology and applications*, Technometrics, 34 (1992) 378–398.
48. Baroni, M., Costantino, G., Cruciani, G., Riganelli, D., Valigi, R. and Clementi, S., *Generating optimal linear PLS estimations (GOLPE): An advanced chemometric tool for handling 3D QSAR problems*, Quant. Struct.-Act. Relat., 12 (1993) 9–20.
49. Little, R.J.A. and Rubin, D.B., *Statistical analysis with missing data*, Wiley, New York, 1987.

Part II

Receptor Models and Other 3D QSAR Approaches

Receptor Surface Models

Mathew Hahn and David Rogers
Molecular Simulations Incorporated, 9685 Scranton Road, San Diego, CA 92121-3752, U.S.A.

1. Introduction

It is common to have measured binding affinities for a set of compounds to a particular protein, but lack knowledge of the three-dimensional structure of the protein active site. A number of methods, called *receptor mapping techniques*, attempt to provide insight about the putative active site and to characterize receptor binding requirements. Often, receptor mapping techniques are used to generate a hypothetical model of the actual receptor site. This is known as a *receptor site model*. In this chapter, we describe a specific type of receptor site model called a *receptor surface model* (RSM) [1,2].

Receptor site models can be distinguished from *pharmacophore models*: pharmacophore models postulate that there is an essential three-dimensional arrangement of functional groups that a molecule must possess to be recognized by the receptor. These models are often generated by finding the chemically important functional groups that are common to the molecules that bind. Receptor site models, in contrast, attempt to postulate and represent the essential features of a receptor site itself, rather than the common features of the molecules that bind to it.

In the absence of direct knowledge of the receptor site, the creation of receptor site models relies on the assumption of an underlying complementarity between the shape and properties of the receptor and the compounds that bind. A molecule and a receptor 'see' each other through characteristics presented on the accessible surface of the other, such as the functional groups exposed and the associated molecular fields of the molecule and receptor. Representations of the receptor-binding surface can contain detailed information relevant to the binding of a wide variety of molecules with differing features and topologies; a single pharmacophore model has difficulty representing this variety of features and topologies. Further, receptor models can easily and directly represent information, such as excluded areas and the shape of hydrophobic regions, that are difficult or impossible to represent using pharmacophore models.

A number of methods for constructing receptor site models have been described. The Hypothetical Active Site Lattice (HASL) [3,4] approach represents the molecules inside an active site as a collection of grid-points. (Strictly speaking, HASL models are not receptor site models, since they characterize molecules and not the active site.) The RECEPS program by Itai and co-workers [5,6] represents the shape around one or more template molecules as a set of grid-points tagged with chemical properties. Crippen and co-workers [7] use voronoi polyhedra to build active site models composed of distinct binding regions. Vedani and co-workers [8] have described the generation of full atomistic models of the active site and refer to these models as pseudo-receptors or mini-receptors. Comparative Molecular Field Analysis (CoMFA) models [9,10] are effectively receptor site models that represent the three-dimensional field properties around a set of superimposed molecules as a set of grid-based probe interaction ener-

H. Kubinyi et al. (eds.), 3D QSAR in Drug Design, Volume 3. 117–133.

gies. Jain and co-workers [11] have developed the Compass program which incorporates the ability to perform some measure of conformational adjustments during the MFA analysis. An interesting new variant is called E-state fields [13], in which atom-based electrotopological indices are reflected out onto a grid, to be followed by PLS analysis. Walters and Hinds [12] described the use of a genetic algorithm to place atoms optimally around a set of superimposed molecules, to arrive at a predictive receptor site model. A novel formalism which derives both the three-dimensional field and the appropriate conformations and alignments of the ligands is presented by Dunn et al. [14].

A critical component of a receptor site model is a representation of the *shape* of the active site surface. Shape can be defined either implicitly or explicitly. Field-based approaches represent shape implicitly; most other techniques represent shape explicitly. Atomistic van der Waals surfaces are the most common explicit representation. Solvent-accessible surfaces can be used to represent the shape of both small and large molecules [15,16]. Molecular surfaces can be constructed from electron density data [17]. Splined surfaces have been used to define both rigid and malleable surfaces [18]. Surface shape has also been described in terms of spherical harmonics [19]. Molecular shape has been variously represented by fields [20], geometrical points [15], surfaces [21–23], volumes [24], indices [25] and three-dimensional topology [26,27].

2. Receptor Surface Models (RSM)

A receptor surface model is generated from a set of one or more aligned structures, usually some subset of the most active. If possible, the conformations of the structures should reflect any knowledge of their active conformations in the actual receptor site. Using the set of aligned structures, a receptor surface model is generated over all or some subregion of the structures.

Selecting the appropriate conformations and obtaining an alignment is a complex matter. While there are a number of techniques for aligning molecules [29–35], arriving at an alignment model is often not trivial. Errors in the alignment model can lead to models that are incorrect or poorly predictive.

Once the alignment model is generated for the chosen subset of compounds, a surface is generated to represent their aggregate molecular shape. The surface encloses a volume common to all the aligned molecules. The approach is conceptually similar to the active analog approach [36], where the union volume is constructed over a set of the most active structures. The shape mapped out by the active structures is assumed to be complementary to the shape of the receptor site itself.

To generate the surface, a volumetric field, characterizing molecular shape, is constructed for each aligned structure. These fields are known as *shape fields*, based on work in the computer graphics world of ‘soft objects’ [37]. The shape fields from each individual structure are combined to produce a final volumetric shape field from which an explicit surface is generated. (The shape fields described here differ from the steric fields generated by probe-based approaches like CoMFA or GRID [38], in which each point in the field corresponds to the steric energy of a probe atom at that point interacting with the structure.)

Once a combined shape field has been created, an *isosurface* of the field can be computed to create an explicit object with well-defined shape ([17], [39], [40]). The isosurface algorithm produces a set of triangulated surface points. The generated surface points have a consistent average point density over all regions of the model, though neighboring points are not necessarily evenly spaced. The point density is determined by the initial grid spacing of the field volume. A grid spacing of 0.5 Å yields an average surface density of 6 points per Å^2.

A receptor surface contains information besides molecular shape. After a surface is created, information corresponding to putative chemical properties of the receptor are associated with each surface point. These properties include partial charge, electrostatic potential, hydrogen-bonding propensity and hydrophobicity. A scalar value for each of these properties is calculated and stored with every surface point in the model. This information serves two purposes: first, it is used during display to convey visually active site characteristics in an intuitive fashion; and second, it is used when calculating interaction energies between a molecule and a surface model.

Receptor site information is conveyed visually by mapping properties onto the surface. Regions of the surface are color-coded to indicate particular chemical properties. The intensity of the color on the surface corresponds to the magnitude of the property. For example, assume that a receptor surface model is constructed from six aligned molecules and each of the molecules position a hydrogen acceptor in the same location. Three of the molecules position a second hydrogen-bond acceptor in a different location. If hydrogen-bonding propensity is mapped onto the surface, the region adjacent to the six acceptors will show a full-intensity color, indicating a strong likelihood of a hydrogen-bond donor existing at that location. The region adjacent to the three hydrogen-bond acceptors will show the same color at half the intensity. Since the receptor surface model is hypothetical, it must be remembered that the property characteristics mapped may not always reflect properties of the actual receptor. Color mapping only displays a single property at one time.

Receptor surface models can be displayed semi-transparently. This allows one to see inside the surface and facilitates docking or modifying a structure within the context of the model. The surface model can be either closed or open: a closed model completely encloses some region of space; and an open model has ‘holes’ in the surface. These openings may represent solvent-accessible regions, or regions about which nothing is known. In fact, the receptor surface model may not even be continuous; instead, it could be composed of a number of smaller surface patches which represent information about known regions, while leaving unknown regions open and undefined.

The receptor surface model supports computations that are analogous to those which can be performed with an atomistic model of a receptor site. A structure can be docked into the model. Energetics calculations can be performed to minimize the structure with respect to the model. Energetic information like the strain energy of the structure in the ‘bound’ state and the interaction energy between the structure and the model is available for evaluation. This information can be used in a qualitative fashion to rank potential test compounds, or used quantitatively as descriptors for a QSAR analysis [2].

A unique feature of the receptor surface model is that a molecule can be energy minimized in the context of the model, where the molecule 'feels' the surface of the model. The energetics calculations rely on a fast, approximate force field, termed *Clean*. The force field quickly calculates reasonable geometries and energies of drug size molecules, either in the presence or absence of a receptor surface model.

The Clean process models a flexible ligand inside a rigid receptor site. This process is analogous to minimizing a structure in an actual receptor, holding the receptor atoms fixed. The assumption that the receptor site remains fixed in geometry is a limitation, but is often a reasonable assumption. Studies of HIV-1 protease bound to a set of inhibitors indicates that the geometry of the receptor remains relatively constant, even when there is significant structural diversity in the inhibitors [41].The structure being minimized, therefore, may be perturbed significantly by the procedure, since the geometry of the structure will adopt a conformation consistent with the shape of the surface.

For example, if a surface is created over a chair cyclohexane, and a boat conformation structure is minimized against the surface, the boat conformation can be flipped to chair in the process. Sometimes a structure will assume a geometry lower in energy than the starting structure. Often, however, a structure will be forced to adopt a geometry higher in energy than the initial geometry because of the shape of the surface. The van der Waals term can induce bond and angle distortions. To detect conformation strain introduced by the minimization, a second minimization is performed on the structure in the absence of the surface. This second minimization will bring the structure to a nearby minimum energy conformation.

The minimizations produce three energy values. The first value is the non-bonded interaction energy between the structure and the surface; this value is termed $\boldsymbol{E}_{\mathbf{interact}}$. The second value is the internal strain energy of the structure with respect to the surface. This is the energy of the 'bound' conformation and is the sum of all bond, angle, torsion, inversion and intra-molecular non-bonded energies; this value is termed $\boldsymbol{E}_{\mathbf{inside}}$. The third value is the internal energy of the structure, after it has been allowed to relax without feeling the surface; this value is termed $\boldsymbol{E}_{\mathbf{relax}}$ and will always be less than or equal to $\boldsymbol{E}_{\mathbf{inside}}$.

The $\boldsymbol{E}_{\mathbf{interact}}$, $\boldsymbol{E}_{\mathbf{inside}}$ and $\boldsymbol{E}_{\mathbf{relax}}$ values can be quickly inspected to facilitate an evaluation of goodness of fit. Evaluation is typically based upon two criteria: $\boldsymbol{E}_{\mathbf{interact}}$ and the difference between $\boldsymbol{E}_{\mathbf{inside}}$ and $\boldsymbol{E}_{\mathbf{relax}}$. The more negative is $\boldsymbol{E}_{\mathbf{interact}}$, the better the complementarity between the molecule and the model.

The difference between $\boldsymbol{E}_{\mathbf{inside}}$ and $\boldsymbol{E}_{\mathbf{relax}}$ is a measure of strain energy between the bound conformation and a nearby relaxed conformation. The smaller the value, the less strain introduced by the minimization within the model. This strain estimate indicates nothing about the difference between the bound conformation and the global energy minimum. If a conformational search has previously been performed on the structure, then $\boldsymbol{E}_{\mathbf{relax}}$ can be replaced with the global energy minimum (or lowest minimum found) to give a better estimate of strain energy.

These energies can be used as three-dimensional descriptors in QSAR studies. Hopfinger advocates using binding energetics as QSAR descriptors when the receptor is known [42,43]. Even when the receptor is unknown, using binding energetics from a hypothetical receptor surface model can be a useful predictive tool.

The energetic results can also be visualized by mapping energy of interaction onto the surface. This allows the user to see where favorable and unfavorable interactions are present. Van der Waals energies can be mapped to see where steric groups 'bump' into the receptor surface model. Electrostatic energies can be mapped to see good and bad charge interactions. After the minimization of a molecule, information about location-specific van der Waals and electrostatic interactions is maintained.

Because a structure can be minimized quickly, with the results displayed in color on the surface, a user can quickly test a hypothesis by editing the molecule to see if changes can be made that strengthen the interaction energy without introducing significant strain in the structure. In addition, because the user can always map the initial receptor properties (charge, H-bonding, hydrophobicity), the user can be guided in terms of what editing changes to make in various regions of the model.

2.1. Strengths of receptor surface models

Receptor surface models provide an intuitive, quantitative description which captures three-dimensional information about receptor–ligand interactions. A number of advantageous features of this representation will be discussed:

1. A receptor surface model is *conservative* as compared to a pharmacophore model. A molecule fits a pharmacophore model if the appropriate functional groups can be assigned to the pharmacophores; a receptor surface model includes information on the steric extent of the training molecules, and so can penalize or eliminate molecules that cannot also assume the appropriate steric shape. This conservativeness can be of great benefit in focusing *de novo* construction or database search to the most likely molecules. (Recent work on 'shrink-wrapped' surfaces are an attempt to compensate for this limitation of pharmacophore models [28].)
2. A receptor surface model is a natural representation for the receptor site information, and so is *visually intuitive*, and can be graphically manipulated in real time.
3. Structures can be *energy minimized* within the receptor surface model to arrive at conformations that are consistent with the model. The interaction energies between the surface and the ligand can be estimated.
4. A receptor surface model can be used in database search, to rapidly find compounds similar in shape and consistent in electrostatics to a given receptor surface model query.
5. The total interaction energies are a compact 3D respresentation that can be used within quantitative structure–activity relationship (QSAR) studies to provide a novel form of 3D QSAR.
6. Local surface interaction energies can be captured to provide a table of localized 3D QSAR descriptors. This table can be analyzed similarly to the analysis of CoMFA probe energies, though with the difference that the sample points are localized to be within the likely interaction regions suggested by the model.

3. Applications of Receptor Surface Models

3.1. 3D QSAR with receptor surface models

An assumption behind the appropriate construction and use of receptor surface models is that the template molecules are appropriately aligned and in their putative active conformations. Otherwise, manipulations and applications of the model may be uninformative or even misleading. This is a similar set of restrictions to those applied to CoMFA-like models ([9], [11], [12]). (Unlike CoMFA studies, however, only the molecules used to generate the receptor surface model need to be so aligned and conformed; the evaluation of other molecules use an alignment and conformation provided by minimizing the molecule inside the RSM.)

Our original work on receptor surface models in 3D QSAR demonstrated that for rigid and semi-rigid molecules, the global interaction energies provide a useful, compact 3D descriptor that can be used to build a 3D QSAR equation [2]. The ability of the RSM to 'fit' new molecules within its surface frees the user from having to specify a detailed conformation beforehand. Still, of more interest is the case where the training and test molecules have significant flexibility.

Recently, technologies have been developed to generate likely alignments of flexible molecules. Examples of such technologies are Catalyst/HipHop (for series with no activity data or when all molecules have similar activities) [35], Catalyst/HypoGen (when many orders of magnitude of activity data are available) or DISCO [33]. These programs can provide possible alignments and conformations, which can then be used by the chemist to generate a receptor surface model.

An example of this is shown by a series of 15 highly flexible peptoids which are known antagonists for the human cholecystokinin B (CCK-B) receptor [44]. Using HipHop, these molecules were aligned into a specific conformation. The aligned molecules are shown in Fig. 1.

Note that while the alignment and conformations of the molecules is an improvement over the original minimized conformations, there is still too much randomness to use techniques such as molecular field analysis (MFA) against this dataset. However, it is possible to use the alignments and conformations of the three most active molecules to construct a receptor surface model; the remaining molecules can then be minimized within the RSM to obtain quantitative fit information. The receptor surface model generated using the top three molecules (and with the hydrogen-bonding characteristics mapped onto the surface) is shown in Fig. 2.

The final question is whether this RSM can be used to obtain quantitative information about the entire series of peptoids. Genetic Function Approximation [45] was used to generated possible QSARs. The QSARs were allowed to use both linear terms and nonlinear spline terms; the use of splines allows the negative effect of bad interactions to be limited in their effect. (And unlike neural networks, spline-based models are still easily interpretable.)

The top QSAR and its statistics are shown in Fig. 3. This simple 3D QSAR shows moderate predictivity ($q^2 = 0.55$); it is encouraging that some level of predictivity is shown in the face of the complexity of the problem, which includes a small dataset, flexible molecules and lack of known receptor information. At the least, it should be a useful guide for future experiments or database searching for possible alternate lead compounds. (Such a 3D search using receptor surface models is described in the next section.)

3.2. Shape-based searching of flexible molecules

This section explores using a receptor surface model as a database query to search a database for hits that fit a particular query's shape. Such a method is useful in a number of contexts, including database screening, database mining and combinatorial library diversity analysis [46].

In order to allow the evaluation of databases of potentially millions of compounds, a two-phase approach is used. Those candidates passing a rough shape similarity filter are then evaluated with a fitting procedure for a more rigorous steric and electrostatic analysis. Such a two-phase approach works for large databases, since the first phase (shape

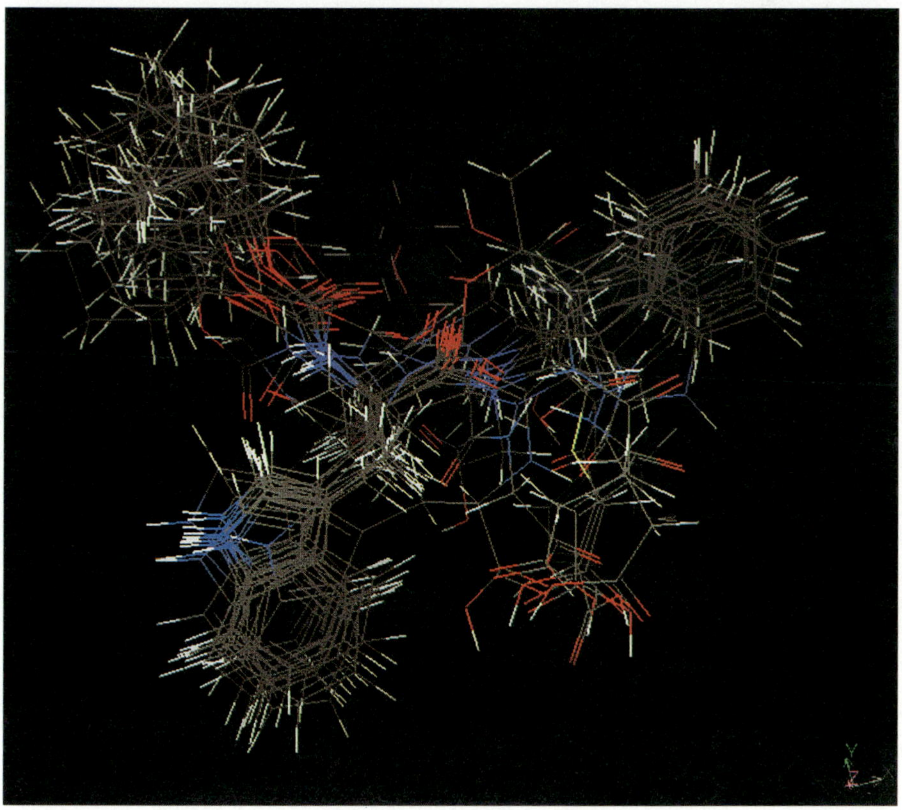

Fig. 1. Overlapped peptoids. The conformations and alignments were generated by the Catalyst/HipHop program.

similarity screening) is both fast and significantly reduces the number of potential candidates. This screening approach is analogous to 2D substructure searches which use topological bit screens before undertaking the algorithmically time-consuming atom-by-atom comparison.

This approach to shape-based searching first requires the creation of a compound database containing multiple 3D conformations per compound. Compounds and their associated conformations are stored in a Catalyst database. After the compound database has been created, a shape filter database is then created. The shape filter database contains information for rapidly screening the database for shape candidates. The shape filter database is constructed by retrieving each conformer from the compound database,

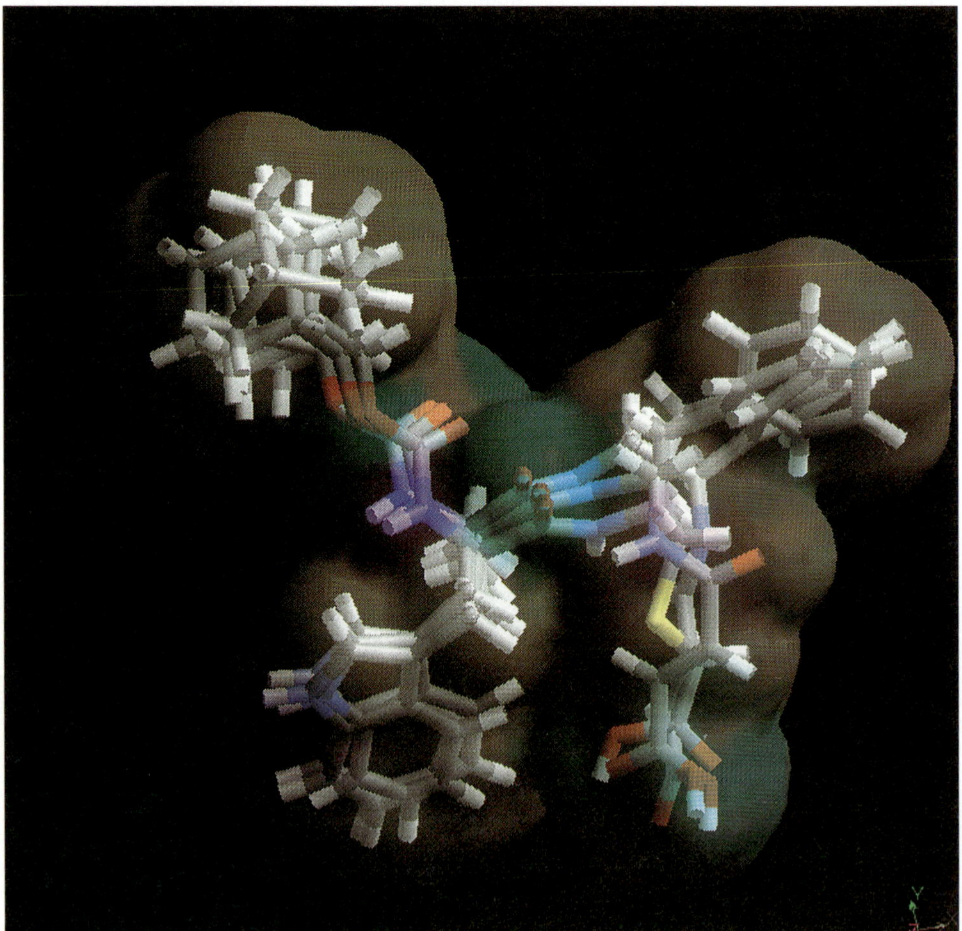

Fig. 2. The receptor surface model generated using the three most-active peptoids. Hydrogen-bonding characteristics are mapped onto the model, with the acceptor/donor status marked by different colors.

computing a set of volume and shape indices and storing these per conformer shape indices in the filter database. Shape filter database creation is fast relative to database creation, and typically takes less than 30 min per million conformations processed.

A shape query is represented as an RSM. The surface encloses a defined volume, which is represented as a grid (0.5 to 1.0 Å spacing). Using the RSM surface points, shape indices are derived.

First, the geometric center and three principal component vectors of the set of points are computed. No special weighting (either VDW radius or atomic mass) is used in the centroid calculation. Next, the maximum extents along each principal axis are found. M0 and NM0 are the extent lengths along the positive (longest) and negative (shortest) direction of the major axis, respectively. M1 and NM1 are the positive and negative extents along the minor axis. In three dimensions, the third axis contains M2 and NM2 components. In addition to these six indices, the total volume of the query (or conformer) is computed from the total number of surface interior grid-points and the grid resolution. These seven indices are stored per conformer in the shape filter database when constructing the database. The same indices generated for a query are used in the screening process. The indices provide a simple and compact way of representing the gross overall size and shape of a query.

The database screening process for a given query is as follows. The volume and six shape indices are computed for the query. These indices are then compared with the corresponding indices for each conformation in the shape filter database. The filter database is actually sorted on the first index, so that only a subset of the indices need be compared. This process quickly eliminates conformations that do not have similar shape, as defined by these indices. A user-settable tolerance on the indices defines what is possibly 'similar'. This tolerance specifies the plus and minus variation allowed for the extents and volume indices.

The database screening phase results in a list of candidate conformations that have shape indices similar to the query. Next, the query and candidate structures are aligned based upon their principal axis. Clearly, if the query or target molecule have any symmetry or near-symmetries, aligning on only the principal axis may not be adequate. After trying all symmetry-equivalent permutations, the alignment yielding the best volume similarity is retained. Finally, a descent optimization algorithm can be executed to improve the volume overlap of axis-based alignment.

```
Activity = 3.301 - 0.204 * <1.976 - InteractionEnergy>
```

SD:	42.633
r^2:	0.658

Fig. 3. The top QSAR and its statistics against the 15-compound peptoid dataset. The angle brackets denote a spline term; it is equal to zero if the value of its contents are negative, and is equal to the value of the contents otherwise.

The grid volumes of the query and target are then compared to determine shape similarity using a Tanimoto score (the intersection divided by union volumes of the query and target) to estimate similarity. This score can be used as a secondary screen to the indices-based screen. The hit list, sorted by similarity, can be saved and browsed, or can be passed on for the final phase of the search procedure.

The final stage is flexible fitting into the receptor surface model. Up till now, electrostatic features of the query (i.e. H-bonding, hydrophobic and charged groups) have not been taken into account, and so each hit may or may not have electrostatic similarity to the query. This evaluation procedure minimizes each hit into the RSM, flexibly fitting each geometry to be consistent with the shape and electrostatics of the model. The evaluation procedure estimates both intramolecular strain energy and intermolecular interaction energy between the hit and the surface model.

To arrive at a final set of shape matches, the evaluated structures are sorted by strain energy and all structures with a strain energy greater than a specified threshold are discarded. The default threshold is 20 kcal/mol. To measure electrostatic similarity, the remaining candidate list is resorted on increasing interaction energy. The user is then presented with the sorted hit compounds.

3.3. Receptor surface analysis (RSA)

As previously described, the surface representation used by a receptor surface model is based on a set of locally defined mesh-points in 3D space. The combined interaction effects of these points can be calculated and used in 3D QSAR modelling as a small set of information-rich descriptors ($\boldsymbol{E}_{\mathbf{interact}}$, $\boldsymbol{E}_{\mathbf{inside}}$, etc.). However, it is also possible to use these points directly and their interaction values in 3D QSAR [47]. This may be useful if the user suspects that only a few local regions of interaction within the site are important, or if the user wishes to identify and view those regions. This approach is analogous to MFA and is termed Receptor Surface Analysis (RSA). RSA is performed as follows. A receptor surface model is generated around some number of aligned active molecules in the putative active conformation. For example, a series of 22 inhibitors of rat-liver squalene epoxidase ([34], [48]) can be aligned with HipHop and the three most-active used to generate a receptor surface model. Such a model is shown in Fig. 4.

The RSM is composed of thousands of localized points which store a local measure of the quality of the interaction during evaluation. It is these points and their VDW and electrostatic interaction energies which can be unpacked, analyzed and viewed. When unpacked, each point provides three columns in a table: the VDW interaction energy, the electrostatic interaction energy and the combined interaction energy.

Many of these points will be uninteresting as there will be little variation in interaction energy across the compounds. A variance filter can be used to remove these points; one rule-of-thumb is to accept only the 5% most-variant columns for further analysis. This reduces the table to a few hundred columns.

In this example, partial least squares (PLS) was used to analyze the data table but the cross-validated r^2 (q^2) showed the model to be non-predictive. Upon inspection of the table, the cause of this could be inferred. Since the RSM points are often quite close to

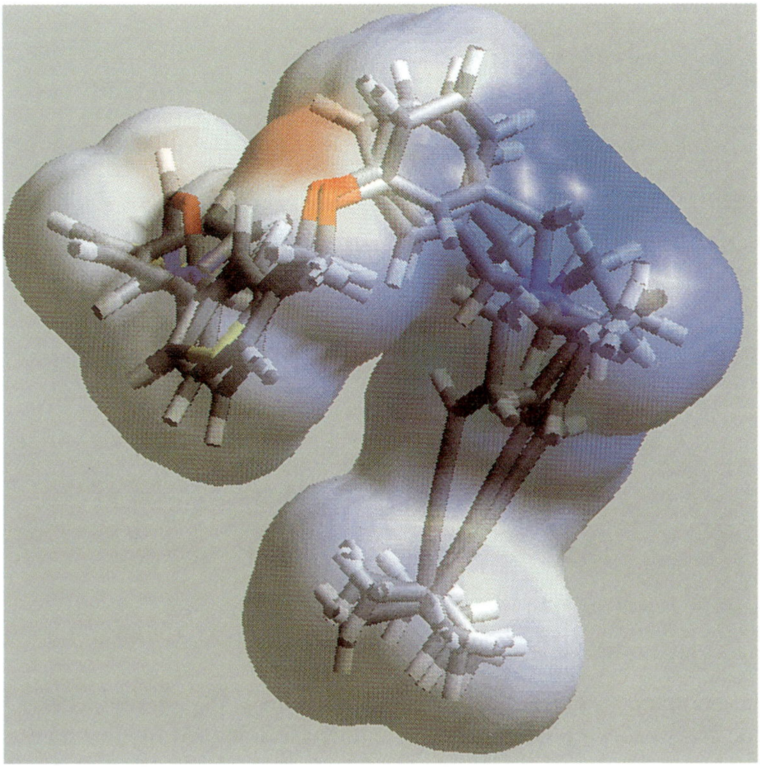

Fig. 4. Squalene epoxidase inhibitors used to create a receptor surface model. Electrostatic potential of the molecules is mapped on RSM. This can suggest the electrostatic environment within a given receptor site.

the test compounds, the interaction energies measured can grow rapidly, since interaction energy is a nonlinear function. This nonlinear effect made it difficult for linear methods such as PLS to find useful patterns in the data. (This suggests one reason why models based upon linear PLS, such as CoMFA models, might overreact to changes in molecular structure near highly loaded grid-points.)

Instead, we used nonlinear genetic partial least squares (G/PLS) [49–51]. This selects a subset of the points, adds them to a model as either linear or spline terms, and fits the generated model with PLS. Many such models are created, and the population of G/PLS models is evolved to discover better models. Using a population of 300 models, 14-term models, 5000 evolution steps and fitting using 4-component PLS, the best-rated model is shown in Fig. 5.

The fitness function used during the evolution was a penalized least-squares error measure called Friedman's lack-of-fit (LOF) function [49]. Cross-validated r^2 (q^2) was not used during training. q^2 is a useful posterior estimator of the significance of a model if it is not previously used during training.

```
Activity = 3.095
- 0.0438 * TOT851
- 0.5194 * <-34.221 - InteractionEnergy>
- 0.0334 * TOT2148
- 0.0366 * TOT1449
+ 0.1334 * TOT1374
- 0.0918 * TOT1023
+ 0.0862 * <-12.524 - TOT1163>
+ 0.0316 * <-8.651 - TOT1264>
- 0.0052 * <-0.323 - TOT1433>
+ 0.1032 * TOT1744
+ 0.0742 * <-8.50286 - TOT859>
+ 0.0489 * <TOT2724 - 0.0977>
- 0.0447 * TOT107
```

SD:	18.639
r^2:	0.987
PRESS:	7.503
Cross-validated r^2 (q^2):	0.60

Fig. 5. The best-rated model for squalene epoxidase binding.

Note the common use of spline terms of the form <A – energy>; these terms are nonzero for positive interactions (with the cutoff level defined by the value of A), and are zero for bad interactions. Again, we see a restriction on the range of energy used to reduce the effect of the nonlinearities in the energy function.

It is also possible to view the points used by the QSAR in 3D space, showing their placement around the given molecule. Such a figure for the subset of linear points in the QSAR is shown in Fig. 6. The small number of points in a nonlinear G/PLS model can focus the user on important details in a receptor–ligand interaction that may be missed in viewing the more diffuse PLS loading maps.

4. Summary

A novel form of receptor site model, called a receptor surface model, has been described. A *receptor surface model* is generated from a series of aligned molecules with associated binding activities. A steric surface is generated to enclose the aggregate aligned molecules, and scalar properties corresponding to putative receptor properties are associated with each surface point. Regions of the receptor surface model can be removed to reflect corresponding openings in the receptor site, or areas of the receptor site about which nothing is known.

The receptor surface model has characteristics that make it a desirable representation for receptor site hypotheses. The models are intuitive and visually appealing. The receptor surface model supports energetics calculations for the interactions of molecules with the model. The model uses theClean force field, which is optimized for speed and accu-

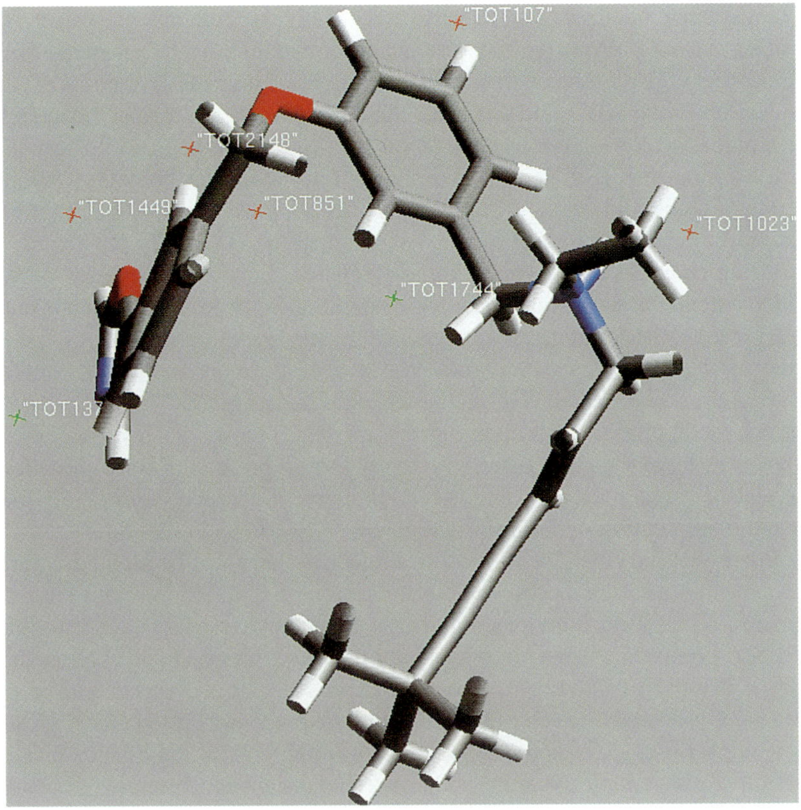

Fig. 6. Some of the interaction points in the QSAR displayed around the most-active molecule. Green crosses are points where positive interactions are negatively correlated with activity; red crosses are where positive interactions are positively correlated with greater activity.

racy when used with the receptor surface model representation. The model provides interactive and qualitative feedback for evaluating and testing new structures. The models are easily modified as the active site hypothesis is refined.

Receptor surface models differ from pharmacophore models, in that the former try to capture essential information about the receptor, while the latter capture information about the commonality of compounds that bind. Pharmacophore models generally represent some minimal set of features present in the actives and postulate that those features, in some configuration, are required for binding. Since these models do not usually represent the receptor boundary, molecules that fit the model can still be inactive because of additional regions of the molecule that are sterically unfavorable. Pharmacophore models, therefore, tend to be geometrically under-constrained (while topologically over-constrained); this steric under-constraint leads to false positives, that is compounds that are deemed active by the model but which are inactive when tested.

Receptor surface models, on the other hand, tend to be geometrically over-constrained (and topologically neutral), since in the absence of steric variation in a region, they assume the tightest steric surface which fits all training compounds. This may be significantly more restrictive than the actual boundaries of the receptor. This means they are prone to false negatives: new actives (not used in creating the model) may map out new regions of the active site and, thus, may evaluate poorly against the model. This is illustrated by the opiate analgetics. Generation of a receptor surface model from molecules such as morphine, meperidine and levorphanol (all having an N-methyl group) would indicate that a meperidine analog where the N-methyl is extended by a phenyl butyl side chain would be inactive. In fact, this analog has 100 to 1000 times the activity of morphine. In such cases (as new information is obtained), the receptor surface model can be modified to extend the surface into new regions; pharmacophore models, since they do not directly represent steric boundaries, are less suitable for such modification.

As the number of ligands increases, it can become increasingly difficult to build models or to overlap the ligands in such a way that their essential commonalties and differences are made obvious. Receptor surface models directly display the commonalties and differences by associating them with the natural representation for the information: a 3D model of a receptor site. The use of modern, high-speed computers makes the display and manipulation of this information easy to perform in real time.

Once the model is constructed, new test molecules need not be aligned or conformed precisely: the model itself is responsible for generating the appropriate alignment and conformation. This is most obvious in the case of molecules which have an initial, rough conformation proposed by matching against a pharmacophore model such as those generated by HipHop; this initial set of conformations may be too variable to be used in a grid-based analysis method such as CoMFA, but the receptor surface model is able to optimize the conformations to approximate the conformations of the ligands chosen in the construction of the model. (Note that other methods, such as Compass [11] or the work of Dunn et al. [14], are also designed to deal with conformational variability.)

Most companies have an internal database of molecules, and many public or commercial databases are also available. Receptor surface models provide a direct way to search for molecules that can be conformed to a given shape, and then can be used to order the hit by the quality of their electrostatic match.

Receptor surface models provide compact, quantitative descriptors which capture three-dimensional information about a putative receptor site. These descriptors may be used alone, or in combination with more traditional 2D descriptors. Such combined QSAR models may better reflect the combination of mechanisms (transport, binding, absorption, etc.) responsible for drug activity.

Receptor surface models and their descriptors are generated quickly. Numerous alternate receptor surface models can be constructed with varying combinations of active structures, surface fit tolerances and alignments. A variable selection technique like GFA can be used to suggest which receptor surface model(s) are likely most informative. GFA also facilitates the discovery of nonlinear relationships by allowing spline models; this makes explicit the location of the discontinuity in the relationship between energy-derived terms and activity. Such relationships are not easily discovered using linear modelling tools such as PLS.

The RSM shape indices can be used to characterize the 3D shape of molecules. By taking averages and ranges of the shape indices of all conformations for a given compound, whole molecule descriptors can be derived which represent shape and size variability. Such descriptors should be useful in diversity and similarity analysis.

Finally, we report on ongoing work that uses local interaction energies to build a 3D QSAR. This is useful when the user wishes to isolate local effects that may be important in the activity of molecules. Unlike grid-based approaches, all the sample points are on a surface where the presumed interactions of interest would be happening at ligand–receptor contact regions.

References

1. Hahn, M., *Receptor surface models: 1. Definition and construction*, J. Med. Chem., 38 (1995) 2080–2090.
2. Hahn, M.A. and Rogers, D., *Receptor surface models: 2. Application to quantitative structure–activity relationship studies*, J. Med. Chem., 38 (1995) 2091–2102.
3. Doweyko, A.M., *The hypothetical active site lattice: An approach to modeling active sites from data on inhibitor molecules*, J. Med. Chem., 31 (1988) 1396–1406.
4. Wiese, M., *The hypothetical active-site lattice*, In Kubinyi, H. (Ed.) 3D QSAR in drug design: Theory, Methods and Applications, ESCOM, Leiden, The Netherlands, 1993, pp. 80–116.
5. Kato Y., Inoue A., Yamada, M., Tomioka, N. and Itai, A., *Automatic superposition of drug molecules based on their common receptor site*, J. Comput. Assist. Mol. Design, 6 (1992) 475–486.
6. Kato, Y., Itai, A. and Iitaka, Y., *A novel method for superimposing molecules and receptor mapping*, Tetrahedron, 43 (1987) 5229–5236.
7. Srivastava, S., Richardson, W.W., Bradley, M.P. and Crippen, G.M., *Three-dimensional receptor modeling using distance geometry and voronoi polyhedra*, In Kubinyi, H. (Ed.), 3D QSAR in drug design: Theory, methods and applications, ESCOM, Leiden, The Netherlands, 1993, pp. 80–116.
8. Snyder, J.P., Rao, S.N., Koehler, K.F. and Vedani, A., *Minireceptors and pseudoreceptors*, In Kubinyi, H. (Ed.), 3D QSAR in drug design: Theory, methods and applications, ESCOM, Leiden, The Netherlands, 1993, pp. 336–354.
9. Cramer, R.D., Patterson, D.E. and Bunce, J.D., *Comparative molecular field analysis (CoMFA): 1. Effect of shape on binding of steroids to carrier proteins*, J. Am. Chem. Soc., 110 (1988) 5959–5967.
10. Cramer, R.D., DePriest, S.A., Patterson, D.E. and Hecht, D.E., *The developing practice of comparative molecular field analysis*, In Kubinyi, H. (Ed.), 3D QSAR in drug design: Theory, methods and applications, ESCOM, Leiden, The Netherlands, 1993, pp. 443–485.
11. Jain, A., Koile, K. and Chapman., D., *Compass: Predicting biological activities from molecular surface properties — performance comparisons on a steroid benchmark*, J. Med. Chem., 37 (1994) 2315–2327.
12. Walters, D.E. and Hinds, R.M., *Genetically evolved receptor models: A computational approach to construction of receptor models*, J. Med. Chem., 37 (1944) 2527–2535.
13. Kellogg, G.E., Kier, L.B., Gaillard, P. and Hall, L.H., *E-state fields: Applications to 3D QSAR*, J. Comput.-Aided Mol. Design, 10 (1996) 513–520.
14. Dunn III, W.J., Hopfinger, A.J., Catana, C. and Duraiswami, C., *Solution of the conformation and alignment tensors for the binding of trimethoprim and its analogs to dihydrofolate reductase: 3D-quantitative structure–activity relationship study using molecular shape analysis — 3-way partial least-squares regression and 3-way factor analysis*, J. Med. Chem., 39 (1996) 4825–4832.
15. Connolly, M.L., *Analytical molecular surface calculation*, J. Appl. Crystallogr., 16 (1983) 548–558.
16. Connolly, M.L., *Solvent-accessible surface of proteins and nucleic acids*, Science, 221 (1983) 709–713.
17. Purvis, G.D., *On the use of isovalued surfaces to determine molecule shape and reaction pathways*, J. Comput.-Aided Mol. Design, 5 (1991) 55–80.
18. Klein, T.E., Huang, C.C., Pettersen, E.F., Couch, G.S., Ferrin, T.E. and Langridge, R., *A real-time malleable surface*, J. Mol. Graphics, 8 (1990) 16–24.
19. Leicester, S.E., Finney, J.L. and Bywater, R.P., *Description of molecular surface shape using Fourier descriptors*, J. Mol. Graphics, 6 (1988) 104–108.

20. Grant, J. and Pickup, D., *A Gaussian description of molecular shape*, J. Phys. Chem., 99 (1995) 3503–3510.
21. Masek, B., Marchant, A. and Matthew, J., *Molecular skins: A new concept for quantitative shape matching of a protein with its small molecule mimics*, Proteins, 17 (1993) 193–202.
22. Masek, D., Marchant, A. and Matthew, J., *Molecular shape comparison of angiotensin II antagonists*, J. Med. Chem. Proteins, 36 (1993) 1230–1238.
23. Bohaceck, R. and McMartin, C., *Definition and display of steric, hydrophobic, and hydrogen-bonding properties of ligand binding sites in proteins using Lee and Richards' accessible surface: Validation of a high-resolution graphical tool for drug design*, J. Med. Chem., 35 (1992) 1671–1684.
24. Perkins, T., Mills, J. and Dean, P., *Molecular surface–volume and property matching to superimpose flexible dissimilar molecules*, J. Comput.-Aided Mol. Design, 9 (1995) 479–490.
25. Todeschini, R., Lasagni, M. and Marengo, E., *New molecular descriptors for 2D and 3D structures, theory*, J. Chemometrics, 8 (1994) 263–272.
26. Mezey, P., *Three-dimensional topological aspects of molecular similarity*, In Johnson, M. and Maggiora, G. (Eds.) Concepts and applications of molecular similarity, John Wiley, New York, 1990, 321–368.
27. Mezey, P., *Shape in chemistry*, VCH, New York, 1993.
28. VanDrie, J.H., *'Shrink-wrap' surfaces: A new method for incorporating shape into pharmacophore 3D database searching*, J. chem. Inf. Comp. Sci., 37 (1997) 38–42.
29. Kearsely, S.K. and Smith, G.M., *An alternative method for the alignment of molecular structures: Maximizing electrostatic and steric overlap*, Tetrahedron Comput. Method., 3 (1990) 615–633.
30. Dammkoehler, R.A., Karasak, S.F., Berkely Shands, E.F. and Marshall, G.R., *Constrained search of conformational hyperspace*, J. Comput.-Aided Mol. Design, 3 (1989) 3–21.
31. Perkins, T.D. and Dean, P.M., *An exploration of a novel strategy of superimposing several flexible molecules*, J. Comput.-Aided Mol. Design, 7 (1993) 155–172.
32. Blaney, J.M. and Dixon, J.S., *A good ligand is hard to find: Automatic docking methods*, Perspectives in Drug Discovery and Design, 1 (1993) 301–319.
33. Martin, Y.C. and Bures, M.G., Danahar, E.A., DeLazzar, J., Lico, I. and Pavlik, P.A., *A fast new approach to pharmacophore mapping and its application to dopaminergic and benzodiazepine agonists*, J. Comput.-Aided Mol. Design, 7 (1993) 83.
34. Hoffmann, R. and Langer, T., *Use of the CATALYST program as a new alignment tool for 3D QSAR*, In Proceedings of the 10th European Symposium on Structure–Activity Relationships: QSAR and molecular modeling, Prous Science Publishers, Barcelona, Spain, 1995, pp. 466–469.
35. Barnum, D., Greene, J. and Smelie, A., *Identification of common functional configurations*, J. Chem. Inf. Comp. Sci., 36 (1996) 563–571.
36. Marshall, G.R., *Binding site modeling of unknown receptors*, In Kubinyi, H. (Ed.), 3D QSAR in drug design: Theory, methods and applications, ESCOM, Leiden, The Netherlands, 1993, pp. 80–116.
37. Wyvill, G., McPheeters, C. and Wyvill, B., *Data structures for soft objects*, The Visual Computer, 2 (1986) 227–234.
38. Goodford, P.J., *A computational procedure for determining energetically favorable binding sites on biologically important macromolecules*, J. Med. Chem., 28 (1985) 849–857.
39. Lorensen, W.E. and Cline, H.E., *Marching cubes: A high resolution 3D surface construction algorithm*, Computer Graphics (Proc. SIGGRAPH), 21 (1987) 163–169.
40. Heiden, W., Schlenkrich, M. and Brickman, J., *Triangulation algorithms for the representation of molecular surface properties*, J. Comput.-Aided Mol. Design, 4 (1990) 225–269.
41. Appelt, K., *Cyrstal structures of HIV-1 protease-inhibitor complexes*, Perspect. Drug Discov. Design, 1 (1993) 23–48.
42. Hopfinger, A.J., Nakata, Y. and Max, N., *Quantitative structure–activity relationship of anthracycline antitumor activity and cardiac toxicity based upon intercalation calculations*, In Pullman, B. (Ed.) Intermolecular forces, Reidel, Dordrecht, The Netherlands, 1981, p. 431.
43. Hopfinger, A.J. and Kawakami, Y., *QSAR analysis of a set of benzothiopyranoindazole anti-cancer analogs based on their DNA intercalation properties as determined by molecular dynamics simulation*, Anti-Cancer Drug Design, 7 (1992) 203–217.

44. Hoffmann, R. and Bourguignon, J.-J., *Building a hypothesis for CCK-B antagonists using the CATALYST program*, In Proceedings of the 10th European Symposium on Structure–Activity Relationships: QSAR and molecular modeling, Prous Science Publishers, Barcelona, Spain, 1995, 298–300.
45. Rogers, D. and Hopfinger, A.J., *Application of genetic function approximation to quantitative structure–activity relationships and quantitative structure–property relationships*, J. Chem. Inf. Comput. Sci., 34 (1994) 854–866.
46. Hahn, M., *Three dimensional shape-based searching of conformationally flexible compounds*, J. Chem. Inf. Comput. Sci., 37 (1997) 80–86.
47. This is ongoing work done by ourselves, Dr. Remy Hoffmann and Dr. Max Muir.
48. Hoffmann, R. and Sprague, P., *Building a hypothesis for competitive inhibition of rat liver squalene expoxidase*, CATALYST Application Note, 1995.
49. Rogers, D., *Genetic function approximation: A genetic approach to developing quantitative structure–activity relationships models*, In Proceedings of the 10th European Symposium on Structure–Activity Relationships: QSAR and molecular modeling, Prous Science Publishers, Barcelona, Spain, 1995, pp. 420–426.
50. Dunn III, W.J. and Rogers, D., *Genetic partial least-squares in QSAR*, In Devillers, J. (Ed.) Genetic Algorithms in Molecular Modeling, Academic Press, London, 1996, pp. 109–130.
51. Rogers, D. and Dunn III, W.J., *Genetic partial least-squares*, J. Comput.-Aided Mol. Design, (1997) (accepted).

Pseudoreceptor Modelling in Drug Design: Applications of Yak and PrGen

Marion Gurrath[a*], Gerhard Müller[b] and Hans-Dieter Höltje[a]
[a] *Heinrich Heine University-Düsseldorf, Institute for Pharmaceutical Chemistry, Universitätsstr. 1, D-40225 Düsseldorf, Germany*
[b] *Bayer AG, IM-FA, Computational Chemistry, Q18, D-51368 Leverkusen, Germany*

1. Introduction

Structure-based drug design comprises two methodologically different strategies in the identification of new drug candidates, commonly termed 'direct' and 'indirect' design (see e.g. [1,2]). The common aim of both strategies is to understand structure–activity relationships and to employ this knowledge for proposing new compounds with enhanced activity and selectivity profiles for a specified therapeutic target. For a direct design strategy, the 3D structure of e.g. a target enzyme or even a receptor–effector complex is required with atomic resolution, generally determined by either high-resolution crystallography or multidimensional and multinuclear NMR spectroscopy [3]. Unfortunately, most receptor systems of current pharmaceutical interest are membrane-bound multidomain proteins, the 3D structure of which are unknown at present, thereby restricting molecular modelling studies to an indirect approach. Thus, the indirect approach is based on comparative analyses of structural features of known active and inactive low-molecular weight compounds, which are interpreted in terms of steric and physico-chemical complementarity with a fictional receptor binding site of unknown structure, typically termed 'receptor mapping'.

The 3D QSAR techniques are the most prominent computational means to support chemistry within indirect drug-design projects [4,5]. The primary aim of these techniques is to establish a correlation of biological activities of a series of structurally and biologically characterized compounds with the spatial 'fingerprint' of numerous field properties for each molecule, such as steric demand, lipophilicity and electrostatics. Typically, a 3D QSAR study allows identifying the pharmacophoric arrangement of molecular fragments in space, and provides guidelines for the design of next-generation compounds with enhanced biological performance.

In practice, the experience from several projects in converting 3D QSAR-derived recommendations into new chemical entities teaches us that non-atomistic models as provided by e.g. CoMFA studies are not always intuitive for synthetic chemists. Atomistic receptor models, in contrast, allow us to gain detailed insights into the key interactions between macromolecular target and ligand in a straightforward fashion, which definitely helps to facilitate the design process and synthesis of new compounds.

In this contribution, we report on the pseudoreceptor modelling concept exemplified by recent molecular modelling studies on different classes of receptor agonists and antagonists from our own laboratories and from the literature. We mainly restrict ourselves to the discussion of the latest developments and applications of the software package Yak and its successor program PrGen [6–8]. Special emphasis will be placed on the

H. Kubinyi et al. (eds.), 3D QSAR in Drug Design, Volume 3. 135–157.

opportunity of the pseudoreceptor modelling concept to combine the receptor mapping philosophy, indicative for the indirect design approach, with the receptor fitting aspects derived from the direct design approaches. It is this conceptual combination that ascribes more transparency to the drug-design process which, as a consequence, is appreciated more easily by the synthetic community in pharmaceutical research.

2. Methodology

The pseudoreceptor modelling approach attempts to generate a 3D model of the binding site of a structurally unknown target protein (enzyme, receptor) based on the superimposed structures of known ligand molecules in their bioactive conformation, together with the experimentally determined binding affinities towards the target protein. The goal of the pseudoreceptor modelling is to engage these superimposed molecules in specific non-covalent ligand–target interactions so as to mimic the receptor-bound state for each ligand. In general, type and spatial arrangement of the pseudoreceptor building blocks surrounding the ensemble of superimposed ligands will bear no structural resemblance to the 'true' biological target protein. Instead of reproducing the complex structure of the ligand-binding protein of interest, the receptor surrogate should be envisioned as a purely hypothetical model of the binding pocket, accommodating a series of structurally related ligands in a similar binding mode, thus allowing a semi-quantitative prediction of binding affinities. The estimation of binding affinities relies on the evaluation of ligand–pseudoreceptor interaction energies, ligand desolvation energies and changes in ligand internal energy and entropy upon the receptor binding event [9]; the mathematical details of the energy evaluations are given below.

Although various pseudoreceptor concepts have been developed by e.g. Frühbeis et al. [10], Snyder and Rao [11,12], Momamy et al. [13], Hong et al. [14], Snyder et al. [15,16], Höltje and Anzali [17], Walters and Hinds [18], Doweyko [19] and Hahn et al. [20,21], we focus mainly on the methodology and applications of Yak and the follow-up program PrGen, developed by Vedani et al. [6–8].

The entire pseudoreceptor modelling procedure employed by PrGen can be split into the following distinct steps:

1. Generation of ligand alignment.
2. Identification of receptor nucleation sites.
3. Construction of the pseudoreceptor.
4. Energetic equilibration.
5. Validation — pseudoreceptor analysis.

2.1. *Generation of ligand alignments*

In the initial step of pseudoreceptor modelling, the 'molecular probes' utilized for reconstructing a hypothetical binding pocket (training set) need to be aligned according to molecular fragments, common to the entire ensemble of ligand molecules, thus constituting the potential pharmacophore. Obtaining a meaningful superposition for a series of ligand molecules is by no means a straightforward task, since the bioactive con-

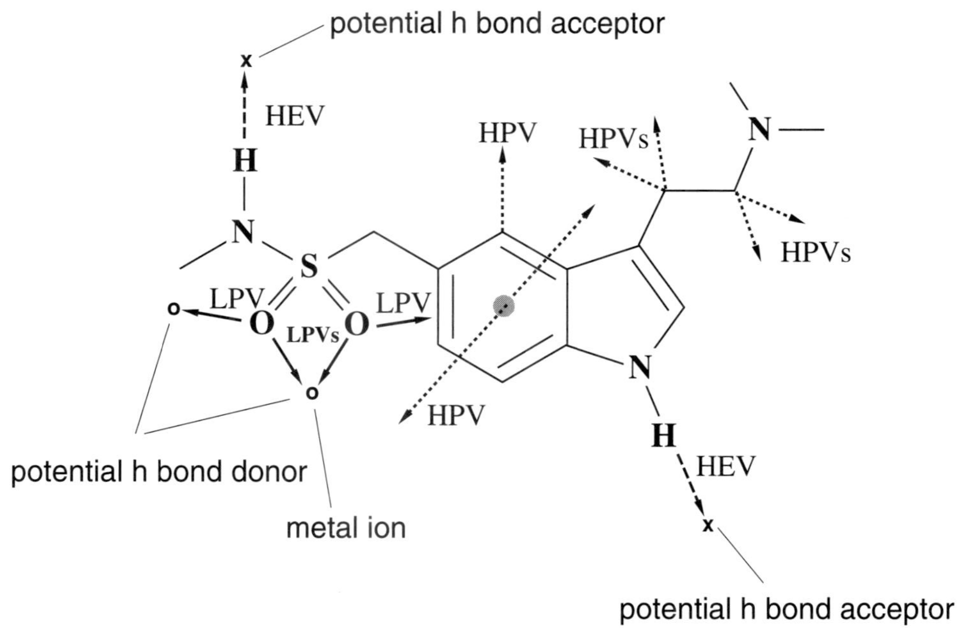

Fig. 1. Different vector types utilized by PrGen for identification of potential interaction sites on ligand molecules exemplified with the drug molecule sumatriptan. For reasons of clarity, only selected HPVs are shown: HEVs, hydrogen extension vectors — mark the ideal position of hydrogen-bond acceptor sites; LPVs, lone pair vectors — mark the ideal position of hydrogen-bond donor sites; and HPVs, hydrophobicity vectors — indicate sites for hydrophobic interactions.

formations and relative positions and orientations within the binding pocket of the native target protein cannot be deduced solely from the molecular structures of the ligands. In this context, PrGen offers a procedure termed 'receptor-mediated pharmacophore alignment' that especially addresses the superposition problem. Within this technique, a primordial receptor model is generated only based on a single ligand molecule that preferably exhibits the highest intrinsic affinity towards the biological receptor of interest among all training set molecules. Only this root molecule serves as molecular probe to map the steric and physico-chemical demand of the receptor surrogate. After refinement of the resulting model complex, the remaining ligands of the training set are added to the model and allowed to relax within the receptor environment.

2.2. *Identification of receptor nucleation sites*

After structural superpositon of all ligand molecules constituting the training set, the ligand groups capable of interacting with receptor residues are identified. For that purpose, three different types of vector, originating on ligand functionalities, associated with different types of directional interaction, are generated (Fig. 1) [22–29]:

1. HEVs, hydrogen extension vectors: mark the ideal position of hydrogen-bond acceptor sites.
2. LPVs, lone pair vectors: mark the ideal position of hydrogen-bond donor sites.
3. HPVs, hydrophobicity vectors: indicate sites for hydrophobic interactions.

After vector generation, a cluster analysis identifies for each vector type spatial areas of high vector density as potential anchor points for receptor residues in space. Dense clusters comprised of a single vector type are interpreted as indications for interaction sites relevant for molecular recognition — i.e. being complementary to the postulated pharmacophore. Dense clusters comprised of different vector types can be envisioned as diagnostic sites for specific discrimination — i.e. for ligand selectivity.

2.3. *Construction of the pseudoreceptor*

Identified anchor points are 'saturated' with receptor fragments (amino acids, metal ions, predefined protein substructures) according to the directionality of the corresponding interaction type involved [22–29]. The pseudoreceptor modelling is an iterative procedure based on successive addition of receptor fragments, unless all potential anchor points are engaged in intermolecular interactions, or, more likely, unless the spatial conditions prevent the addition of any further receptor residue. One of the major advantages of such an atomistic approach over 'classical' 3D QSAR techniques consists in the opportunity to include available biological information other than the binding affinities of ligands within the pseudoreceptor generation process. Results from various investigations on the target protein, such as secondary structure predictions, identification of common folding motifs within a protein homology family, site-directed mutagenesis or cross-linking studies with affinity labels, can specifically tailor the pseudoreceptor generation protocol.

After generation of a truncated protein core consisting of only a few residues or fragments surrounding the ensemble of superimposed ligands, it turned out to be advantageous to augment the atomistic part of the receptor surrogate by virtual particles, mimicking hydrophobic interactions and accounting for the electrostatic field of the residual protein. The virtual particles used in PrGen are spherical Lennard-Jones particles that may vary in size and polarizability [30]. Initially, these are uncharged entities, but during correlation-coupled minimization (see below) finite charge values are assigned in order to improve the correlation between experimental and predicted binding affinities within the training set.

2.4. Energetic equilibration

The ligand training set is not only used for the positioning of receptor residues in space, but also for calibrating the resulting pseudoreceptor model. Based on the 3D model of the generated ligand–receptor complex, the experimentally obtained binding energies relate to the calculated ligand–pseudoreceptor interaction energy according to the following equations [31–33]:

$$\Delta(\Delta G^0_{\mathrm{calc}}) \approx \Delta(\Delta E_{\mathrm{calc}}) - \Delta(T\Delta S_{\mathrm{binding}}) - \Delta(\Delta G_{\mathrm{solvation,\ ligand}}) + \Delta(\Delta E_{\mathrm{int.,\ ligand}}) \qquad (1)$$

where $\Delta(\Delta E_{\mathrm{calc}})$ is the calculated interaction energy between ligand and pseudoreceptor; $\Delta(T\Delta S_{\mathrm{binding}})$ is the loss of conformational entropy upon binding of ligands; $\Delta(\Delta G_{\mathrm{solvation,\ ligand}})$ is the solvation energy of ligands; and $\Delta(\Delta E_{\mathrm{int.,\ ligand}})$ is the difference of the internal energy for ligands upon binding from a strain-free reference conformation.

The following linear regression can be applied to optimize the pseudoreceptor in the field of the training set ligand molecules and to predict binding energies for ligands included in the test set:

$$\Delta(\Delta G^0_{\mathrm{predicted}}) = |a| \cdot \Delta(\Delta G^0_{\mathrm{calculated}}) + b \qquad (2)$$

where $|a|$ is the absolute value of the slope, and b is the intercept.

Equation 1 assumes that all ligands are 'equally buried' within the receptor and that differences in the solvation energy of the different ligand–receptor complexes become negligible. After completion of residue addition, the pseudoreceptor is generally submitted to a multi-step minimization and calibration procedure which cannot be summarized in a generic protocol applicable to any type of pseudoreceptor projects. Furthermore, for each different pseudoreceptor modelling approach a specifically fine-tuned protocol has to be established.

However, each initial model is usually minimized to remove internal strain due to the receptor-building procedure [7,8]. The receptor residues are minimized keeping the ligands of the training set fixed, generally resulting in a model that will rarely show a satisfactory correlation between experimental and predicted binding energies. To obtain a better correlation, a correlation-coupled minimization of all receptor residues can be performed, while all ligands are kept at their initial position. A subsequent minimization of the ligands allows the removing of unfavorable contacts while the receptor residues are kept fixed, again leading to a decreased correlation. This procedure is repeated iteratively until a highly correlated model is obtained for the relaxed state [8]. A further advantage of PrGen is the possibility to alter position, orientation and conformation of all ligand molecules during the refinement, which helps to diminish the user-bias imposed in the superposition strategy in the initial set-up of the pseudoreceptor modelling approach. Additionally, PrGen offers the application of a Monte Carlo procedure after ligand relaxation in order to explore the pseudoreceptor cavity for alternative binding modes. Within this protocol, the position, orientation and conformation of each ligand is altered using the Metropolis criterion for acceptance. This procedure is not only applicable to the ligand and receptor equilibration protocols based on the training set-derived pharmacophore, but also for an efficient 'docking' of the ligand molecules of the test set, the activities of which are predicted [8].

2.5. Validation — pseudoreceptor analysis

After completion of the pseudoreceptor construction and energetic equilibration, it is necessary to analyze the model for its biophysical relevance. Typically, a pseudoreceptor model can be validated by replacing the training set with a series of test ligands. These have to be minimized in combination with the Monte Carlo driven protocol (mentioned above) within the pseudoreceptor model. Thereafter, free energies of binding can be predicted for these ligands using the linear regression obtained with the training set molecules (Eq. 2). Further criteria to assess the quality of a pseudoreceptor include the analysis of secondary structure elements within the receptor surrogate, the distribution of hydrophobic and hydrophilic residues, and the solvent accessibility of the binding site.

3. Case Studies

The pseudoreceptor modelling studies discussed in this chapter attempted to establish structure–activity relationships on receptor agonists and antagonists targeted at distinct members of two receptor superfamilies, namely the G protein-coupled receptors [34] and the integrins [35] (Fig. 2). Both receptor types are transmembrane proteins and mediate signal transduction across the cellular membrane.

The G protein-coupled receptors represent a prominent class of drug targets, exemplified in this contribution with two biogenic amine and the cannabinoid receptor. The potential of integrins as valid targets of considerable pharmaceutical interest became apparent with the finding that RGD (Arg-Gly-Asp) peptides and RGD-derived peptidomimetics interfere in the adhesive mechanisms associated with platelet aggregation, thus preventing clot formation by selective binding to the $\alpha_{IIb}\beta_3$ (gpIIb/IIIa) integrin on platelets [36]. Apart from the platelet-associated receptor, further members of the integrin family emerged as promising drug targets, such as $\alpha_v\beta_3$ and $\alpha_v\beta_5$ for treatment of cancer and osteoporosis [37]. In this context, the fidelity of the pseudoreceptor modelling approach will be demonstrated on rationally designed and conformationally restricted cyclic peptides, the 3D structures of which were experimentally determined by 2D NMR in solution [38,39].

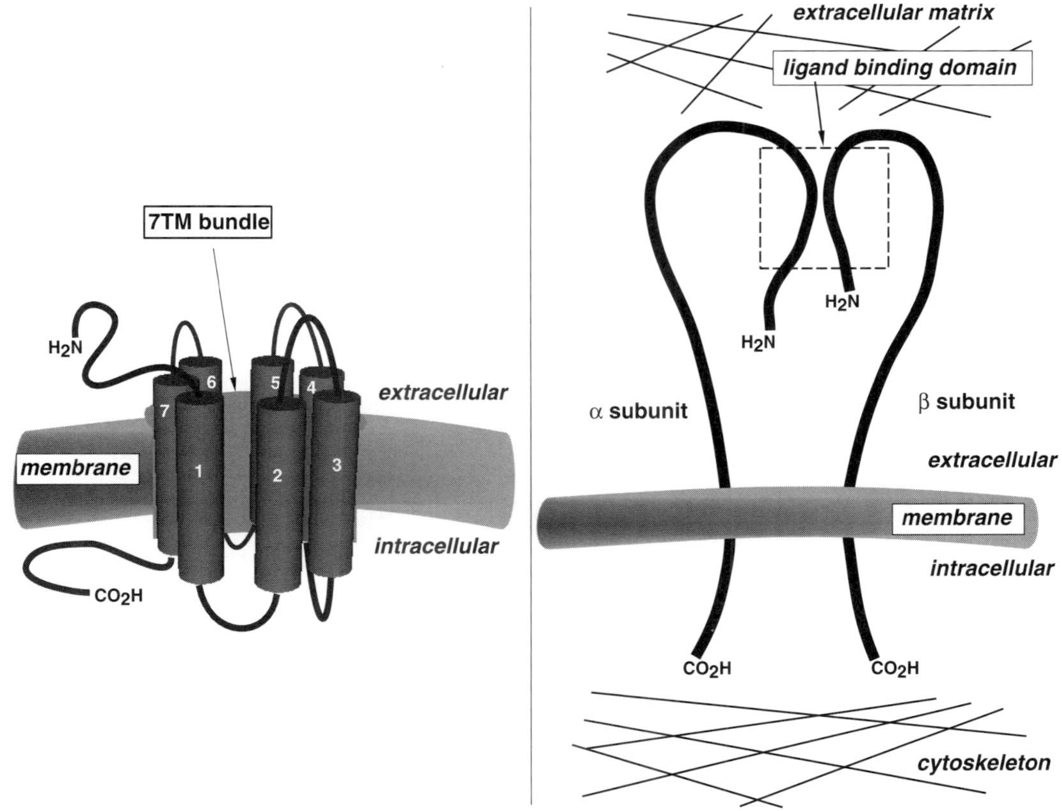

Fig. 2. *Schematic representation of the molecular topology of G protein-coupled receptors (left) and integrins (right). Members of the G protein-coupled receptor family are composed of a single polypeptide strand that folds into a seven-transmembrane helix domain. The integrins are heterodimeric transmembrane proteins with large extracellular domains. The ligand binding domain is constituted by the interface of the α and β subunit.*

3.1. Binding site of the cannabinoid receptor, reference [8]

The pharmaceutical interest in the cannabinoid receptor modulation is not mainly focused on the psychotropic effects elicited by the cannabis preparations marihuana and hashish, containing cannabinoids, but predominantly aimed to exploit the more beneficial pharmacological potential, such as anti-emetic, analgetic, muscle-relaxing or bronchodilatory effects [40–42]. The pseudoreceptor modelling approach carried out by Folkers et al. [8] is based on 28 cannabinoid antagonists, 14 of which are assigned to the training set and the remaining 10 compounds used as a test set for predicting the binding affinity. These 28 antagonists comprised classical **1** and non-classical **2** cannabinoids, the most active molecule being **1a** (DMH: R^1 = CH_2OH; R^2 = 1,1-dimethylheptyl; ring C: 8-en) in the series of classical cannabinoids and **2a** (CP55: R^2 = 1,1 dimethylheptyl; R^3 = 4-hydroxybutyl; stereochemistry: 1*R*,3*R*,4*R*) in the series of non-classical cannabinoids, respectively.

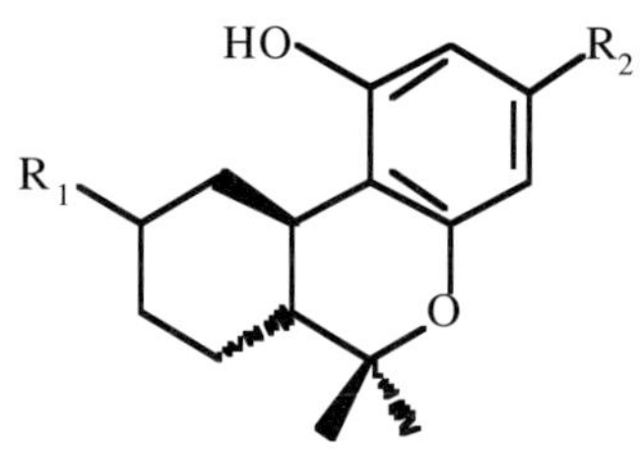

1

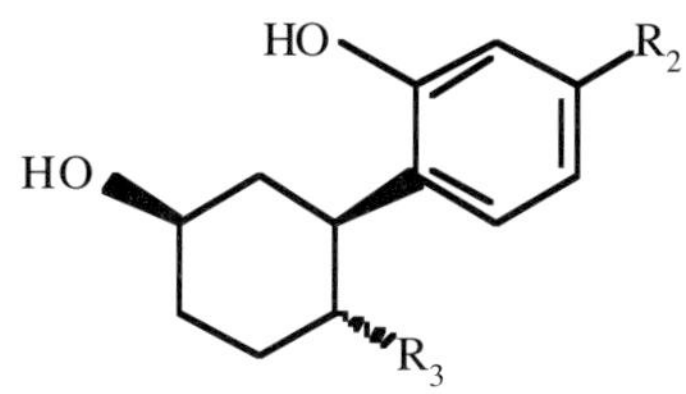

2

The authors followed a receptor-mediated pharmacophore alignment approach by restricting only on 4 compounds for the construction procedure of a primordial receptor model. It is noteworthy that the receptor fragments consisted of small helical fragments bearing key residues for ligand interactions, thus inherently accounting for the fact that the cannabinoid receptor is comprised of 7 transmembrane sequence stretches adopting α helical conformations, the so-called 7TM domain common to all G protein-coupled receptors [34]. The resulting pseudoreceptor was composed of 7 helical rods accommodating the 4 'root'-compounds (Fig. 3).

After equilibration, the 14 remaining antagonists of the training set were docked into the binding pocket and minimized within the static cavity. Finally, a ligand equilibration protocol including the Monte Carlo procedure was performed. The obtained receptor surrogate converged to a correlation coefficient of 0.94. This model (Fig. 3) was used to predict the binding affinities of the 10 test set compounds that were docked into the cavity and subjected to 25 rounds of free Monte Carlo minimizations, thereby ensuring a sufficient spatial exploration of the cavity by the ligands.

The receptor model reproduces the experimentally derived binding data with an RMS error in prediction of about 0.8 kcal/mol, corresponding to an uncertainty factor of 4.1 in the dissociation constant. Apart from this semi-quantitative evaluation, the model reveals atomistic details refering to the spatial distribution of interacting receptor

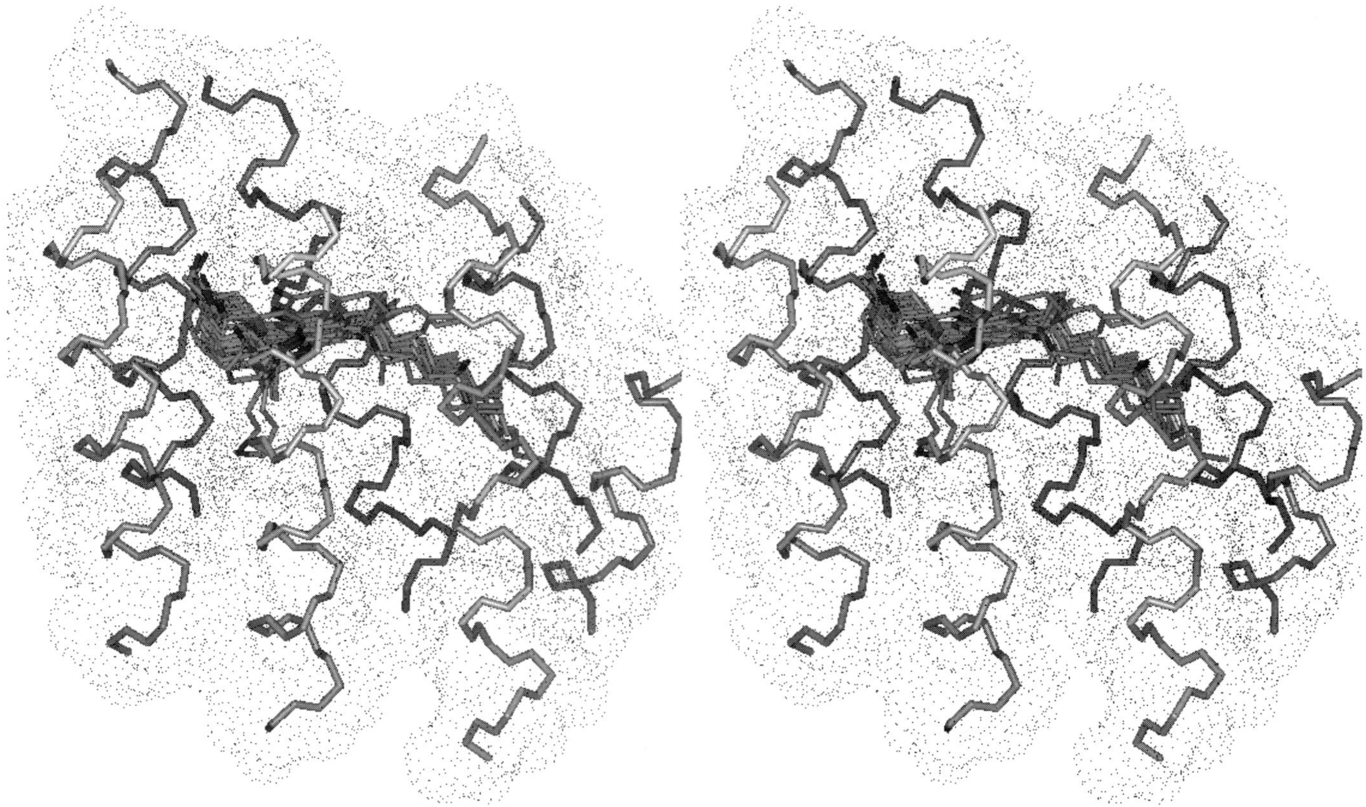

Fig. 3. *Stereoview of the cannabinoid pseudoreceptor model consisting of seven α-helical rods. The superimposed training set compounds are depicted within the cavity. For reasons of clarity, only heavy atoms of the receptor residues are shown. The coordinates of the model were kindly provided by A. Vedani and P. Zbinden.*

residues within a '7-helix mini-bundle' which can be exploited for *de novo* design of new or derivatization of known analogs [8].

3.2. *Binding site of the β_2 adrenergic receptor, references [7,8]*

The β_2 adrenergic receptor, a further member of the G protein-coupled receptor family [43], was studied by the same group by means of pseudoreceptor modelling employing PrGen. From a pharmaceutical point of view, a 3D model reflecting the binding characteristics of selective agonists would be beneficial for the design of drugs for e.g. the clinical treatment of asthma [43].

The study relies on 15 β_2 adrenergic antagonists of the common generic structure **3**.

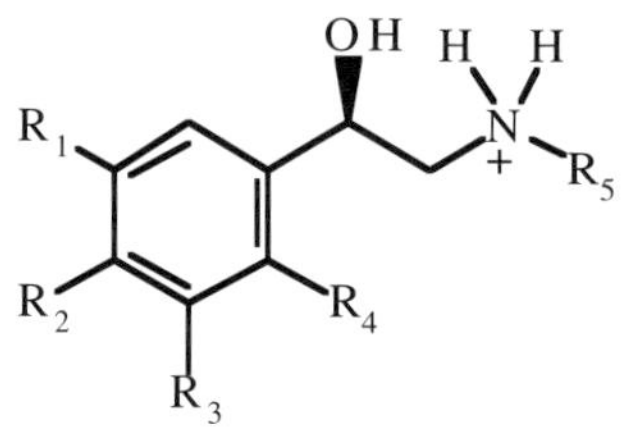

3

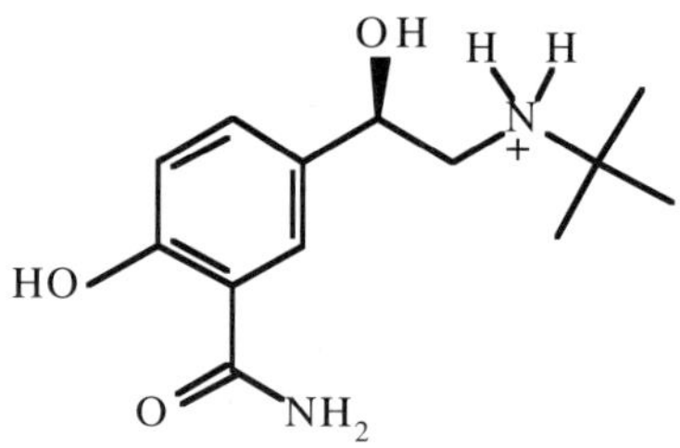

3a

The 15 adrenalin derivatives exhibit different substitution patterns at their ring positions R^1 to R^5. Ring positions R^1, R^2 and R^4 vary only moderately (H, OH, NH_2, Cl), whereas R^3 represents $CONH_2$, OH, Cl, CH_2OH, NH_2 or $NHCH_3$. The ammonium functionality bears either a further H atom, or CH_3, *iso*-propyl, *tert.*-butyl groups. The most active compound **3a** is shown explicitly. Nine of the 15 receptor antagonists were selected as the training set for pseudoreceptor generation, whereas the remaining 6 ligands served as test set for receptor analysis. Within this study, 3 different types of receptors were constructed, a completely atomistic model, a purely virtual model and a mixed model (Fig. 4).

This enabled the authors to judge comparatively the reliability of the different receptor model types with respect to their predictive power. Common to the atomistic and the mixed model (Fig. 4) is a series of key amino acids engaging the adrenalin derivatives in highly conserved interactions, already proposed by protein modelling studies on G protein-coupled receptors [44–46]. The hydrogen-bonding capabilities of the distinct ligand molecules essentially governed the pseudoreceptor construction process, in that the spatial positions of complementary functionalities encoded in amino acid residues were assigned according to the directionality of the corresponding interaction. The predictive quality was assessed by the same procedure described for the cannabinoid receptor model and turned out to be in a comparable range, as mentioned in section 3.2.

However, the authors conclude that receptors composed purely of virtual Lennard-Jones particles are not suited to mimic stereochemically demanding environments as

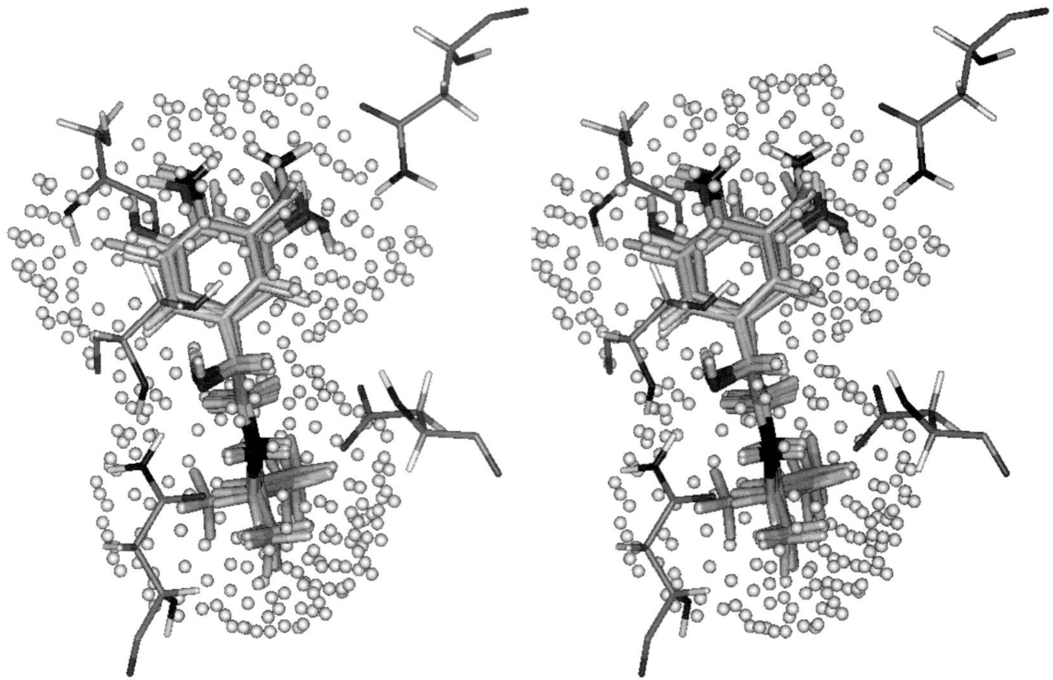

Fig. 4. *Stereoview of the β2-adrenergic pseudoreceptor model composed of a truncated protein core and augmented by inclusion of virtual Lennard-Jones particles (small spheres), together with the superimposed training set molecules. For reasons of clarity, only heavy atoms of the receptor residues are shown. The coordinates of the model were kindly provided by A. Vedani and P. Zbinden.*

found in proteins which are indeed capable of chiral discrimination. In contrast, as shown with the mixed model consisting of 5 key amino acids saturating the predominant directional ligand–receptor interactions, the utilization of virtual particles to augment a truncated protein core worked out satisfactorily [7,8].

3.3. The histaminergic H_3 binding site

Histaminergic H_3 receptors [47] were found to act as auto- as well as hetero-receptors and, therefore, are of broad importance in many physiological processes. They do not only regulate the biosynthesis and liberation of histamine, but also influence cholinergic, adrenergic, serotoninergic and several peptidergic neurons. Even in the brain, where the H_3 receptor density is maximal, the quantity observed amounts only to 1% compared to the H_1 and H_2 subtypes. The extremely low receptor density explains why so little is known about the H_3 receptor structure. On the basis of conformational correspondences for structurally rather diverse histaminergic H_3-receptor agonists **4** to **15** [48–52], we have been able to define a H_3-specific pharmacophore. The proposed pharmacophore [53] correctly describes the stereoselectivity of the H_3-receptor and illustrates that the methyl groups of e.g. Immepyr (Sch 49648) and Sch 50971 can occupy the same region of space as the R-α-methyl group of R-α,S-β-dimethylhistamine, while the pyrolidine rings overlap with the S-β-methyl group.

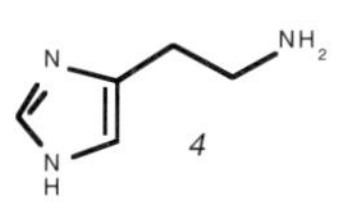

Histamine

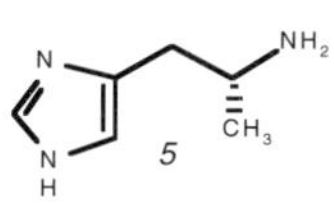

R-α-Methylhistamine

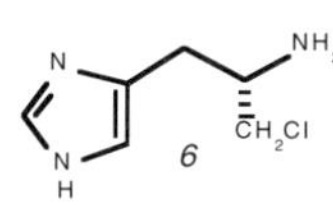

S-α-Chloromethyl-histamine

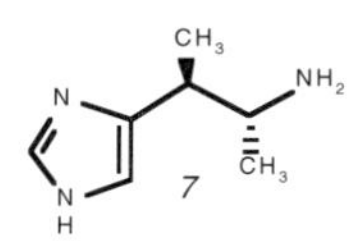

R-α,S-β-Dimethyl-histamine

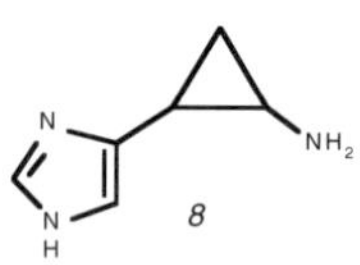

Cyclopropylhistamine

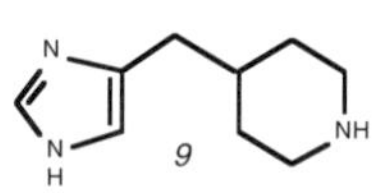

Immepip

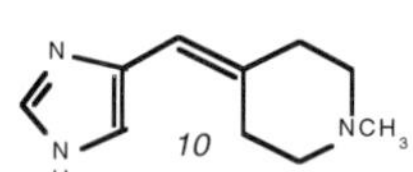

4-(1*H*-4-imidazolylmethylene) 1-methylpiperidine

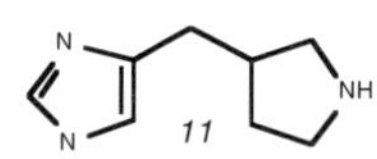

VUF 4864

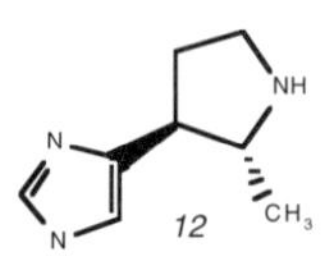

Immepyr (Sch 49648)

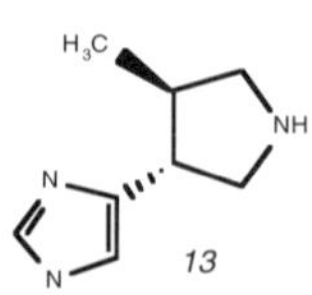

Sch 50971

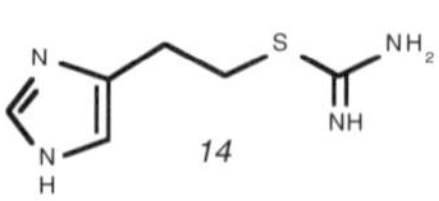

Imetit (SKF 91105)

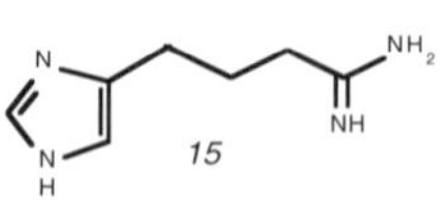

SKF 91606

Investigations of corresponding molecular interaction fields derived from GRID computations 54 using hydroxyl and methyl probes show very similar distributions and suggest that the H_3-agonists may interact with a common binding site. The comparable localizations and intensities of the hydrophobic interaction patterns are remarkable and indicate that, in addition to hydrogen donor and acceptor sites, hydrophobic amino acids may act as potential selectivity-producing binding regions for H_3-receptor agonists.

Using the H_3-receptor pharmacophore as a template, a Yak pseudoreceptor model for the H_3-receptor agonist binding was constructed as well. The model consists of 6 amino acid residues (Fig. 5) suggested in the course of the Yak procedure as the ones with highest probability. Because the amino acid sequence of the H_3 receptor is hitherto not known, the selection cannot be supported by alignment or mutation experiments.

The imidazole moiety of H_3-agonists is involved in two hydrogen bonds: a tyrosine residue donates a proton to the ring system, whereas an asparagine residue serves as proton acceptor. The positively charged side chain nitrogens interact with a negatively charged aspartate. The other pseudoreceptor binding sites are hydrophobic in character: a phenylalanine is involved in dispersion interactions with the imidazole ring system,

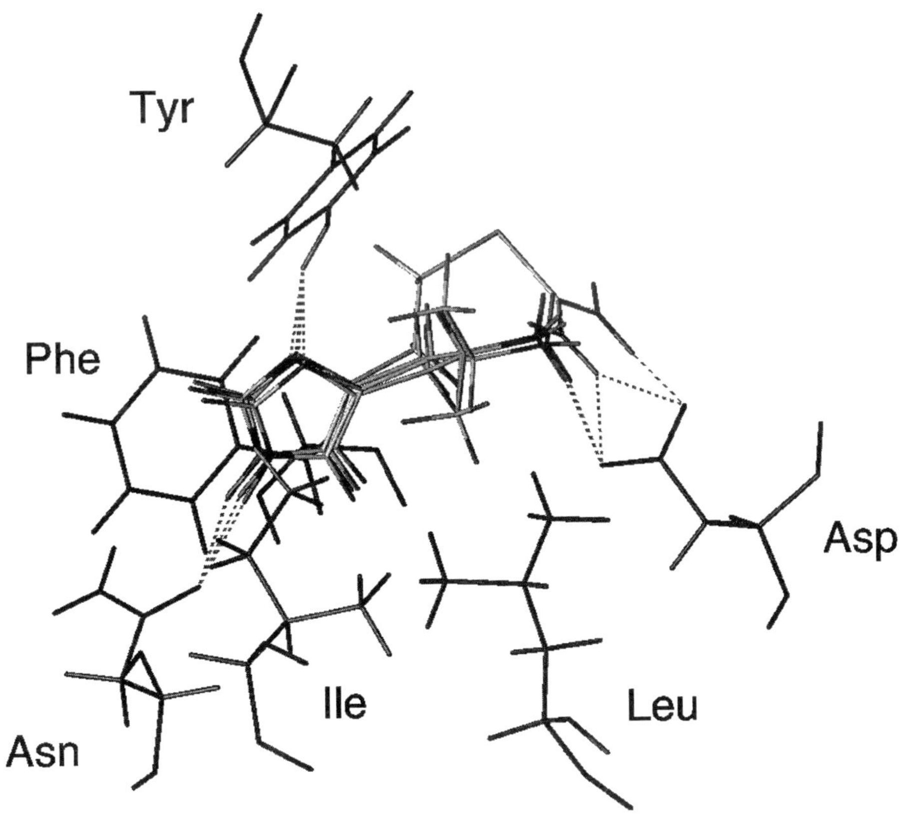

Fig. 5. Pseudoreceptor model for histaminergic H_3-agonists. The four predicted ligands are shown inside the binding region.

```
                      X   Y   Z  -Y  -X    -Z
                  90
EF (3CLN.BRK):    -R-V-F-D-K-D-G-N-G-Y-I-S-A-A-E-L-R-H-Y-
                  363
αIIb:             -P-L-G-D-L-D-R-D-G-Y-N-D-I-A-V-A-A-P-Y-
                  346
αV:               -P-L-G-D-L-D-Q-D-G-F-N-D-I-A-I-A-A-P-Y-
```

Fig. 6. Multiple sequence alignment of an EF-hand region of calmodulin (upper lane) with assumed EF-hand regions of the α subunits of two integrins. The bold letters above the alignment mark the residues that constitute the octahedral coordination polyhedron of the Ca^{2+} ion in the calmodulin structure (PDB3CLN.BRK).

whereas a leucine and an isoleucine fragment are located in close contact to the hydrophobic part of the side chains. At least some of the hydrophobic contacts have been recently found in the crystal structure of the histidine-binding protein 1HSL [55], where a tyrosine residue is located in the same position relative to the ring system of the bound L-histidine as the phenylalanine in this model. Using the 12 ligands **4** to **15** as a training set, the correlation coefficient for experimental versus calculated free energies of binding is 0.99. The RMS deviation for the training set was found to be 0.21 kcal/mol. Subsequently, the pseudoreceptor model was tested by predicting biological binding data for 4 ligand molecules not considered in model construction (histamine, R-α-methylhistamine, S-α-methylhistamine and imetit). The RMS deviation for this test set amounts to 0.66 kcal/mol, which underlines the significance of the model.

Comparing the Yak model with the GRID interaction fields yields a very high correspondence not only of type, polar or hydrophobic, but also of relative spatial positions and sizes of the common fields. The good agreement between the results obtained from two absolutely independent techniques led us to believe that the developed H_3-pseudoreceptor might be successfully used for prediction purposes.

Concluding the G protein-coupled receptor related studies, the receptor model of the dopaminergic receptor, based on a series of 3-pyridylalkyl indoles, constructed by Vedani et al., should only be mentioned for the sake of completeness [7].

3.4. Binding sites of the $\alpha_v\beta_3$ and $\alpha_{IIb}\beta_3$ integrins

The integrins are a superfamily of heterodimeric transmembrane proteins (Fig. 2) which interact extracellularly with numerous adhesion proteins, thus mediating various adhesion phenomena, such as platelet aggregation, tumor metastasis, angiogenesis, and osteoclast and osteoblast anchorage on bone tissue [35]. At the beginning of the 1990s, the tripeptide sequence RGD (Arg-Gly-Asp) was identified in numerous integrin ligands and termed as the universal cell recognition sequence which served as lead structure for the rational structure-based design of adhesion antagonists [56]. This finding offered new perspectives in the development of antithrombotic, antimetastatic and anti-osteoporose drugs [36,56]. Several RGD-derived non-peptidic compounds have entered phase III of clinical trials for the prevention of clot formation by competitively antagonizing the fibrinogen-$\alpha_{IIb}\beta_3$ integrin interaction on platelets [57].

Stimulated by the progress made in this particular research area of peptide-based drug design, several research groups currently seek for selective $\alpha_v\beta_3$ antagonists, thereby attempting to establish new anticancer and osteoporose therapies. In this context, we report on a pseudoreceptor modelling study based on NMR-derived and MD-refined conformations of a series of rationally designed cyclic peptides **16** to **19** (Table 1), which inhibit competitively tumor cell adhesion and platelet aggregation by binding to the integrins $\alpha_v\beta_3$ and $\alpha_{IIb}\beta_3$, respectively.

Comparable to the cannabinoid receptor modelling study introduced in section 3.2, structural information available from protein sequence comparisons was used as external boundary condition for the pseudoreceptor generation process. Sequence homology studies uncovered significant similarities between the integrin binding regions

Table 1 *Comparison of experimentally derived binding data (estimated from IC_{50} values and solvation enthalpies), with interaction energies of the four RGD peptides with both hypothetical binding pockets. The scaling factors are obtained by defining the binding energy of the most active peptide as 100% and scaling all remaining peptides according to it*

Ligand	Experimental scaling factor	Scaling factor from Yak
• $\alpha_v\beta_3$		
cyclo(-Arg-Gly-Asp-*D*-Phe-Val-) **16**	1.000	0.971
cyclo(-Arg-Gly-Asp-Phe-*D*-Val-) **17**	0.987	0.962
cyclo(-Arg-Gly-Asp-Phe-Leu-*D*-Ala-) **18**	0.939	0.971
cyclo(-Arg-Gly-Asp-*D*-Phe-Val-Gly-) **19**	0.979	1.000
• $\alpha_{IIb}\beta_3$		
cyclo(-Arg-Gly-Asp-*D*-Phe-Val-Gly-) **19**	0.978	0.938
cyclo(-Arg-Gly-Asp-Phe-Leu-*D*-Ala-) **18**	0.973	1.000
cyclo(-Arg-Gly-Asp-Phe-*D*-Val-) **17**	1.000	0.887
cyclo(-Arg-Gly-Asp-*D*-Phe-Val-) **16**	0.960	0.873

(α subunit) and certain Ca^{2+}-binding EF-hand motifs as present in e.g. calmodulin [58] (Fig. 6).

It is assumed that in RGD-sensitive integrins the Ca^{2+}-coordination polyhedron is formed by 5 receptor functionalities and the carboxylate group of Asp from the RGD sequence of the ligands, thus initiating electrostatically the RGD–integrin interaction [58]. Therefore, the Asp-carboxylate–Ca^{2+} interaction was chosen as the primary anchor point for pseudoreceptor construction. In both modelling studies, generating the $\alpha_v\beta_3$ and the $\alpha_{IIb}\beta_3$ binding sites, a $Ca^{2+}\cdot 4H_2O$ cluster was docked to both *syn*-electron pairs of the Asp-carboxylate oxygen atoms, resulting in a bidentate metal–ligand interaction.

The hypothetical binding pocket for the tumor cell-associated receptor ($\alpha_v\beta_3$) consists of 22 amino acid residues linked to 6 peptide fragments, together with the $Ca^{2+}\cdot 4H_2O$ cluster (Fig. 7).

The Phe^4 side chains of the peptide ligands could be embedded in a tight and coherent hydrophobic binding pocket comprised of 8 pseudoreceptor residues (Fig. 7). The model for the platelet-associated receptor ($\alpha_{IIb}\beta_3$) comprised 21 amino acid residues and the metal ion–water cluster (Fig. 8).

Since the side chains of the Phe^4 residues populate a more extended spatial area within the superimposed ligand set, no tight binding pocket could be generated (Fig. 8). However, a narrow binding cleft resulted around Gly^2C^{α} formed by the side chain of a Val in $Gly^2H^{\alpha proR}$ direction and by the aromatic ring of a Tyr in $Gly^2H^{\alpha proS}$ orientation, respectively. The Tyr simultaneously acts as hydrogen bond donor to an *anti*-electron

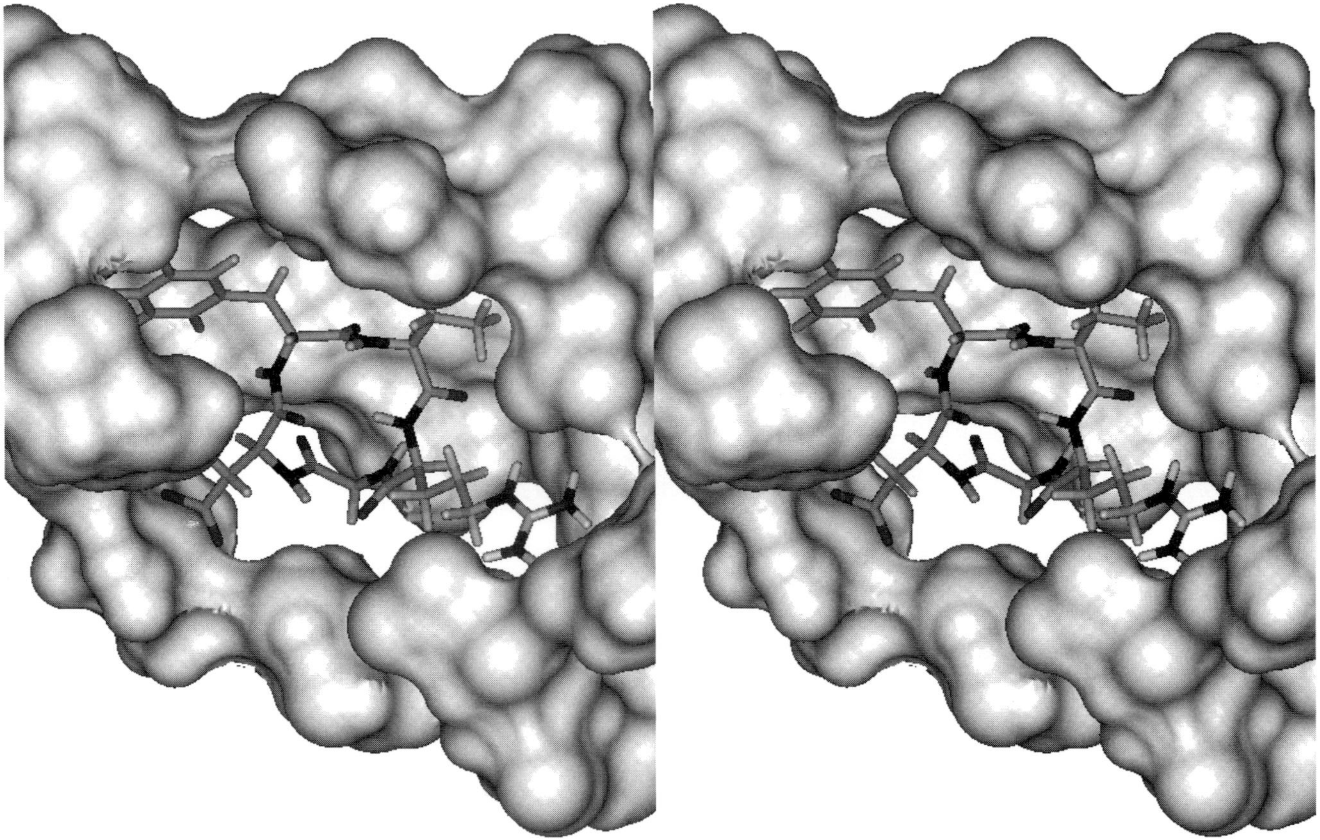

Fig. 7. *Stereoview of the pseudoreceptor model of the $\alpha_v\beta_3$ integrin. Only the primary ligand* ***16*** *with highest affinity (Table 1) is shown. The receptor residues are surrounded by a solid solvent-accessible surface.*

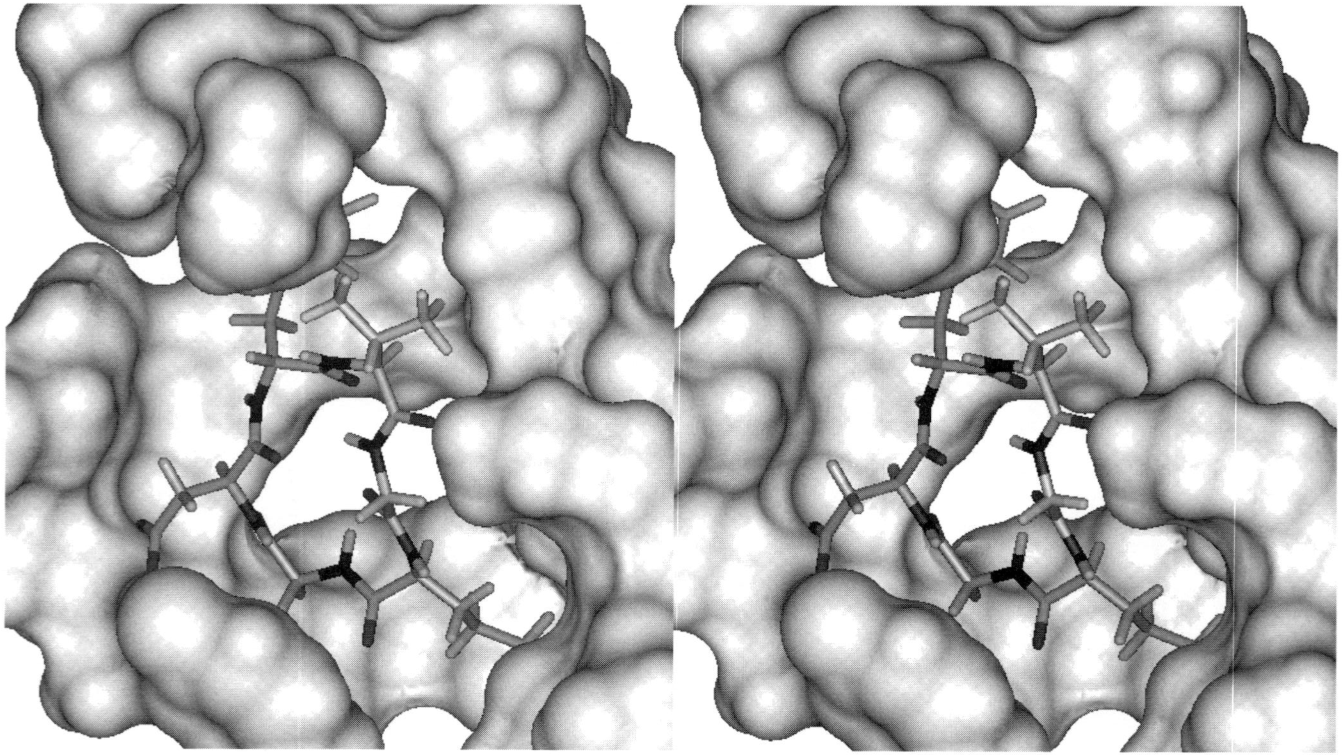

Fig. 8. *Stereoview of the most active RGD peptide* ***19*** *(Table 1) docked into the pseudoreceptor model of the* $\alpha_{IIb}\beta_3$ *integrin. The receptor residues are surrounded by a solid solvent-accessible surface.*

pair of Asp^3 carboxylate of the ligands, thereby reinforcing the torsional orientation of the carboxylate group for optimal interaction with the calcium ion (Fig. 9).

Both pseudoreceptor models qualitatively reproduce the experimentally derived antagonist activities of the antiadhesive peptides **16** to **19** used as ligand set (Table 1). In both models, the Arg^1 and Asp^3 side chains, the potential pharmacophore, are engaged in a network of attractive interactions (Figs. 7 and 8). While $\alpha_v\beta_3$ seems to exhibit a high steric demand for binding the Phe^4 side chain, the $\alpha_{IIb}\beta_3$ turned out to be less restrictive.

The most striking difference is found around the Gly^2 residue within the recognition sequence RGD. No sterically demanding binding cleft was obtained in the $\alpha_v\beta_3$ model which is supported by the finding that 2 RAD peptides, notably *cyclo*(R$\underline{A}$DFV) **20** and *cyclo*(RAD$\underline{F}$V) **21**, exhibit inhibitory activities of 63 μM and 1.0 μM, respectively [38,39]. These peptides are almost inactive in the $\alpha_{IIb}\beta_3$ assay, which is rationalized by the generated narrow binding cleft shielding Gly^2 in the corresponding pseudoreceptor model (Fig. 9). A methyl substituent in *proR* orientation would clash with the *iso*-propyl group of a Val residue, a methyl group in *proS* orientation would create major steric conflicts with the aromatic ring of a Tyr residue (Fig. 9). Concluding, the 3D pseudoreceptor models retrospectively verified structure–activity relationships already elaborated from comparative analyses embedded in a classical indirect molecular design strategy [59] by means of an atomistic blueprint of a hypothetical receptor-binding cavity. With these models in hand, it is possible to switch from an indirect to a direct molecular design strategy, applying a *de novo* ligand design approach. Additionally, the receptor models allow defining more precisely geometric profiles, suitable for mining 3D structure databases [60].

Again, it should be emphasized that these pseudoreceptor structures certainly bear little structural resemblance with their natural counterparts. They were designed to accommodate a series of ligands in a similar binding mode, thus representing the physico-chemical and sterical surface properties of the 'true' binding pocket, rather than reproducing the real receptor binding cavity with atomic accuracy.

4. Conclusion

The pseudoreceptor modelling approach discussed in this chapter tries to take advantage of the receptor fitting methodologies applied in a direct drug-design scenario for property-based receptor mapping projects, indicative for indirect drug design. A major advantage of the techniques implemented in Yak and PrGen lies in the combination of an atomistic receptor model, being represented by a truncated protein-binding cleft, and a directional force field [61–63] that is capable of treating ligand-metal ion–protein interactions, frequently found to be of prime importance for the docking event in various pharmaceutically targeted receptors and enzymes. Expanding the precursor program Yak by including pharmacophore relaxation, equilibration, receptor-mediated pharmacophore alignment, correlation-coupled minimization and the options to explore ligand and receptor space by Monte Carlo simulations certainly accounts for a more

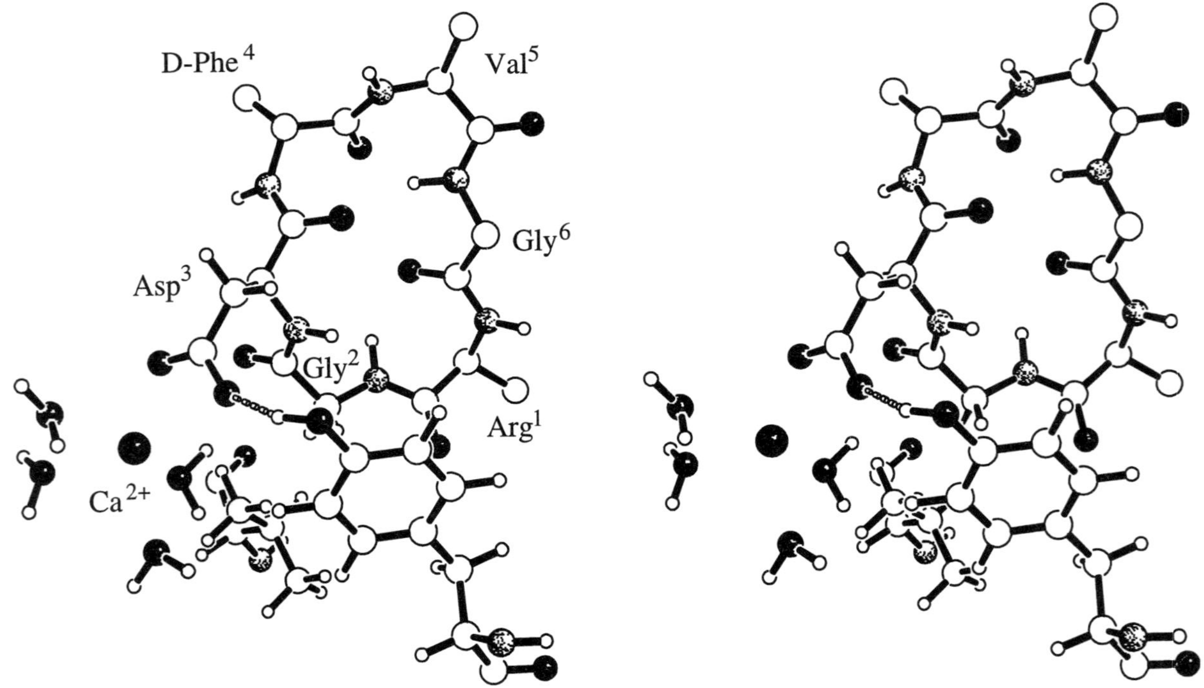

Fig. 9. *Stereoview of a fragment from the pseudoreceptor model of the the* $\alpha_{IIb}\beta_3$ *integrin, highlighting the narrow binding cleft around the Gly*[2] *residue of the universal recognition sequence. From the most active peptide cyclo(-Arg-Gly-Asp-D-Phe-Val-Gly-)* ***19****, which adopts a typical conformation comprising two* β *turns, only the backbone, the* C^{β} *atoms and the Asp*[3] *side chain atoms are shown. The Gly*$^{2}C^{\alpha}$ *is shielded by a Tyr residue (front face) and a Val residue (back face), respectively. The hydrogen bond of the Tyr-OH function to Asp*[3] *carboxylate is depicted as broken line.*

realistic approach treating pharmacophore–receptor interactions by computational means.

From our experience, we strongly believe that atomistic models help to increase the apprehension of the structure-based drug-design approach by chemists, thereby facilitating the chemical realization of proposed compounds that emerged from modelling studies.

References

1. Kuntz, I.D., *Structure-based strategies for drug design and discovery*, Science, 257 (1992) 1078–1082.
2. Höltje, H.-D. and Folkers, G., In Mannhold, R., Kubinyi, H. and Timmerman, H. (Eds.) Methods and principles in medicinal chemistry: Vol. 5. Molecular modeling — basic principles and applications, VCH Verlagsgesellschaft, Weinheim, Germany, 1997.
3. Müller, G., Feriani, A., Capelli, A.M. and Tedesco, G., Multidimensional NMR for macromolecular structure determination, La Chimica e l'Industria, 77 (1995) 937–957.
4. Kubinyi, H. (Ed.), 3D QSAR in drug design: Theory, methods and applications, ESCOM, Leiden, The Netherlands, 1993.
5. van de Waterbeemd, H., Testa, B. and Folkers, G. (Eds.), Computer-assisted lead finding and optimization: Current tools for medicinal chemistry, Verlag Helvetica Chimica Acta, Basel, Switzerland, 1997.
6. Vedani, A., Zbinden, P. and Snyder, J.P., *Pseudo-receptor modeling: A new concept for the three-dimensional construction of receptor binding sites*, J. Receptor Res., 13 (1993) 163–177.
7. Vedani, A., Zbinden, P., Snyder, J.P. and Greenidge, P.A., *Pseudoreceptor modeling: The construction of three-dimensional receptor surrogates*, J. Am. Chem. Soc., 117 (1995) 4987–4994.
8. Zbinden, P., Dobler, M., Folkers, G. and Vedani, A., *PrGen: Pseudoreceptor Modeling using receptor-mediated ligand alignment and pharmacophore equilibration*, J. Comput.-Aided Mol. Design (in press).
9. Murcho, A. and Murcko, M.A., *Computational methods to predict free energy in ligand–receptor complexes*, J. Med. Chem., 38 (1995) 4953–4967.
10. Frühbeis, H., Klein, R. and Wallmeier, H., *Computer-assisted molecular design: An overview*, Angew. Chem. Int. Ed. Engl., 26 (1987) 403–418.
11. Snyder, J.P. and Rao, S.N., *Pseudoreceptors: A bridge between receptor fitting and receptor mapping in drug design*, Chem. Design Automation News, 4 (1989) 13–15.
12. Snyder, J.P. and Rao, S.N., *Pseudoreceptor modeling: An experiment in large scale computation*, Cray Channels, 11 (1990) 4–12.
13. Momamy, F., Pitha, R., Klimkowsky, V.J. and Venkatachalam, C.M., *Drug design using a protein pseudoreceptor*, In Hohne, B.A. and Pierce, T.H. (Eds.) Expert systems applications in chemistry, ACS Symp. Ser. 408, 1989, pp. 82–91.
14. Hong, J.-L., Namgoong, S.K., Bernardi, A. and Still, W.C., *Highly selective binding of simple peptides by a C3-macrotricyclic receptor*, J. Am. Chem. Soc., 113 (1990) 5111–5112.
15. Snyder, J.P., Rao, S.N., Koehler, K.F. and Pellicciari, R., *Drug modeling at cell membrane receptors: The concept of pseudoreceptors*, In Angeli, P., Gulini, U. and Quaglia, W. (Eds.) Trends in Receptor Research, Elsevier Science Publishers, Amsterdam, The Netherlands, 1992, pp. 367–403.
16. Snyder, J.P., Rao, S.N., Koehler, K.F. and Vedani, A., *Minireceptors and pseudoreceptors*, In Kubinyi, H. (Ed.), 3D QSAR in drug design: Theory, methods and applications, ESCOM, Leiden, The Netherlands, 1993, pp. 336–354.
17. Höltje, H.-D. and Anzali, S., *Molecular modeling studies on the digitalis binding site of the* Na^+/K^+*-ATPase*, Pharmazie, 47 (1992) 691–698.
18. Walters, D.E. and Hinds, R.M., *Genetically evolved receptor models: A computational approach to construction of receptor models*, J. Med. Chem., 37 (1994) 2527–2536.
19. Doweyko, A.M., *Three-dimensional pharmacophores from binding data*, J. Med. Chem., 37 (1994), 1769–1778.

20. Hahn, M., *Receptor surface models: 1. Definition and construction*, J. Med. Chem., 38 (1995) 2080–2090.
21. Hahn, M. and Rogers, D., *Receptor surface models: 2. Application to quantitative structure–activity studies*, J. Med. Chem., 38 (1995) 2091–2102.
22. Murray-Rust, P. and Glusker, J.P., *Directional hydrogen bonding to sp^2- and sp^3-hybridized O atoms and its relevance to ligand–macromolecule interactions*, J. Am. Chem. Soc., 106 (1984) 1018–1025.
23. Taylor, R. and Kennard, O., *Hydrogen bonding geometry in organic crystals*, Acc. Chem. Res., 17 (1984) 320–326.
24. Baker, E.N. and Hubbard, R.E., *Hydrogen bonding in globular proteins*, Prog. Biophys. Molec. Biol., 44 (1984) 97–179.
25. Vedani, A. and Dunitz, J.D., *Lone-pair directionality of H-bond potential functions for molecular mechanics calculations: The inhibition of human carbonic anhydrase II by sulfonamides*, J. Am. Chem. Soc., 107 (1985) 7653–7658.
26. Tintelnot, M. and Andrews, P., *Geometries of functional group interactions in enzyme–ligand complexes: Guides for receptor modeling*, J. Comput.-Aided Mol. Design, 3 (1989) 67–84.
27. Alexander, R.S., Kanyo, Z.F., Chirlian, L.E. and Christianson, D.W., *The stereochemistry of phosphate–lewis acid interactions for nucleic acid structure and recognition*, J. Am. Soc., 112 (1990) 933–937.
28. Klebe, G. and Diederich, F.A., *A comparison of the crystal packing in benzene with the geometry seen in crystalline cyclophane–benzene complexes: Guidelines for rational design*, Phil. Trans. Roy. Soc., London, ser. A, 345 (1993) 37–48.
29. Klebe, G., *The use of composite crystal-field environments in molecular recognition and the de novo design of protein ligands*, J. Mol. Biol., 237 (1994) 212–235.
30. Kern, P., Brunne, R.M., Rognan, D. and Folkers, G., *A pseudo-particle approach for studying protein–ligand models truncated to their active site*, Biopolymers, 38 (1996) 619–637.
31. Blaney, J.M., Weiner, P.K., Dearing, A., Kollman, P.A., Jorgensen, E.C., Oatley, S.J., Burridge, J.M. and Blake, J.F., *Molecular mechanics simulation of protein–ligand interactions: Binding of thyroid analogues to prealbumin*, J. Am. Chem. Soc., 104 (1982) 6424–6434.
32. Still, W.C., Tempczyk, A., Hawley, R.C. and Hendrickson, T., *Semianalytical treatment of solvation of molecular mechanics and dynamics*, J. Am. Chem. Soc., 112 (1990) 6127–6129.
33. Searle, M.S. and Williams, D.H., *The cost of conformational order: Entropy changes in molecular associations*, J. Am. Chem. Soc., 114 (1992) 10690–10697.
34. Iismaa, T.P., Biden, T.J. and Shine, J. (Eds.), G Protein-coupled receptors, Springer-Verlag, Heidelberg, Germany, 1995.
35. Heavner, G.A., *Active sequences in cell adhesion molecules: Targets for therapeutic intervention*, Drug Discovery Today, 1 (1997) 295–304.
36. D'Souza, S.E., Ginsberg, M.H. and Plow, E.F., *Arginyl-glycyl-aspartic acid (RGD): A cell adhesion motif*, Trends Biochem. Sci., 16 (1991) 246–250.
37. Engleman, V.W., Kellogg, M.S. and Rogers, T.E., *Cell adhesion integrins as pharmaceutical targets*, Annu. Rep. Med. Chem., 31 (1996) 191–200.
38. Gurrath, M., Müller, G., Kessler, H., Aumailley, M. and Timpl, R., *Conformation/activity studies of rationally designed potent anti-adhesive RGD peptides*, Eur. J. Biochem., 210 (1992) 911–921.
39. Pfaff, M., Tangemann, K., Müller, B., Gurrath, M., Müller, G., Kessler, H., Timpl, R. and Engel, J., *Selective recognition of cyclic RGD peptides of NMR defined conformation by $\alpha_{IIb}\beta_3$, $\alpha_v\beta_3$, and $\alpha_5\beta_1$ integrins*, J. Biol. Chem., 296 (1994) 20233–20238.
40. Johnson, M.R., Melvin, L.S., Althuis, T.H., Bindra, J.S., Harbert, C.A., Milne, G.M. and Weissman, A., *Selective and potent analgesics derived from cannabinoids*, J. Clin. Pharmacol., 21 (1981) 271–282.
41. Johnson, M.R. and Melvin, L.S., *The discovery of non-classical cannabinoids*, In Mechoulam, R. (Ed.) Cannabinoids as therapeutic agents, CRC Press, Boca Raton, FL, 1986, pp. 121–146.
42. Razdan, R.K., *Structure–activity relationships in cannabinoids*, Pharmacol. Rev., 38 (1986) 75–149.
43. Main, B.G., *β-Adrenergic receptors*, In Emmett, J.C. (Ed.) Comprehensive medicinal chemistry, Volume 3. Membranes and receptors, Pergamon Press, Oxford, U.K., 1990, pp. 187–228.

44. Kontoyianni, M., DeWeese, C., Penzotti, J.E. and Lybrand T.P., *Three-dimensional models for agonist and antagonist complexes with $\beta 2$ adrenergic receptor*, J. Med. Chem., 39 (1996) 4406–4420.
45. Nederkoorn, P.H., van Lenthe, J.H., van der Goot, H., Donné-Op den Kelder, G.M. and Timmerman, J., *The agonistic binding site at the histamine H_2* receptor: 1. *Theoretical investigations of histamine binding to an oligopeptide mimicking a part of the fifth transmembrane α helix*, Comput.-Aided Mol. Design, 10 (1996) 461–478.
46. Nederkoorn, P.H.J., van Gelder, E.M., Donné-Op den Kelder, G. and Timmerman, J., *The agonistic binding site at the histamine H_2 receptor: 2. Theoretical investigations of histamine binding to receptor models of the seven α-helical transmembrane domain*, Comput.-Aided Mol. Design, 10 (1996) 479–489.
47. Arrang, J.M., Garbarg, M. and Schwartz, J.-C., *Auto-inhibition of brain histamine release by a novel class of histamine receptors*, Nature, 302 (1983) 832–837.
48. Lipp, R., Stark, H. and Schunack, W., *Absolute configuration, stereochemistry and receptor selectivity of (R)α,(S)β-dimethylhistamine, a novel highly potent histamine H_3-receptor agonist*, In Schwartz, J.-C. and Haas, H.L. (Eds.) The histamine receptor: Vol. 16, Wiley-Liss Inc., New York, 1992, pp. 57–72.
49. Shih, N.-Y., Aslanian, R., Lupo, A.T., Duguma, L., Orlando, S., Piwinski, J.J., Green, M.J., Gangluy, A.K., Clark, M., Tozzi, S., Kreutner, W. and Hey, J.A., *A novel pyrrolidine analog of histamine as potent, highly selective histamine H_3-receptor agonist*, J. Med. Chem., 38 (1995) 1593–1599.
50. Vollinga, R.C., de Koning, P., Jansen, F.P., Leurs, R., Menge, W.M.P.B. and Timmerman, H., *A new potent and selective histamine H_3-receptor agonist: 4-(1H-imidazol-4yl-methyl)-piperidine*, J. Med. Chem., 37 (1994) 332–333.
51. Howson, W., Parson, M.E., Raval, P. and Swayne, G.T.G., *Two novel potent and selective histamine H_3-receptor agonists*, Bioorg. Med. Chem. Lett., 2 (1992) 77–78.
52. Ganellin, C.R., Bang-Andersen, B., Khalaf, Y.S., Tertiuk, W., Arrang, J.M., Garbarg, M., Ligneau, X., Rouleau, A. and Schwartz, J.C., *Imetit and N-methyl derivatives: The transition from potent agonists to antagonists at histamine H_3-receptors*, Bioorg. med. Chem. Lett., 2 (1992) 1231–1234.
53. Sippl, W., Stark, H. and Höltje, H.-D., *Computer-assisted analysis of histamine H_2- and H_3-receptor agonists*, Quant. Struct.-Act. Relat., 1 (1995) 121–125.
54. Goodford, P.J., *A computational procedure for determining energetically favourable binding sties on biologically important macromolecules*, J. Med. Chem., 27 (1985) 849–857.
55. Yao, N., Trakhanow, S. and Quiocho, F.A., *Refine structure of the histamine binding protein complexed with histamine and its relationship with many other active transport/chemosensory proteins*, Biochemistry, 33 (1994) 4769–4775.
56. See e.g. Cox, D., Aoki, T., Seki, J., Motoyama, Y. and Yoshida, K., *The pharmacology of the integrins*, Med. Res. Rev., 14 (1994) 195–228.
57. Samanen, J., *GPIIb/IIIa antagonists*, Annu. Rep. Med. Chem., 31 (1996) 91–100.
58. Smith, J.W. and Cheresh, D.A., *Integrin ($\alpha_v\beta_3$)–ligand interaction*, J. Biol. Chem., 265 (1990) 2168–2172.
59. Müller, G., Gurrath, M. and Kessler, H., *Pharmacophore refinement of gpIIb/IIIa antagonists based on comparative studies of antiadhesive cyclic and acyclic RGD peptides*, J. Comput.-Aided Mol. Design, 8 (1994) 709–730.
60. Manallack, D.T., *Getting that hit: 3D database searching in drug discovery*, Drug Design Today, 1 (1997) 231–238.
61. Vedani, A., Dobler, M. and Dunitz, J.D., *An empirical potential function for metal centers: Application to molecular mechanics calculations on metalloproteins*, J. Comput. Chem., 7 (1986) 701–710.
62. Vedani, A., *YETI: An interactive molecular mechanics program for small-molecule protein complexes*, J. Comput. Chem., 9 (1988) 269–280.
63. Vedani, A. and Huhta, D.W., *A new force field for modeling metalloproteins*, J. Am. Chem. Soc., 112 (1990) 4759–4767.

Genetically Evolved Receptor Models (GERM) as a 3D QSAR Tool

D. Eric Walters

Department of Biological Chemistry, Finch University of Health Sciences/The Chicago Medical School, 3333 Green Bay Road, North Chicago, IL 60064-3095, U.S.A.

1. What is GERM?

Genetically Evolved Receptor Models (GERM) [1,2] is a procedure for construction of three-dimensional models of receptor sites in the absence of a crystallographically determined structure of the real receptor. Most biological receptors have not yet been crystallized and X-rayed; many will be quite difficult to study experimentally (for example, if they are membrane bound or have not yet been isolated). Very often, we have only a structure–activity series, and from this we would like to infer the three-dimensional requirements of the receptor site. This can be viewed either as a receptor modelling task or a 3D QSAR task. In either case, GERM is a method for constructing quantitative 3D models.

2. How Does GERM Work?

The starting point for a GERM analysis is a structure–activity series for which a 'reasonable' alignment of 'reasonable' conformers has been determined. The conformational analysis and alignment problems are beyond the scope of this review.

Conceptually, it is quite straightforward to take a superimposed set of compounds, surround the compounds with a shell of atoms (corresponding to the first layer of atoms in the receptor site) and assign to these atoms specific atom types (aliphatic H, polar H, etc.) which correspond to the types of atoms which would be found in proteins. The practical limitation is this: suppose we use a set of 15 different atom types (which may be typical of a protein-oriented molecular mechanics force field); with a shell of 60 atoms surrounding our superimposed ligands, the number of possible combinations is 15^{60} (about 4×10^{70}), so that we have no hope of systematically finding the 'best' possible model. Certainly, we could look at one position at a time and find the model which binds most tightly to our set of ligands (or to the one with highest potency), but real receptors are not necessarily designed for maximum possible affinity. We do not want the model with the best affinity, but the model with the best correlation between calculated affinity and experimentally determined bioactivity. Thus, we have encountered a very highly multi-dimensional search problem.

One very fruitful approach to such multi-dimensional search problems has been the genetic algorithm (GA) method [3]. GA does not guarantee that the global 'best' solution will ever be found, but it very rapidly finds a large number of 'very good' solutions. It does this by mimicking biology — specifically, by using recombination and mutation. The first step is to encode each solution to the problem (in this case, a shell of atoms and their corresponding atom types) into a linear string of numbers; these strings are the

H. Kubinyi et al. (eds.), 3D QSAR in Drug Design, Volume 3. 159–166.

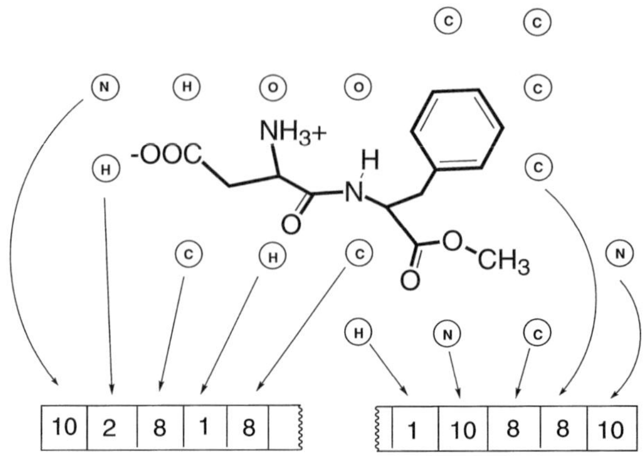

Fig. 1. Procedure for encoding a receptor model into a string of numerical values. Position in the string corresponds to position in three-dimensional space; the numerical value at a given position of the string corresponds to a particular atom type at that position in space.

'genes'. We have implemented this as shown in Fig. 1: the position in the string of numbers corresponds to a specific position in three-dimensional space, and the numerical value at that position corresponds to a specific atom type. Table 1 lists our 'genetic code' which is based on atom types from the CHARMm protein force field [4]. Since we begin the GERM procedure with a closed shell of atoms, and we know that some receptors have an open (solvent-exposed) face, we wanted to allow for the possibility of having no atom at all in some positions. We included in our genetic code the possibility for a 'zero' or null atom type. Any given model can thus be expressed as a string of numbers. The second step is numerically to score each model. We have chosen to do this in the following way. The ligands in the training set are placed, one at a time, in the model (Fig. 2); using a force field, the intermolecular van der Waals and electrostatic interaction energies between the ligand and the model are calculated; finally, we calculate the correlation coefficient for 1/exp(energy) versus log(bioactivity).

With procedures in hand to (1) encode models into strings of numbers, and (2) numerically evaluate any given model, GA can be applied. An initial population of models is generated by assigning random atom types to each position of each model. Each of these models is evaluated. Since fitness scores are correlation coefficients, scores can range from –1.0 (completely inverse correlation) to +1.0 (perfect correlation). In practice, most models are quite mediocre, and an initial score of 0.2–0.3 is quite common, with some models scoring higher and some lower. Now, pairs of models are selected at random from the population to serve as 'parents'. At a randomly chosen point, the 'genes' are cut and recombined — the tail end of gene 2 is added to the head of gene 1, and vice versa, generating two new 'offspring' models. Each new gene is evaluated with the scoring function. If an offspring model has a higher score than one or both parents, it is added to the population and the weaker parent is eliminated. If the offspring model is worse than the parents, it is allowed to die. Recombination allows good

Table 1 *The 'genetic code' and parameters used in this work consisted of a list of atom types and their corresponding force-field parameters. Atom types were chosen from the CHARMM force field (reference [4]) and charges are values which approximate those found in the standard 20 amino acids in the commercially distributed version of the CHARMM force field (Molecular Simulations, Inc., San Diego, CA 92121, U.S.A.). Type 0 corresponds to having no atom at all in a given position*

Atom type code	CHARMm type	E_{min} (kcal/m)	R_{min} (Å)	Partial atomic charge
0	–	–	–	–
1	H (H on polar atom)	–0.0498	0.800	0.25
2	HC (H on charged N)	–0.0498	0.600	0.35
3	HA (aliphatic H)	–0.0450	1.468	0.00
4	C (carbonyl C)	–0.1410	1.870	0.35
5	CH1E (CH group)	–0.0486	2.365	0.00
6	CH2E (CH_2 group)	–0.1142	2.235	0.00
7	CH3E (CH_3 group)	–0.1811	2.165	0.00
8	CT (aliphatic C)	–0.0903	1.800	0.00
9	NP (amide N)	–0.0900	1.830	–0.40
10	NT (amine N)	–0.0900	1.830	–0.30
11	O (carbonyl O)	–0.2000	1.560	–0.50
12	OT (hydroxyl O)	–0.2000	1.540	–0.60
13	OC (carboxylate O)	–0.1591	1.560	–0.55
14	S	–0.0430	1.890	–0.20

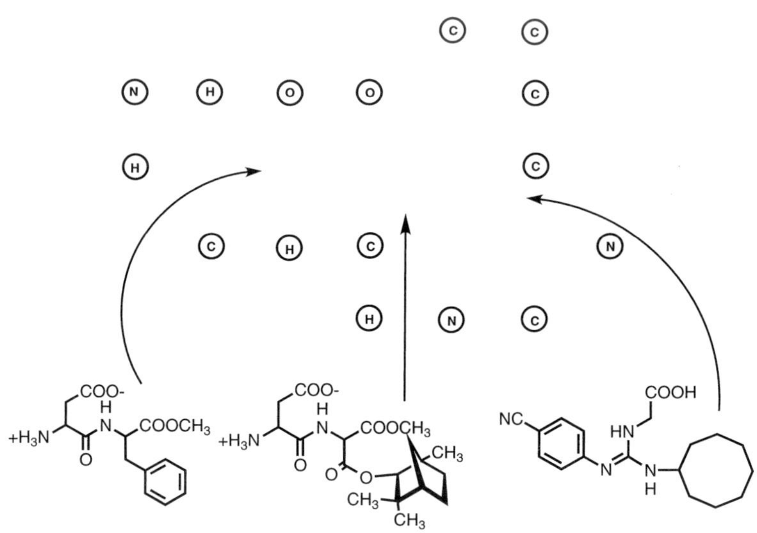

Fig. 2. Fitness scoring procedure. Ligands in the training set are docked into a model, one at a time, and an intermolecular binding energy is calculated using a molecular mechanics force field (van der Waals and electrostatic terms). These calculated binding energies are then correlated with measured bioactivities.

features from many different models to come together, survive and reproduce in the population, while bad features (bad choices of atom types) tend to die off. A mutation operator can be added to the procedure, to add to the 'genetic diversity' of the gene pool. At some user-selected frequency, a randomly chosen atom is assigned a randomly chosen atom type. Genetic diversity is an important consideration; if there is not sufficient diversity, the models become 'inbred', and the population converges too quickly to a lower average fitness score. To guard against inbreeding, we do not allow identical twins in our population.

In setting up calculations with the GERM method, there are several parameters for which the user must choose values. These include the number of atoms to use in making the model, the population size and the mutation rate. Each of these variables has an impact on the length and ultimate success of the calculations.

The number of atoms constituting a model and the size of the population are most important in determining how good the results will be and how long the calculations will take. Models with larger numbers of atoms are more likely to come close to the important functional groups on the ligands. However, the calculations will take longer since energy terms must be calculated between each ligand atom and each model atom. We have used 50 or 60 atom models for ligands of the size of dipeptides, and 75 atoms for larger ligands. The GERM program has a procedure which spaces the model atoms as evenly as possible over the surface of the ligands.

Larger populations will contain more genetic diversity and, in the long run, provide higher fitness scores. But increasing the population size also increases the length of time it takes to reach those higher scores. Figure 3 illustrates typical results. Smaller populations (bold line) rise more rapidly to their maximal scores; but those scores are

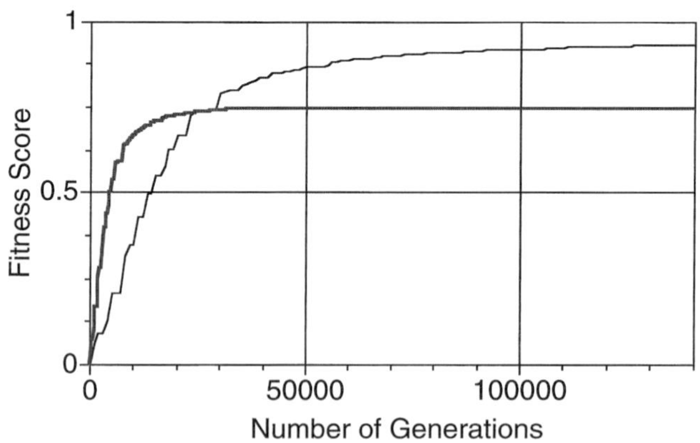

Fig. 3. Evolution of the mean fitness scores of populations of receptor models using the GERM method. The bold line illustrates that a smaller population will initially evolve more rapidly, but will level off at a lower fitness score. The narrow line illustrates a larger population which evolves more slowly, but reaches a higher mean score, because it starts with greater genetic diversity.

lower because of the more limited genetic diversity. We have typically used 500 to 1000 models. Larger models (75 atoms or more) demand larger populations.

We have used a mutation rate of 1 per generation, using a Poisson distribution function, so that in any particular generation there may be 0, 1, 2 or occasionally more mutations, and the average rate is 1 per generation. Higher mutation rates tend to be detrimental, particularly late in the evolutionary process. When the models contain many good features, random changes are more likely to be harmful than beneficial.

3. Results

The initial result of the calculation is a large set of 'very good' models, where 'very good' means a very high correlation (r-squared = 0.9 or better) between calculated binding energy and experimentally measured bioactivity. These models have a number of possible applications. For example, a new structure can be docked into the models, the binding energy calculated and, from the correlation, a bioactivity is calculated. Since there are hundreds of good models available, many estimates can be averaged; a mean and standard deviation can be calculated.

Most of our results, to date, have involved a series of high-potency sweeteners [1,2]. Conformational analysis and superposition of these compounds has been carried out in previous modelling studies [5]. Biological activity data for these compounds were determined by trained taste panelists, who identified concentrations of the test compounds equivalent in sweetness to reference solutions of sucrose [6]. Three structural families of compounds were studied: L-aspartic acid derivatives, arylureas and arylguanidinium-acetic acids. These compounds are considered likely to act at a common receptor site because they have several structural features in common: (1) a carboxylate group; (2) two or more polar N–H hydrogens; (3) a large hydrophobic substituent; and (4), in many cases, an aryl ring with a strongly electron-withdrawing substituent. Furthermore, all of these families of compounds have low-energy conformers which permit good superposition of these features.

First, it was found that good models could be generated for the 8 aspartic derivatives studied (correlation coefficient > 0.979), for the 8 arylureas (correlation coefficient > 0.947) and for the 8 arylguanidinium-acetic acid derivatives (correlation coefficient > 0.943).

Next we investigated the possibility of overfitting by doing leave-*n*-out cross-validation. For the 8 aspartic derivatives, 2 compounds were left out of the model evolution; bioactivities of these 2 compounds were then calculated from the models evolved around the other six structures. This procedure was repeated until all 8 compounds had been predicted on the basis of models for which they were not templates. Average error for the omitted compounds was 0.44. This procedure was repeated for the 8 arylureas (average error = 0.41) and for the arylguanidines (average error = 0.36).

An alternative test for overfitting involves scrambling the bioactivity data; if the method is overfitting, then it should be able to make 'good' models even for meaningless input data. When the log(potency) numbers were randomized 10 different times for the series of 8 aspartic derivatives, the average final r-squared for the models was 0.344, far worse than the 0.96–0.99 usually obtained for these compounds.

A more rigorous test of any QSAR method comes when we go beyond a homologous series to sets encompassing diverse structure types. In the 3 series of high potency sweeteners, we combined all 22 compounds (2 of the compounds are both aspartic derivatives and arylureas). Eleven representative compounds were used as the training set, models were evolved around these and potencies calculated for the remaining 11 from these models. Mean error was 0.44, and the worst case prediction erred by 0.75. Such predictions are well within useful limits for such practical purposes as deciding which new compounds would be worth the effort and expense of synthesis and testing.

The final population of models provides other useful results as well [2]. The final population may contain 1000 different 'good genes', all of which are at least slightly different since we allow no duplicates in the population; furthermore, these gene sequences are all aligned. Visual examination of the population listing shows that there are some positions in the model for which a single atom type is highly conserved; other positions are quite variable. In the case of sweet receptor models, we found that the most highly conserved positions and atom types corresponded to the main structural requirements for sweet taste. Adjacent to the carboxylate groups of the sweeteners were 2 sites with high frequency of positively charged hydrogen atoms. Near the primary cluster of NH groups, the models have a site with highly conserved negative charge. Several sites around the hydrophobic pocket have highly conserved hydrophobic atom types.

We examined the models for sites with a high occurrence of the null atom type, to see if there might be a tendency for some part of the receptor model to have an open face. There is a band of 6 sites across the back face of the model site which has a very strong preference for 'small' atom types (either no atom or a hydrogen atom, regardless of charge). This suggests a region on the ligand structures where it might be possible to add further functionality without sterically preventing binding, and with the possibility of gaining additional interaction sites. Certainly, such insights are an important outcome from any successful QSAR/modelling study.

One unexpected result came out of the sequence analysis. In the region occupied by the methyl ester group of aspartame and the methyl substituent of alitame (Fig. 4), there was consistently found a highly conserved site with negative charge. It seemed odd that an atom with partial negative charge should consistently appear near the oxygen atoms

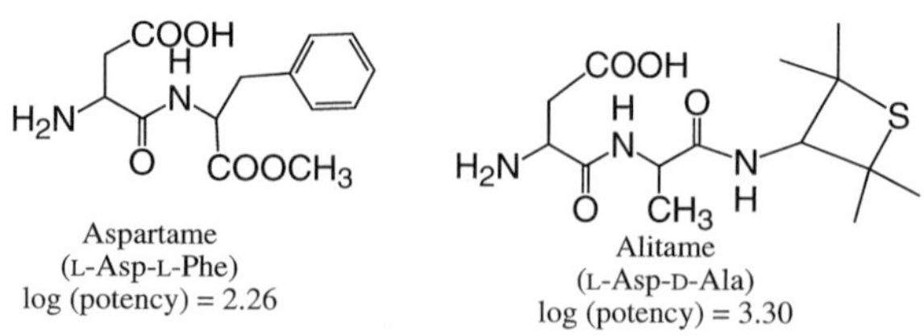

Fig. 4. Structures of aspartame and alitame.

of the ester since this should produce a repulsive interaction. We (and most other workers in this field) had always considered that the order-of-magnitude higher potency of alitame was due to its highly branched hydrophobic substituent (tetramethyl thietane versus phenyl in aspartame). The modelling result suggests another possibility — perhaps aspartame has a repulsive interaction which alitame circumvents? Again, further experiments are suggested: could potency be increased by replacing the methyl ester or methyl sidechain with an appropriate hydrogen bond donor?

A further test of the GERM method is currently in progress [7]. Numerous X-ray crystallographic structures of HIV protease complexed to inhibitors have been published. We have superimposed twelve of these structures, and have used the superimposed inhibitor structures (with the protein removed) as templates for GERM calculations. Comparison of the calculated models with the actual protein structure reveals that many of the important features of the real protein are captured in the computed models. A detailed comparison of the calculated and experimental structures is in press.

4. What Are the Underlying Assumptions and Possible Limitations of the Method?

It is important when using any procedure to understand the underlying assumptions of the method. Here, we wish to point out explicitly the assumptions which go into the GERM method. We also consider some of the likely limitations of the procedure.

The first consideration is that useful three-dimensional models are dependent on the conformational analysis and alignment used as input. This is, of course, true for any 3D QSAR method. The GERM method is not, in its current implementation, able to automate the alignment process. We have observed empirically, however, that the method sometimes points out molecules which are not well aligned. After a population of models is evolved, we use the models to calculate potencies for the training set, to see which structures may be outliers. In several cases, we have found that an outlier is not as well aligned as the other structures in the training set, and with improved alignment, both the models and the predictions for this compound can be improved substantially. We anticipate that future generations of the program may be able to co-evolve the alignments with the models.

There are two other implicit assumptions in GERM model generation which should be stated. We deal only with a single conformation of each ligand in the training set; we know from crystallography that ligands can occasionally bind in related conformations. Similarly, we deal with a single orientation of each ligand in the binding site; again, we know from crystallographic studies that ligands may bind in more than one orientation, or in an unexpected orientation. As an aside, it is possible after models have been generated to dock ligands in different conformations and in different orientations to see if calculated binding energies might improve.

Clearly, we are assuming that receptor binding is directly proportional to bioactivity; we do not take into account differential effects on second messengers or other signaling steps which occur between receptor binding and experimentally observed response.

It is important to keep in mind that we are using very simple force-field calculations (non-bonded terms only) in calculating ligand–receptor binding. We take no steps to account for solvent effects, conformational strain induced in ligands or flexibility of the receptor molecule.

As stated previously, we start with a completely closed receptor site. Our current implementation does not give us a means to leave an open face on the receptor binding site. We can only infer possible open regions on the basis of frequency of null or small atom types, or on the occurrence of regions which have no discernible preference for any particular atom type.

5. Conclusion

The GERM method shows considerable promise as a procedure for 3D QSAR and for making useful models of receptor sites, particularly for problems where a crystallographic or homology-modelled receptor structure is not available. Further applications of the models have yet to be explored, such as screening 3D structure databases to find novel leads, or using the models in conjunction with *de novo* ligand-design programs.

Program Availability

The GERM program is available through Finch University of Health Sciences/ The Chicago Medical School; contact the author: walterse@mis.finchcms.edu or http://www.finchcms.edu/biochem/Walters/germ.html for further information.

References

1. Walters, D.E. and Hinds, R.M., *Genetically evolved receptor models: A computational approach to construction of receptor models*, J. Med. Chem., 37 (1994) 2527–2536.
2. Walters, D.E. and Muhammad, T.D., *Genetically evolved receptor models (GERM): A procedure for construction of atomic-level receptor site models in the absence of a receptor crystal structure*, In Devillers, J. (Ed.) Genetic algorithms in drug design, Academic Press, London, 1996, pp. 193–210.
3. Holland, J.H., *Adaptation in natural and artificial systems*, University of Michigan Press, Ann Arbor, MI, 1975.
4. Brooks, B.R., Bruccoleri, R.E., Olafson, B.D., States, D.J., Swaminathan, S. and Karplus, M., *CHARMM: A program for macromolecular energy minimization and dynamics calculations*, J. Comput. Chem., 4 (1983) 187–217.
5. Culberson, J.C. and Walters, D.E., *Development and utilization of a three-dimensional model for the sweet taste receptor*, In Walters, D.E., Orthoefer, F.T. and DuBois, G.E. (Eds.) Sweeteners: Discovery, molecular design and chemoreception, American Chemical Society, Washington, DC, 1991, pp. 214–223.
6. DuBois, G.E., Walters, D.E., Schiffman, S.S., Warwick, Z.S., Booth, B.J., Pecore, S.D., Gibes, K., Carr, B.T. and Brands, L.M., *A systematic study of concentration–response relationships of sweeteners*, In Walters, D.E., Orthoefer, F.T., and Dubois, G.E. (Eds.) *Sweeteners: Discovery molecular design and chemoreception*, American Chemical Society, Washington, DC, 1991, pp. 261–276.
7. Walters, D.E. and Muhammad, T.D., *Genetically evolved receptor models (GERM): A comparison of evolved models with crystallographically determined binding sites.* In Liljefors, T., Jørgensen, F.S., and Krogsgaard-Larsen, P. (Eds.) Rational molecular design in drug research, Munksgaard, Copenhagen, 1998 (in press).

3D QSAR of Flexible Molecules Using Tensor Representation

William J. Dunn III and Antony J. Hopfinger[a]
Department of Medicinal Chemistry and Pharmacognosy, College of Pharmacy, University of Illinois at Chicago, Chicago, IL 60612, U.S.A.

1. Introduction

The process by which a biologically active compound in an *in vitro* or an *in vivo* system is transported and binds to its receptor is poorly understood. This process is an example of molecular recognition [1], and understanding it is a major goal of drug discovery and development research. Computer-aided efforts to understand the process have their beginnings in the early work of Hansch [2], who extended the principles of physical organic chemistry to the study of biological structure–activity relationships. Hansch's work evolved into the field of quantitative structure–activity relationships, or QSAR, which treated drug–receptor interactions as an equilibrium or pseudo-equilibrium process in the same way that substituent effects on the ionization of weak organic acids and bases were treated. The active compounds were quantitatively described by features determined from a consideration of their 2-dimensional structures and these features were correlated with changes in activity. As the appreciation of the role of 3-dimensional structure in biological activity became more acute in the early 1980s, methods of 3-dimensional QSAR, or 3D QSAR, began to emerge. As a note, QSAR studies are a special case of quantitative structure–property relationships, QSPR studies.

In an effort to provide the discussion of 3D QSAR methods with more focus, Hopfinger and Tokarski [3] have recently reviewed this topic and divided the methods into (a) receptor independent and (b) receptor dependent. Receptor-independent methods are developed with little or no prior knowledge of the receptor geometry, while receptor-dependent methods use knowledge of receptor geometry in their derivation. The tensor treatment of structure–activity data to derive 3D QSAR models is a receptor-independent method and is designed to provide information indirectly about the receptor geometry.

By way of introduction to our work, the more important receptor-independent 3D QSAR methods are briefly mentioned here. The reader is referred to the work of Hopfinger and Tokarski [3] for a more in-depth and timely discussion of this topic, and other relevant chapters in this volume.

Tensor analysis has only recently been applied to problems in chemistry. Before its discussion, some definitions and conventions are introduced in order to avoid confusion with terminology. Initially, it is important to distinguish between structural dimensionality and the spatial dimensionality in which the data analysis is carried out. When discussing structural dimensionality, upper-case notation will be used (e.g. 2-Dimensional descriptors or 3D QSAR). Structural Dimensionality is not limited to 3-Dimensions. As

[a] Chem21 Group, Inc., Lake Forest, Illinois, U.S.A.

H. Kubinyi et al. (eds.), 3D QSAR in Drug Design, Volume 3. 167–182.

will be pointed out later in this chapter, the tensor approach encompasses higher structural Dimensions (e.g. time).

The dimensionality of descriptor space will be indicated by lower-case *d* and is determined by the product of the number of descriptors and the number of elements considered in each structural Dimension. For example, if 4 descriptors are evaluated for 10 conformers (conformation is one element of structural Dimensionality) and 15 receptor alignments (alignment is another element of structural Dimensionality), the dimensionality of descriptor space is $4 \times 10 \times 15$.

Tensors are not commonly referred to in computer-aided drug design, even though they are dealt with routinely. For example, a scalar is a zero-order tensor and a vector is a first-order tensor. A first-order tensor is a quantity that has magnitude and direction, while a second-order tensor has magnitude and two directions. Here, column vectors are designated by lower-case, bold characters, $\mathbf{u}$. A row or transpose vector is indicated by prime, $\mathbf{u}'$. A matrix, or 2-way array, is a second-order tensor and a 3-way array of data is third-order tensor. Matrices are designated as upper-case bold characters, $\mathbf{X}$, while 3-way arrays are designated by upper-case, bold italic, $\boldsymbol{X}$. Higher-order arrays can be represented as *N*-way arrays, where *N* is the order to the tensor. In the social science literature, where tensor analysis is used more extensively, the terminology 2-mode and 3-mode analysis is used. The use of the terminology, *N*-way, is consistent with current usage in the physical science literature and will be used here.

Since a major thrust of the approach presented here is treating structure–activity data of molecules which are conformationally flexible and can assume numerous possible receptor alignments, definitions of conformation and alignment are necessary. Regarding the former, the definition of Eliel et al. [4] is taken: ‘By “conformations” are meant the non-identical arrangements of the atoms in a molecule obtainable by rotation about one or more single bonds’ [4]. An alignment is the arrangement of two or more molecules in which a common set of atoms, substructures or features is approximately superimposed. In the example presented in this chapter, only pair-wise alignments are used, but the approach presented is not limited to the use of pair-wise alignment rules. The assumption of a reference compound for the pair-wise alignment rule, while a good starting assumption, has limitations. For one, it introduces a bias into the alignment process, and if an error is contained in the reference alignment rule, this error is amplified in the analysis. There would be an advantage, in some cases, in using a ‘consensus’ alignment rule which is not based on a reference, but gives each compound in the dataset equal weight in the alignment rule. There has been one reference to the use of a consensus alignment rule in structure–activity studies [5], but the method uses an annealing method which is computationally not practical for a large series of compounds.

2. Receptor-Independent 3D QSAR Analysis

Having the 3-dimensional structure of the receptor available to the medicinal chemist reduces drug-design problem to fitting ligands into the receptor site in sterically allowed geometries. While the number of X-ray and nmr determined structures is increasing rapidly, the majority of drug-design problems require designing ligands for receptors of unknown structure. In such cases, geometric information about the receptor can then be obtained in indirect ways and a number of receptor-independent methods of 3D QSAR have been developed to provide this information.

An underlying assumption of all currently used receptor-independent 3D QSAR methods is that the members of series of bioactive compounds bind to their respective receptor in a common conformation and alignment that allows optimal interaction of the functional groups of the pharmacophore with their complements in the active site.

Comparative molecular field analysis [6,7,8], or CoMFA, is one of the more powerful and frequently used receptor-independent methods. Several other 3D QSAR methods have been proposed and these include molecular shape analysis, or MSA [3], molecular similarity matrices [9], distance geometry techniques [10], the hypothetical active site lattice, HASL, model [11], genetically evolved receptor models, GERM [12], grid analysis [13] and CATALYST [14]. Reference [15] is a good current review of 3D QSAR analysis, and reference [3] provides a focused update and analysis of current work in 3D QSAR. Again, there is no current 3D QSAR approach which is capable of handling the general 3D QSAR problem for flexible molecules for which variable alignment rules can be simultaneously considered. This is the subject of the remainder of this review.

3. The General 3D QSAR Formalism

By relaxing the conformation and alignment constraints imposed by most currently used methods of 3D QSAR, a general formalism for 3D QSAR can be proposed in terms of tensor analysis of the resulting structure–activity data [16]. This formalism is presented here in terms of MSA descriptors. However, in the most general case, it can be applied to any conformation/alignment-dependent descriptor set. The model, in terms of MSA descriptors, is:

$$Y_u = T_u * [V_u(s,m,n), F_u(p,r_{i,j,k},m,n), H_u(h_p,m,n), E_u(e_p,m,n)] \tag{1}$$

where Y is the activity, or dependent variable; conformation is noted by m and alignment by n; and u states that the relationship is absolute rather than relative — i.e. based on a reference compound. In order to use the absolute form of the model, a consensus alignment rule is necessary. The variables, V, F, H and E are four tensors, of which V and F have their roots in MSA. V incorporates shape, s, in molecular description and contains the intrinsic molecular shape, IMS, features of the compounds. It is a measure of the effect of molecular shape within the steric contact surface of the molecule. It is highly dependent on conformation and alignment. F is the molecular field, MF, tensor computed with the set of field probes, p, at spatial positions r_{ijk} from the molecular surface and measures the effect of molecular shape outside the steric contact surface of the molecule. It, too, is highly dependent on conformation and alignment. The H tensor incorporates the physico-chemical descriptors which may or may not be conformation and alignment dependent. Examples are lipophilicity, pK_a, solubility, etc. The E are largely experimentally determined descriptors for which the conformational dependence is expressed only as a function of the Boltzmann average in the experimental result. The H and E are the basis of 2-Dimensional QSAR or traditional Hansch analysis and can enter the analysis independently of conformation and alignment. If only information about the geometry of the ligand–receptor complex is of interest, the H and E may not directly enter into the analysis.

The relative MSA 3D QSAR model is:

$$Y_{u,v} = T_{u,v} * [V_{u,v}(s,a,b),\ F_{u,v}(p,r_{i,j,k},a,b),\ H_{u,v}(h_p,a,b),\ E_{u,v}(e_p,a,b)] \qquad (2)$$

Where the subscript v indicates that the tensor is evaluated relative to a reference compound.

The application of the method involves solution for the transformation tensors, T_u and $T_{u,v}$, in Eqs. 1 and 2. The transformation tensors project the descriptors onto the **Y** and can be obtained with a number of data analytical methods. Due to the unique nature of the structure–activity data generally encountered in 3D QSAR, data reduction methods are necessary. Two methods, 3-way factor analysis and 3-way PLS [16] have been applied to this problem and these are discussed below.

3.1. 3D QSAR data structure

The data structure for the 3D QSAR problem with conformation and alignment fixed is shown in Fig. 1. It is identical to the 2-Dimensional QSAR data structure and the data are treated identically. The biological activity measure is **Y**, which is a vector for a single activity or a matrix for more than one measured response. The descriptors, or independent variables, are **X**, and comprise the V, F, H and E tensors, as discussed above. In the case of a CoMFA problem, the descriptors are the respective probe-dependent energies computed at points on the grid for each compound. As usual, there are many more variables than compounds, so that a data reduction method — i.e. PLS regression — is required in the data analysis step.

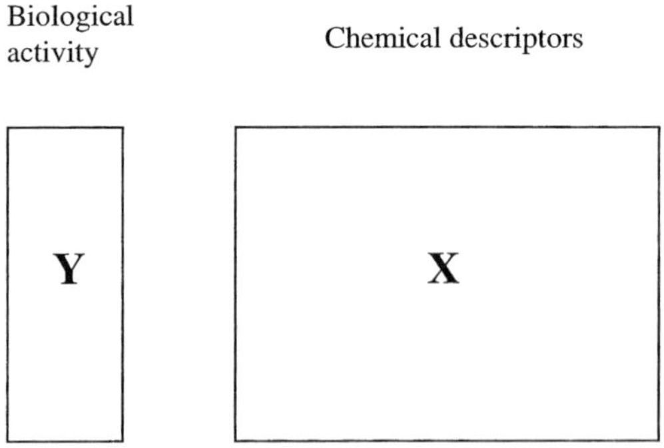

Fig. 1. The standard dataset for a 2D QSAR.

By relaxing the conformation and alignment constraints, the data structure in Fig. 2 results for a single variable. In order to solve the 3D QSAR problem, the resulting 3-way array must be decomposed to yield the transformation tensors, T. This can be done in several ways, but the use of 3-way factor analysis and 3-way PLS is proposed. Both have advantages and disadvantages, as will be seen in the discussion which follows.

The use of factor analysis and PLS regression in this application is quite different from their use in traditional 3D QSAR. It is not the objective of their application here to derive a predictive QSAR model, but to solve for the conformation and alignment most highly correlated with activity. It is assumed that only one conformation and alignment is involved in the ligand–receptor complex. However, by varying the resolution of the conformation/alignment space explored and the number of descriptors considered, the 3-way array in Fig. 2 can be small or as large as computationally feasible. It is of interest to extract and rank the important one or two descriptor vectors. These can then be used with more traditional correlation methods, and with other variables, to derive predictive QSARs. In a way, the methods are used here as a variable selector, or filter, to extract the conformation/alignment information from noise.

3.2. 3-way arrays

The QSAR resulting from decomposition of the 2-way array of chemical descriptor data in Fig. 1 provides the change in biological activity with change in 2-Dimensional structure, or with 3-Dimensional structure with conformation and alignment fixed. In the case in which a structure is unconstrained with respect to conformation and alignment, the objective is to decompose the 3-way array in Fig. 2 to explore how the change in structure with respect to changes in conformation *and* alignment is related to the change in biological response. This information is in the unfolded 3-way arrays, as shown in Fig. 3. The unfolding leads to 3 matrices, **O**, **P** and **Q**, which contain the requisite information. The indices l, m and n refer to compound, conformation and alignment,

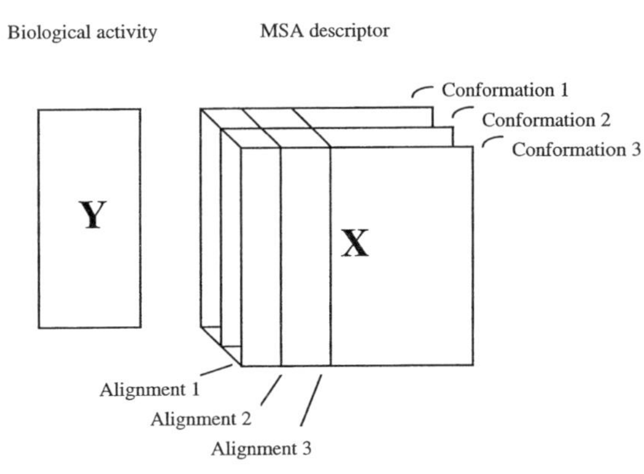

Fig. 2. Data structure for a 3D QSAR problem with one descriptor. Each layer in the x, y-plane of the array represents the variable with conformation fixed and each layer in the y, z-plane represents the layer with alignment fixed. Each column represents the variable with alignment fixed.

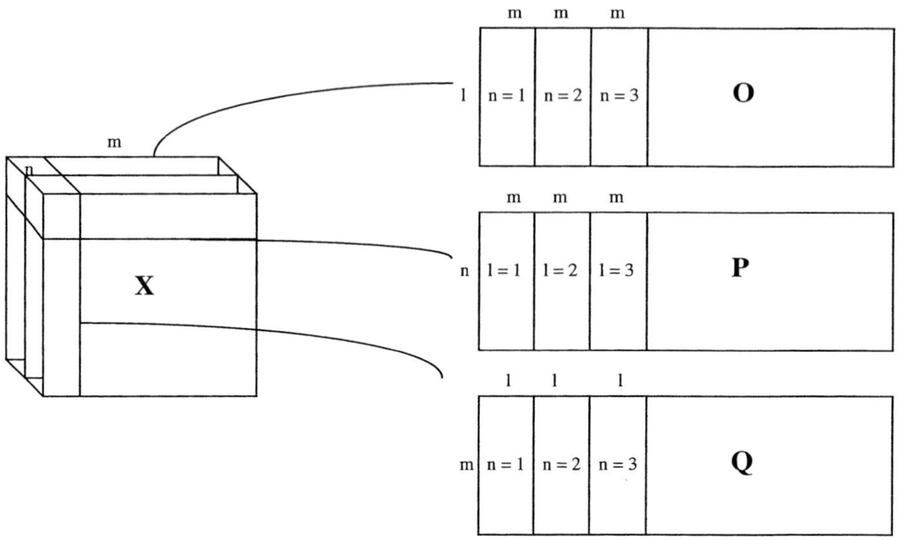

Fig. 3. Unfolding a 3-way array into three 2-way arrays.

respectively, while *o*, *p* and *q* are the number of significant factors or components in the compound, conformation and alignment matrices. 3-Way factor analysis deals with **O**, **P** and **Q**, while 3-way PLS regression deals with **O** from the 3-way array.

3.3. 3-Way factor analysis

3-Way factor analysis was developed first by Tucker [18], and more recently by Kroonenberg [19]. It has also been applied more recently to analysis of analytical [20,21] and environmental chemical [22] data. 3-Way factor analysis decomposes a 3-way array into three factor weight matrices, **A**, **B** and **C**, and a 3-way core matrix, **G** (Fig. 4). The factor weight matrices are associated with compound, conformation and alignment, respectively, with the magnitude of the weights being measures of the variance in the descriptor vectors in the array. The core matrix contains the correlation structure of the 3-way array.

The weight matrices **B** and **C**, which are conformation and alignment specific, are of interest for this application. They indicate the conformation and alignment vectors in the 3-way array which have the greatest systematic variation. The descriptor vectors associated with these heavily loaded conformations and alignments are used in regression to derive the 3D QSAR which is equivalent to principal components regression and subject to the advantages and disadvantages of this method. They are not conditioned to be correlated with **Y**.

The algebraic model for the decomposition is:

$$x_{l,m,n} = \sum_{o}\sum_{p}\sum_{q} a_{l,o} b_{n,p} c_{m,q} g_{o,p,q} \tag{3}$$

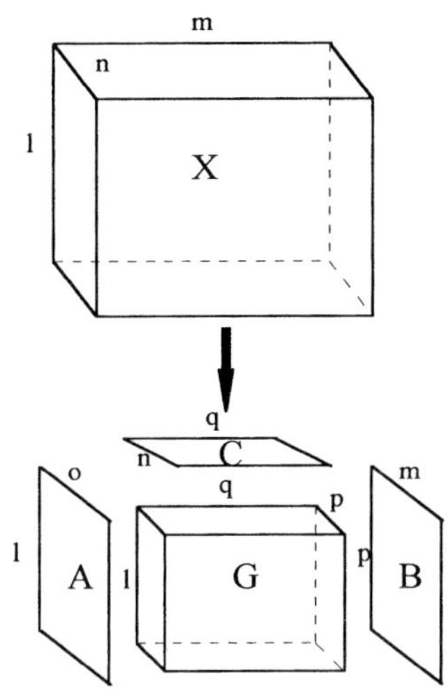

Fig. 4. Decomposition of a 3-way array by 3-way factor analysis.

where a, b and c are the elements of **A**, **B** and **C**, respectively, with o, p and q being the number of significant factors in each. The weights, o, p and q, are not necessarily equivalent. The matrix form is given as:

$$X_{l,m,n} = A_{l,o}G_{o,p,q}(B_{p,n} \otimes C_{q,m}) \tag{4}$$

where the terms are as defined above, and ⊗ indicates the Kronecker product.

3.4. 3-Way PLS regression

Referring to Fig. 5, 3-way PLS regression extracts from $\boldsymbol{X}$ and **Y** the latent variable which are vectors computed along the axes of greatest variation in $\boldsymbol{X}$ and **Y** and are most highly correlated. PLS can be applied to $\boldsymbol{X}$ in terms of a single variable or over a number of variables, J. This is shown in algebraic notation in Eqs. 5–7, below. Here, the usual PLS:

$$x_{l,m,n} = \sum_{j=1}^{J}[\bar{x}_{l,m,n} + \sum_{z=1}^{Z} t_{l,z} \otimes P_{z,m,n}] + e_{l,m,n} \tag{5}$$

$$y_{l,i} = \bar{y}_i + \sum_{z=1}^{Z} u_{l,z} d_{z,i} + e_{l,i} \tag{6}$$

$$\hat{u} = b * t \tag{7}$$

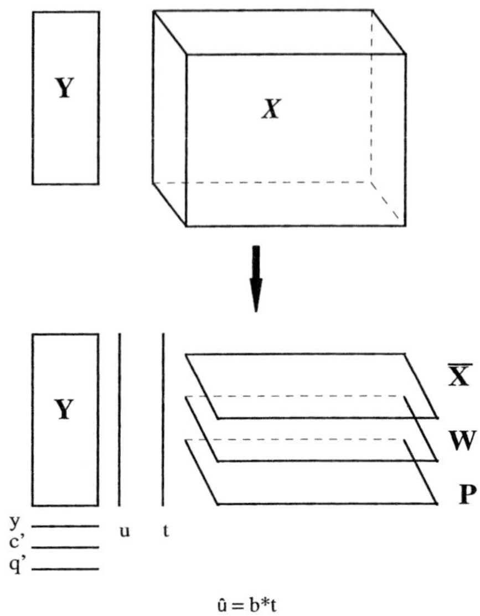

Fig. 5. Decomposition of a 3-way array by PLS regression shown here for one latent variable.

notation is used with l, m and n referring to compound, conformation and alignment, respectively. The latent variables are t from the descriptor data and u from the biological activity data. The ***X***-loadings are **P** and the **Y**-loadings are q. **W** contains the PLS weights. In 3-way PLS, the ***X***-loadings, **P**, are a 2-way array. The number of significant components is Z. The sums of the squares of the residuals, $e_{l,m,n}$ and $e_{l,i,}$ are minima. In the calculation of the ***X***-data from the PLS parameters, ⊗ indicates the Kronecker product. Algorithms for computing the 3-way factor and PLS regression models are presented in the algorithm.

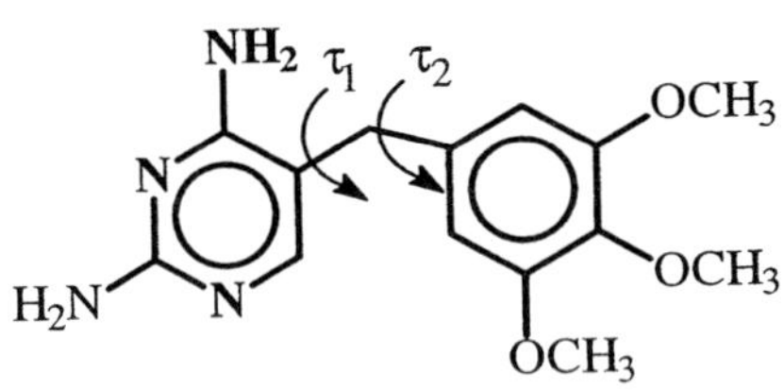

I

3.5. *Conformation–alignment weights*

In order to weight, or rank, the conformations and alignments that result from 3-way PLS, conformation/alignment weights, or CAW, are computed from the ***X***-loadings, **W**; these are computed as below:

$$CAW_{mn} = \sum_{z=1}^{Z} Var_z W_{zmn}^2 \tag{8}$$

Where Var_z is the **Y**-variance explained in component z. A similar statistic can be computed from the 3-way factor analysis results by using the sum of squares of the weights from **B** and **C** to rank the conformations and alignments, respectively.

4. Application of the Methodology

In order to illustrate the utility of the 3D QSAR formalism, it has been applied to structure-binding data for trimethoprim, ***I***, and trimethoprim-like analogs to dihydrofolate reductase, DHFR. The geometry of the binary DHFR–trimethoprim complex has been extensively studied [23], making this an ideal set of data for testing the general 3D QSAR formalism. If there is an active conformation and alignment and the tensor analysis approach can predict its geometry, this would help establish its general utility. An account of this work has been published [17], and a summary of the technique and its results are given here.

4.1. Generation of conformation, alignment and MSA descriptor data

Enzyme-inhibitor binding data were taken from the literature on 20 analogs of structure ***I***. Earlier 3D QSAR studies of 2,4-diaminopyrimidine inhibitors of DHFR have shown that the MSA descriptor, common steric overlap volume, COSV, has been a significant variable [24] which led to its use in this study. The structures were built using bond

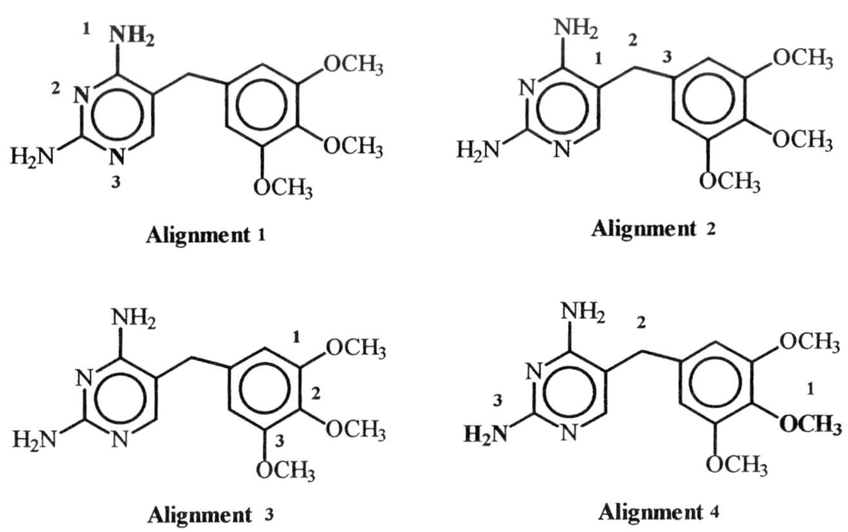

Fig. 6. The four alignment rules used with the key atoms of superimposition noted.

lengths and bond angles from the trimethoprim crystal structure. Partial charges were computed using the MNDO method [25]. Fixed valence conformational analysis was performed for each of the analogs at 10° resolution for the torsion angles, τ_1, and τ_2, as shown in **I**. The MMII non-bonded potential, a Coulomb potential with a dielectric constant of 3.5, and a MMII-scaled hydrogen bonding potential, were used [26]. To be consistent, this force field was used in the study cited above [24]. The conformational profiles of the series of analog inhibitors are defined by the torsion angles τ_1 and τ_2. The conformation of trimethoprim bound in its binary complex with *E. coli* DHFR is defined by torsion angles $\tau_1 = 177°$ and $\tau_2 = 76°$, where $\tau_1 = \tau_2 = 0$ corresponds to the reference conformation in the cis configuration. The active site bound conformation is not the global minimum for any of the analogs. Trimethoprim was used as the shape reference, and 10 trial conformations were considered for each compound. The 10 conformations are operationally equivalent to one another with respect to bonding topology defining the torsion angles, as discussed below.

Trimethoprim is found to have 8 free space minimum energy conformations within 5 kcal/mol of the global intramolecular minimum energy conformation. For each of the other analogs in the dataset, the minimum energy conformations within 5 kcal/mol of the global minimum energy conformation and nearest in τ_1, τ_2 torsion angle space to the minimum energy conformations of trimethoprim were considered; that is, the (10° resolution in τ_1 and τ_2) minimum energy conformations within 5 kcal/mol, closest to the τ_1 and τ_2 values of the selected 8 minima of trimethoprim, were selected. For those compounds that do not have minima for τ_1 and τ_2 values close to those of trimethoprim, the τ_1 and τ_2 values were set to those of the trimethoprim minimum. For the series, overall the τ_1 and τ_2 values vary within a range of ±30° of 177° and 76°, respectively. In total, 10 conformations were selected for each compound, with one conformation being the crystal-bound geometry.

Four alignment rules were selected, as shown in Fig. 6. In each test alignment, 3 key atoms were identified for superposition and all compounds in the dataset are compared pair-wise to trimethoprim using the 3 alignment atoms defining the alignment rule. The COSV for each analog, relative to trimethoprim, for each of the 10 conformations and 4 alignments, was computed. The result was a $20 \times 10 \times 4$ 3-way array. The reader is referred to the original work for further details regarding the structure–activity data. 3-Way factor analysis was applied directly to the 3-way array, and 3-way PLS regression was applied to the data with pIC_{50} as the dependent variable.

4.2. *Results*

The application of 3-way factor analysis to the data resulted in two significant eigenvalues (based on variance explained) from **M**, **P** and **Q**, respectively. Their eigenvectors were used in the construction of **A**, **B** and **C** (Tables 1–3). The factor loadings were largest for conformation 10, alignment 2, conformation 10, alignment 3 and conformation 9, alignment 2. 3-Way PLS gave results (Table 4) consistent with these with CAW values of 0.10, 0.07 and 0.05, respectively, for the same 3 conformation/alignment sets. The bound conformation of trimethoprim is that of conformer 10, so it is

Table 1 *Factor loadings for the compounds*

Compound	Component number 1	2
1	–0.48	–0.26
2	–0.10	0.01
3	–0.22	0.10
4	–0.07	0.03
5	0.01	–0.08
6	–0.27	–0.08
7	–0.23	–0.07
8	–0.14	–0.11
9	–0.22	–0.16
10	0.13	–0.12
11	0.11	–0.11
12	0.07	–0.06
13	0.47	–0.16
14	0.12	0.07
15	0.30	0.07
16	0.05	–0.05
17	0.26	–0.22
18	–0.09	0.66
19	0.03	0.55
20	0.25	–0.01

Table 2 *Factor loadings for alignments*

Alignment	Factor 1	2
1	–0.53	–0.65
2	–0.18	0.69
3	0.82	–0.24
4	–0.10	0.21

satisfying that the two results give consistent results. Alignment rules 2 and 3 are indicated to be significant in binding and are reasonable in light of nmr spectroscopy studies of the solution structure of the enzyme–inhibitor complex.

To this point, the tensor approach has been used as a filter to extract from the 3-way arrays the geometries of the ligands having the most systematic variation and most highly associated with activity. The descriptor vectors associated with these geometries can be used, either alone or in combination with other descriptors, to develop 3D QSARs. If used with 2-Dimensional structural descriptors, hybrid QSARs result; this is shown below.

The MSA descriptor, $COSV^2$, when regressed with activity gave the 3D QSAR below:

$$\log(1/IC_{50}) = 5.51\,COSV^2 + 2.74$$

$$n = 20,\ r^2 = 0.50,\ XV - R^2 = 0.42,\ F = 17.84,\ F_{1,18\alpha\ .01} = 8.28 \qquad (9)$$

Table 3 *Factor loadings for conformations*

Conformation	Factor 1	Factor 2
1	–0.14	–0.40
2	–0.28	0.14
3	–0.24	–0.10
4	–0.32	–0.11
5	0.17	0.52
6	–0.26	–0.09
7	–0.14	–0.31
8	0.21	0.50
9	0.31	0.06
10	0.69	–0.42

Table 4 *PLS loadings for conformation and alignment*

Alignment	Conformation 1	2	3	4	5	6	7	8	9	10
1	–0.02	0.00	–0.01	0.00	–0.05	0.00	–0.01	0.00	0.18	0.29
2	0.08	0.03	0.01	–0.03	0.17	–0.02	0.06	0.20	0.01	0.46
3	0.13	0.17	0.14	0.12	0.15	0.13	0.16	0.19	0.23	0.41
4	–0.02	0.04	–0.04	0.02	–0.05	0.01	–0.03	0.03	0.13	0.22

where $XV - R^2$ is the cross-validated R^2 for the equation. The single variable, $COSV^2$, explains 50% of the variation in activity, and when combined with 2- and other 3-Dimensional variables, the result below is obtained:

$$\log(1/IC_{50}) = 0.36\pi - 0.35\,MR^2 + 17.55\,COSV + 0.05\,NOV + 1.89\,S^2 - 10.28 \tag{10}$$

$$n = 20,\ R^2 = 0.91,\ XV - R^2 = 0.82,\ F = 29.33,\ F_{1,14\alpha\ .01} = 3.70$$

where *NOV* is the nonoverlap volume, *S* is the torsion angle unit entropy and *MR* is the scaled molar refractivity.

The tensor analysis approach to 3D QSAR provides computer-aided drug design with a generalized treatment of structure–activity data within a framework of existing QSAR methods. It is an heuristic approach which is subject to the caveats of such methods. The method is based on the same rules of statistics as are all such methods, and in order to be used successfully, they are highly dependent on a good experimental design.

This application indicates the potential for tensor analysis of 3-Dimensional structure–activity to provide information about the receptor-bound geometry of ligands. The methodology is a correlative one and an extension of the 2D QSAR approach. Further applications are under way to explore the utility of tensor analysis not only in 3D

QSAR studies, but in the more general 3D QSAR arena, where it has the potential for providing the structural basis for fundamental processes which have embedded in them complex molecular ordering and orientation.

5. Appendix 5

5.1. Algorithm for decomposition of 3-way arrays by 3-way factor analysis

A variation of the algorithm of Zeng and Hopke [22] has been programmed and is given below:

Step 1. Unfold $\boldsymbol{X}$ to obtain its 3, 2-way arrays, $\boldsymbol{X}_{(l)mn}$, $\boldsymbol{X}_{(n)mn}$ and $\boldsymbol{X}_{(m)ml}$, as in Fig. 3.

Step 2. Compute:

$$\boldsymbol{O}_l = \boldsymbol{X}'_{(l)mn}\,\boldsymbol{X}_{(l)mn}$$
$$\boldsymbol{P}_m = \boldsymbol{X}'_{l(m)n}\,\boldsymbol{X}_{l(m)n}$$
$$\boldsymbol{Q}_l = \boldsymbol{X}'_{lm(n)}\,\boldsymbol{X}_{lm(n)}$$

Step 3. Construct:

$\boldsymbol{A}_{lo}$ from the o significant eigenvectors of $\boldsymbol{O}_l$
$\boldsymbol{B}_{mp}$ from the p significant eigenvectors of $\boldsymbol{P}_m$
$\boldsymbol{C}_{nq}$ from the q significant eigenvectors of $\boldsymbol{Q}_n$.

Step 4. Compute the unfolded core matrix, $\boldsymbol{G}_{(o)pq}$ as:

$$\boldsymbol{G}_{(o)pq} = \boldsymbol{A}_{lo}\boldsymbol{X}_{(l)mn}(\boldsymbol{B}_{mp} \otimes \boldsymbol{C}_{np})$$

Step 5. In the prediction phase, estimate the 3-way array, $\boldsymbol{X}_{lmn}$, where the estimate is in unfolded form:

$$\hat{\boldsymbol{X}}_{(l)mn} = \boldsymbol{A}_{lo}\boldsymbol{G}_{opg}(\boldsymbol{B}_{pn} \otimes \boldsymbol{C}_{qm})$$

Diagnostic statistics can be computed to determine the number of significant eigenvectors, o, p and q, to include in $\boldsymbol{A}$, $\boldsymbol{B}$ and $\boldsymbol{C}$. For this, cross-validation is the method of choice.

5.2. Algorithm for decomposition of 3-way arrays by PLS regression

An algorithm for PLS regression decomposition of 3-way arrays based on the NIPALS algorithm has been published by Lohmöller and Wold [27]. More recently, a cursory discussion of PLS regression decomposition of *N*-way arrays was published [28], also based on the NIPALS algorithm. Due to the combinatorial problem of treating multiple alignments of flexible molecules, this algorithm is computationally inefficient. Here, a variation of the UNIPALS algorithm [29,30] developed in this laboratory is presented. It differs from the conventional PLS methods, in that it uses a Kronecker product, as does 3-way factor analysis, in the prediction phase. This algorithm has been programmed and, in a limited number of applications, has performed well. Other PLS regression algorithms have been published [31,32] and could possibly be adapted to 3-way array decomposition.

To begin:

Step 1. Compute from $\boldsymbol{X}_{(l)mn}$ and $\boldsymbol{Y}$:
$\boldsymbol{D} = \boldsymbol{X}'_{(l)mn}\,\boldsymbol{Y}$

Step 2. Compute the first eigenvalue, $\boldsymbol{c}$, of $\boldsymbol{D}'\boldsymbol{D}$.

Step 3. Compute the $\boldsymbol{Y}$-scores:
$\boldsymbol{u} = \boldsymbol{Yc}$

Step 4. Compute the $\boldsymbol{X}$-weights, $\boldsymbol{W}$, as:
$\boldsymbol{W}' = \boldsymbol{u}'\boldsymbol{X}_{(l)mn}/\boldsymbol{u}'\boldsymbol{u}$; $\boldsymbol{W}$ is the unfolded form of the 2-way array in Fig. 5. Normalize $\boldsymbol{W}$ to length l.

Step 5. Compute the $\boldsymbol{X}$-scores as:
$\boldsymbol{t} = \boldsymbol{X}_{(l)mn}\,\boldsymbol{W}$

Step 6. Compute the $\boldsymbol{X}$-loadings as:
$\boldsymbol{P} = \boldsymbol{X}'_{(l)mn}\,\boldsymbol{t}/\boldsymbol{t}'\boldsymbol{t}$; $\boldsymbol{P}$ is obtained as the unfolded form of the 2-way array in Fig. 5.

Step 7. Compute the $\boldsymbol{Y}$-loadings as:
$\boldsymbol{q} = \boldsymbol{Y}'\boldsymbol{u}/\boldsymbol{u}'\boldsymbol{u}$

Step 8. Form the inner relation:
$\hat{\boldsymbol{u}} = b^{*}\boldsymbol{t}$

Step 9. Update $\boldsymbol{X}$ and $\boldsymbol{Y}$, respectively, as:
$\boldsymbol{E} = \boldsymbol{X}_{(l)mn} - \boldsymbol{t} \otimes \boldsymbol{P}'$
$\boldsymbol{F} = \boldsymbol{Y} - b\,\boldsymbol{X}_{(l)mn}\,\boldsymbol{Wc}'$

Step 10. To compute the next latent variable, form $\boldsymbol{E}'\boldsymbol{F}$ as the updated $\boldsymbol{X}'\boldsymbol{Y}$ and repeat the algorithm.

In many ways, this algorithm works like regular PLS and the models generated by it can be evaluated in the same way as regular PLS models. In this application, however, the $\boldsymbol{X}$-loadings, $\boldsymbol{P}$, are of interest. The largest elements of $\boldsymbol{P}$ are associated with the receptor-bound conformation and alignment. It may be possible to carry out an orthogonal decomposition of $\boldsymbol{P}$ to obtain the individual conformation and alignment weights but this has not been attempted. Again, cross-validation is the desired method for determining model complexity — i.e. the number of latent variables.

5.3. Kronecker products of matrices

The Kronecker product has not been widely used in the chemical sciences, so that its use may not be familiar to most medicinal chemists. It is used in the prediction phase of both 3-way factor analysis and 3-way PLS. To illustrate its use, consider two matrices $\boldsymbol{Y} = [y_{lm}]$ of order $(i \times j)$ and $\boldsymbol{Z} = [z_{no}]$ of order $(q \times r)$. The Kronecker product, $\boldsymbol{Y} \otimes \boldsymbol{Z}$ will have order $(iq \times jr)$. Unlike the formation of inner and outer products of matrices, the Kronecker product is defined irrespective of the order of the two matrices which are used to form the product. To illustrate the actual operation, consider the two matrices:

$$Y = \begin{bmatrix} a_{11} & a_{12} \\ a_{21} & a_{22} \end{bmatrix} \tag{11}$$

$$Z = \begin{bmatrix} b_{11} & b_{12} \\ b_{21} & b_{22} \end{bmatrix} \tag{12}$$

The Kronecker product, ($\boldsymbol{Y} \otimes \boldsymbol{Z}$) is:

$$(Y \otimes Z) = \begin{bmatrix} a_{11}Z & a_{12}Z \\ a_{21}B & a_{22}B \end{bmatrix} = \begin{bmatrix} a_{11}b_{11} & a_{11}b_{12} & a_{12}b_{11} & a_{12}b_{12} \\ a_{11}b_{21} & a_{11}b_{22} & a_{12}b_{21} & a_{12}b_{22} \\ a_{21}b_{11} & a_{12}b_{12} & a_{22}b_{11} & a_{22}b_{12} \\ a_{21}b_{21} & a_{21}b_{22} & a_{22}b_{21} & a_{22}b_{22} \end{bmatrix} \tag{13}$$

For further reading the works of Graham [33] and Novotny [34] are recommended.

Acknowledgements

The authors wish to acknowledge the support of the National Science Foundation in the form of a Phase I SBIR grant, and Pfizer Corporation, Groton, CT, U.S.A., in the form of a research grant.

References

1. Roberts, S.M. (Ed.), *Molecular recognition: Chemical and biochemical problems II*, Royal Society of Chemistry, Redwood Press, London, U.K., 1993.
2. Hansch, C., *A quantitative approach to biochemical structure–activity relationships*, Accts. Chem. Res., 2 (1968) 232–239.
3. Hopfinger, A.J. and Tokarski, J.S., *3D-QSAR analysis*, In Charifsom, P.S. (Ed.) Practical applications of computer-aided drug design, Marcel Dekker, New York, 1997.
4. Eliel, E.L., Allinger, N.L., Angyal, S.J. and Morrison, G.A., Conformational analysis, The American Chemical Society, Washington, DC, 1981, p. 1.
5. Barakat, M.T. and Dean, P.M., *Molecular structure matching by simulated annealing: II. An exploration of the evolution of configuration landscape problems*, J. Computer-Aided Mol. Design, 4 (1990) 317–330.
6. Cramer III, R.D., Patterson, R.E. and Bunce, J.D., *Comparative molecular field analysis (CoMFA): 1. The effect of shape on binding of steroids to carrier proteins*, J. Am. Chem. Soc., 110 (1988) 5959–5967.
7. Tripos Associates, 1699 Hanley Road, St. Louis, MO 63144, U.S.A.
8. Cramer, R.D., Clark, R.D., Patterson, D.E. and Ferguson, A.M., *Bioisosterism as a molecular diversity descriptor: Steric fields of single 'topomeric' conformers*, J. Med. Chem., 39 (1996) 3060–3069.
9. Good, A.C., Peterson, S.J. and Richards, W.G., *QSARs from similarity matrices: Technique validation and application in the comparison of different similarity evaluation methods*, J. Med. Chem., 36 (1993) 2929–2937.
10. Crippen, G.M., *Distance geometry approach to rationalizing binding data*, J. Med. Chem., 22 (1979) 988–997.

11. Doweyko, A.M., *The hypothetical active site lattice: An approach to modeling active sites from data on inhibitor Molecules*, J. Med. Chem., 31 (1988) 1396–1406.
12. Walters, D.E. and Hinds, R.M., *Genetically evolved receptor models: A computational approach to construction of receptor models*, J. Med. Chem., 37 (1994) 2527–2536.
13. Goodford, P.J., *A computational procedure for determining energetically favorable binding sites on biologically important macromolecules*, J. Med. Chem., 28 (1985) 849–856.
14. CATALYST, Molecular Simulation, Inc., San Diego, CA, U.S.A.
15. Kubinyi, H. (Ed.), 3D-QSAR in drug design: Theory, methods and applications, ESCOM, Leiden, The Netherlands, 1993.
16. Hopfinger, A.J., Burke, B.J. and Dunn III, W.J., *A generalized formalism for three-dimensional quantitative structure–activity relationship using tensor representation*, J. Med. Chem., 37 (1994) 3768–3774.
17. Dunn III, W.J., Hopfinger, A.J.,Catana, C. and Duraiswami, C., *Solution of the conformation and alignment tensors for the binding of trimethoprim and its analogs to dihydrofolate reductase: 3D-quantitative structure–activity relationships study using molecular shape analysis, 3-way partial least squares regression and 3-way factor analysis*, J. Med. Chem., 39 (1996) 4825–4832.
18. Tucker, L.R., *Determination of parameters of a functional relation by factor analysis*, Psychometrika, 23 (1958) 19–23.
19. Kroonenberg, P., Three mode principal component analysis, DSWO Press, Leiden, The Netherlands, 1983.
20. Apellof, C.J. and Davidson, E.R., *Three dimensional rank annihilation for multicomponent determinations*, Anal. Chim. Acta, 146 (1983) 9–14.
21. Sanchez, E. and Kowalski, B.R., *Generalized rank annihilation factor analysis*, Anal. Chem., 58 (1986) 496–499.
22. Zeng, Y. and Hopke, P.K., *The application of three-mode factor analysis (TMFA) to receptor modeling of scenes particle data*, Atmosph. Environ., 26A (1992) 1701–1711.
23. Koetzle, T.F. and Williams, G.J.B., *The crystal and molecular structure of the antifolate drug trimethoprim (2,4-diamino-5-(3,4,5-trimethoxybenzyl)pyrimidine): A neutron diffraction study*, J. Am. Chem. Soc., 98 (1976) 2074–2081.
24. Mabilia, M., Pearlstein, R.A. and Hopfinger, A.J., *Molecular shape analysis and energetics-based intermolecular modeling of benzylpyrimidine dihydrofolate reductase inhibitors*, Eur. J. Med. Chem.-Chim. Thera., 20 (1985) 163–174.
25. Dewar, M.J.S. and Thiel, W., *Ground states of molecules: 38. The MNDO method, approximations and parameters*, J. Am. Chem. Soc., 99 (1977) 4899–4906.
26. Hopfinger, A.J. and Pearlstein, R.A., *Molecular mechanics force-field parameterization procedures*, J. Comput. Chem., 5 (1985) 486–497.
27. Lohmöller, J.B. and Wold, H., *Three-mode path models with latent variables and partial least squares (PLS) parameter estimation*, In Proceedings of the European Meeting of the Psychometric Society, University of Groningen, The Netherlands, 1980, p. 50.
28. Wold, S., Geladi, P., Esbensen, K. and Öhman, J., *Multi-way principal comonents- and PLS-analysis*, J. Chemometrics, 1 (1987) 41–56.
29. Glen, W.G., Dunn III, W.J. and Scott, D.R., *Principal components analysis and partial least squares regression*, Tetrahedron Comput. Method., 2 (1989) 349–376.
30. Glen, W.G., Sarker, M., Dunn III, W.J. and Scott, D.R., *UNIPALS: Software for principal components analysis and partial least squares regression*, Tetrahedron Comput. Method., 2 (1989) 377–396.
31. Lindgren, F., Geladi, P. and Wold, S., *The kernel algorithm for PLS*, J. Chemometrics, 7 (1993) 45–59.
32. Bush, B.L. and Nachbar Jr., R.B., *Sample-distance partial least squares: PLS optimized for many variables, with application to CoMFA*, J. Comput.-Aided Mol. Design, 7 (1993) 587–619.
33. Graham, A., Kronecker products and matrix calculus: With applications, Ellis Horwood, Chichester, U.K., 1981.
34. Novotny, M.A., *Matrix products with application to classical statistical mechanics*, J. Math. Phys., 20 (1979) 1146–1150.

Comparative Molecular Moment Analysis (CoMMA)

B. David Silverman, Daniel E. Platt, Mike Pitman and Isidore Rigoutsos
IBM Thomas J. Watson Research Center, Yorktown Heights, NY 10598, U.S.A.

1. Introduction

The binding of a drug molecule to its targeted receptor site is dependent upon a number of physical and chemical factors. In many instances, this binding is a consequence of non-bonding as opposed to covalent interactions and is, therefore, determined to a large extent by the complementarity of ligand molecular shape and charge to its targeted receptor site. Molecular shape and charge can be characterized in a number of different ways as attested to by chapters in this volume.

Perhaps the most elemental characterization of molecular shape and charge is provided by the moments of the mass (shape) and charge distributions. For those with no prior exposure to the concept of moments of a distribution, such a mass or charge, suitable references might be useful [1,2]. Certain of the lower-order molecular moments — e.g. molecular weight, moments of inertia, net molecular charge and dipole moment — have been used to characterize molecules, and it is perhaps not fully appreciated that these quantities are lower-order terms in a series that extends to infinity. Table 1 lists these commonly used moments and terminology, up to and inclusive of the second order of the molecular mass (shape) and charge. Molecular weight, moments of inertia and dipole moment have been previously used in a number of three-dimensional quantitative structure activity (3D QSAR) studies. Since such lower-order moments had been used to characterize neutral molecules, what captured our interest initially was that quadrupolar moments, the second-order electrostatic analog of the inertial moments, were never mentioned in connection with either discussions of molecular similarity or 3D QSAR procedures. A reason for this became apparent immediately. The comparison of quadrupolar moments between different molecules required the identification of a center — i.e. a center identified in an analogous fashion to the molecular center-of-mass which enables comparison of the moments of inertia of different molecules. Such center had not been identified.

The zero'th-order moment of molecular mass is just the molecular weight, which is obviously independent of a location of the origin of multipolar expansion. The inertial or second-order moments do depend upon the choice of origin about which they are calculated. There is, however, a convenient point or space, namely the center-of-mass,

Table 1 *Molecular moments*

Order	Mass	Charge
0	Molecular weight	Molecular charge
1	0	Dipole
2	Moments of inertia	Quadrupole

H. Kubinyi et al. (eds.), 3D QSAR in Drug Design, Volume 3. 183–196.

about which molecular dynamic rotations and translations separate, and which therefore provides a reference origin for the similarity comparison of the moments of inertia of different molecules. This origin is chosen such that the first-order moment of the mass distribution is zero.

Moments of the molecular charge distribution can be described in a similar manner. The zero'th-order moment of the molecular charge distribution is just the net molecular charge. The first-order or dipole moment of a neutral molecule is not dependent upon the choice of origin about which it is calculated. This independence or invariance is a specific consequence of the more general attribute of molecular electrostatic multipolar expansions, namely the lowest-order non-vanishing moment of such expansion is invariant with respect to the choice of origin. The lowest-order non-vanishing moment, in general, might be the molecular charge (zero'th-order moment), dipole (first-order moment) or quadrupole (second-order moment). The values of all moments of order higher than lowest non-vanishing order are, however, dependent upon the origin one chooses to perform the moment expansion. Therefore, for molecules of zero net charge and dipole moment of finite value, the quadrupole moments will depend upon the choice of origin.

So the question asked was: could one find a reference origin that would enable comparison of the quadrupole moments of different molecules with zero net charge and non-vanishing dipole moment? An answer to this [3] was found within the context of discussion concerning the so-called centers of the various electrostatic multipolar moments [4], namely center-of-charge, center-of-dipole, center-of-quadrupole ..., and a general scheme was developed to enable comparison between the moments of order higher than lowest non-vanishing order. Details of this will be summarized in the next section and can be found in the earlier paper [3].

Enabling comparison between the quadrupole moments of different molecules then provided a 'complete set' of molecular descriptors comprising the molecular moments of mass (shape) and charge up to and including second order. Consequently, the next thought was: having such 'complete set', how would it perform as QSAR descriptors on sets of molecules previously investigated by other 3D QSAR procedures? Our original expectation concerning such performance was not great; however, the results — surprisingly good — formed the basis of a following publication [5]. The original motivation was not to provide a small set of descriptors that would perform well in exclusion of other descriptors — e.g. partition coefficient, substituent constants etc. — but to provide a succinct set of descriptors that would simply characterize the three-dimensional information contained in the moment descriptors of molecular mass and charge up to and inclusive of second order. The 3D QSAR analysis utilizing these moment descriptors exclusively was called Comparative Molecular Moment Analysis (CoMMA) and the concise set of descriptors utilized were referred to as CoMMA descriptors. Such small set of descriptors could easily be amplified to incorporate other molecular features relevant to drug delivery and receptor site binding.

The present chapter will review and summarize some of the issues involved in the development and utilization of CoMMA descriptors in similarity assignments and in 3D QSAR drug recovery.

2. Quadrupolar Moments: Center-of-Dipole

The charge distribution of a molecule can be characterized by its multipolar components [2]. These components are elements in an infinite series whose sum up to some finite order approximates the electrostatic far-field potential — i.e. far field in the sense that the distance at which the potential is sampled is large compared with the extent of the molecular charge distribution.

In general, the partitioning of the far-field electrostatic potential among the various terms in the expansion depends upon the origin chosen to perform the multipolar decomposition. While it is true that the lowest-order non-vanishing moment does not depend upon the choice of origin of expansion, the contribution that this moment makes to the field at any particular location in space does depend upon the location of the expansion center — i.e. the center about which the expansion is performed and the moments calculated.

For example, consider multipolar expansions performed about two different locations in space for a neutral molecule with a dipole moment of significant magnitude. One of the expansion centers will be chosen somewhere in the vicinity of the molecule, while the others will be chosen somewhat distant from the molecule. One might then ask for the dipolar contribution to the electrostatic potential at points distant from both expansion centers. Sampling a number of such 'far-field' points in space, one would find that at a majority of these points the dipolar contribution from the expansion center that is closer to the molecule is a better approximation to the total electrostatic potential than the dipolar contribution from the expansion center distant to the molecule. As one examines all such far-field locations on a sphere of given radius from the expansion center, there will be a unique center of multipolar expansion such that the solid angle average of the squared deviation of the total far-field potential from the dipolar contribution to this potential is a minimum. Formally stated, one minimizes:

$$\frac{1}{\Omega}\oint d\Omega\,(\hat{x})\left|V(\vec{x}) - V_1(\vec{x},\vec{x}_0)\right|^2$$

with respect to the choice of expansion center to find such unique center, where $V(\vec{x})$ is the actual potential, $V_1(\vec{x};\vec{x}_0)$ is the dipolar potential with the center placed at $\vec{x}_0$ and the integral forms the solid angle average at some fixed distance from $\vec{x}_0$.

This center of expansion is then aptly named the center-of-dipole. For multipolar expansions performed about this center, the electrostatic dipolar potential most closely approximates the total far-field potential in an averaged sense. For a dipole moment vector, $\vec{p}$, and a quadrupolar tensor, $\boldsymbol{Q}$, calculated about an arbitrary origin, the displacement from this origin to the center-of-dipole is given by:

$$\vec{x}_0 = \frac{2}{3p^2}\left[\boldsymbol{Q}\cdot\vec{p} - \left(\frac{\vec{p}\cdot\boldsymbol{Q}\cdot\vec{p}}{4p^2}\right)\vec{p}\right]$$

The direction of the dipole and principal quadrupolar axes exhibit an interesting relationship when moments are calculated about the center-of-dipole. The dipole points

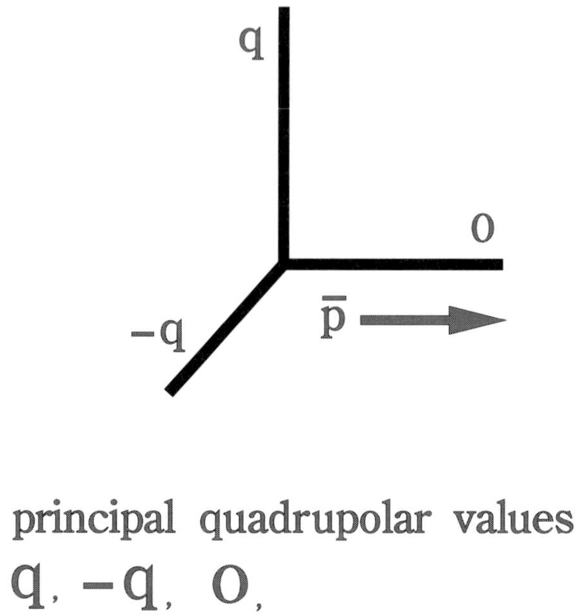

Fig. 1. Dipole direction, principal quadrupolar axes and values about the 'center-of-dipole'.

along the principal axis associated with zero quadrupole moment (Fig. 1). The two remaining principal quadrupolar components are equal in magnitude and opposite in sign as a consequence of the tracelessness or zero sum of the diagonal components of the quadrupolar tensor.

Multipolar expansion with the center-of-dipole as origin and in the frame of the quadrupolar principal axes, therefore, provides molecular electrostatic field descriptors that are independent of the orientation of the molecule in space. Up to and inclusive of second order in the moment expansion, these are the dipole and principal quadrupole magnitude, as well as the orientation of the quadrupolar principal axes with respect to the molecule.

The analogy between center-of-mass and center-of-dipole is not precise. Such analogy is more apt between the center-of-mass and center-of-charge. For ions — i.e. charged molecular species (non-vanishing zero'th-order moment) — one may zero out the dipolar contribution (first-order moment) to the electrostatic field by choice of the expansion center to obtain the more familiar 'center-of-charge' [4]. At this center, the monopolar electrostatic potential most closely approximates the total far-field potential in the averaged sense, as described previously for the dipolar electrostatic potential of neutral molecules. The center-of-mass and center-of-charge are then both defined by zeroing out the first-order moment of their respective distributions.

3. CoMMA Descriptors

Therefore, for neutral polar molecules, we have a set of well-defined molecular descriptors obtained from the moment expansions up to and including second order. The molecular weight, the three moments of inertia, I_x, I_y, I_z, the magnitude of the dipole moment, p, and magnitude of the principal quadrupole moment, Q, comprise six molecular moment descriptors.

The presence of two sets of axes, namely the inertial and principal quadrupolar axes, provides the further possibility of defining descriptors that succinctly describe the relationship between moments of the mass (shape) and charge distributions of the molecule. These additional descriptors may be defined in a number of different ways. In previous work [5], this additional set was defined as follows: the magnitudes of the dipolar components, as well as the magnitudes of the components of displacement between the center-of-mass and center-of-dipole, were calculated with respect to the principal inertial axes. This provides six descriptors, namely p_x, p_y, p_z and d_x, d_y, d_z. Two additional quadrupolar components, Q_{xx} and Q_{yy}, were calculated with respect to a translated inertial reference frame whose origin coincides with the center-of-dipole. The tracelessness (zero sum of the diagonal components of the quadrupolar tensor) precludes use of one of the diagonal tensor components as an independent variable. Use of the magnitudes, as well as a limited number of quadrupolar descriptors, was a consequence of the unsensed nature of the principal inertial axes — ‘unsensed’, in that positive and negative directions are not assigned to the axes. The axes may be sensed by utilizing information from higher-order moments or by reference to common structural molecular features.

The set of CoMMA descriptors, 14 as enumerated, is a set of three-dimensional internal molecular moment descriptors that are independent of molecular rotations and translations in space. Molecular superposition, alignment or registration is, therefore, not essential when comparing the descriptors of different molecules.

While it is formally satisfying to enable the use of molecular moment descriptors inclusive of second order in connection with similarity comparisons between different molecules, the pragmatic value of such analysis with respect to molecular chemical/pharmacological activity remains. This concern motivated the examination of several molecular series that had been previously investigated by other 3D QSAR procedures, namely steroids [6–8], imidazoles [9,10], benzoic acids [9,11], beta-carboline, pyridodiindoles and GGS compounds [9,12] and a set of non-nucleoside HIV inhibitors of current interest, the TIBO series [13].

Comments on the 3D QSAR of these series will be delayed to the next section. However, we will use these results to illustrate the correlations between the descriptors. The five sets of molecules are comprised of 165 molecules. Table 2 shows the correlation matrix for the set of 14 descriptors calculated with *ab initio* results for the combined set of 165 molecules. We have included mass or molecular weight as a descriptor which had not been included in the earlier analysis. Certain of the correlations are apparent, namely between molecular weight and the inertial moments. Some correlations are less apparent, namely between the inertial moments and principal

Table 2 *Correlation matrix for 165 structurally diverse molecules*

	p_x	p_y	p_z	p	q	q_{xx}	q_{yy}
p_x	1.0000	–0.1557	0.0168	0.6728	0.1789	–0.1730	0.3735
p_y	–0.1557	1.0000	–0.0663	0.5820	0.0731	0.5635	–0.1779
p_z	0.0168	–0.0663	1.0000	0.0989	0.2498	–0.2383	–0.0162
p	0.6728	0.5820	0.0989	1.000	0.2317	0.2721	0.1764
q	0.1789	0.0731	0.2498	0.2317	1.0000	–0.0682	0.5795
q_{xx}	–0.1730	0.5635	–0.2383	0.2721	–0.0682	1.0000	–0.1048
q_{yy}	0.3735	–0.1779	–0.0162	0.1764	0.5795	–0.1048	1.0000
I_x	0.0719	0.4615	0.2756	0.4479	0.4338	0.2500	0.0720
I_y	0.3536	–0.0702	0.3361	0.2894	0.5801	–0.2267	0.2623
I_z	0.3374	0.0493	0.3027	0.3607	0.6049	–0.1211	0.2787
d_x	0.0307	–0.2652	0.2064	–0.1445	0.2552	–0.4482	–0.0126
d_y	–0.0838	–0.1743	–0.1171	–0.2057	0.3207	–0.0259	–0.0751
d_z	0.1078	–0.2659	0.3422	–0.0525	0.3808	–0.4353	0.2060
m	0.2459	0.1229	0.3872	0.3454	0.5807	–0.0430	0.1896

	I_x	I_y	I_z	d_x	d_y	d_z	m
p_x	0.0719	0.3536	0.3374	0.0307	–0.0838	0.1078	0.2459
p_y	0.4615	–0.0702	0.0493	–0.2652	–0.1743	–0.2659	0.1229
p_z	0.2756	0.3361	0.3027	0.2064	–0.1171	0.3422	0.3872
p	0.4479	0.2894	0.3607	–0.1445	–0.2057	–0.0525	0.3454
q	0.4338	0.5801	0.6049	0.2552	0.3207	0.3808	0.5807
q_{xx}	0.2500	–0.2267	–0.1211	–0.4482	–0.0259	–0.4353	–0.0430
q_{yy}	0.0720	0.2623	0.2787	–0.0126	–0.0751	0.2060	0.1896
I_x	1.0000	0.6676	0.7690	0.2810	0.3169	0.2726	0.8394
I_y	0.6676	1.0000	0.9833	0.5594	0.3572	0.5762	0.9227
I_z	0.7690	0.9833	1.000	0.5087	0.3751	0.5196	0.9501
d_x	0.2810	0.5594	0.5087	1.0000	0.4603	0.6078	0.4698
d_y	0.3169	0.3572	0.3751	0.4603	1.0000	0.2306	0.3499
d_z	0.2726	0.5762	0.5196	0.6078	0.2306	1.0000	0.4720
m	0.8394	0.9227	0.9501	0.4698	0.3499	0.4720	1.0000

quadrupolar moment. The message, however, is that if one performs a 3D QSAR calculation with such set of descriptors, the analysis should consider the significant correlated nature of the descriptors. Independent of whether the number of data points is larger or smaller than the number of initial descriptors, it is essential to reduce the number of descriptors from the initial number to eliminate collinear descriptor combinations that impact the predictability negatively due to noise or spurious systematic variations. This can be accomplished by principal component regression (PCR) or partial least-squares (PLS) procedures.

4. 3D QSAR

Prior to the examination and discussion of results, an important caveat is in order. It must be appreciated that even though the identification of a center that can be used for the purpose of similarity comparison between the higher-order moments of electrostatic multipole expansions is formally correct, aside from other issues enumerated in the literature — e.g. conformer selection, solution effects, etc. — there is no guarantee that these moments or any other electrostatic moment required for the 3D QSAR can presently be calculated to an accuracy required to suggest chemical/pharmacological predictability. Difficulties in calculated dipole moments have been well documented and computational results that approach experimental values have been achieved only with higher-level quantum chemistry calculations than those performed on the sets of molecules referred to in the previous section. They have also been performed on molecules with many fewer atoms. The calculation of quadrupole moments is of an even higher order of difficulty and, again, has only been partially successful on small molecular species. The difficulties so encountered in calculating electrostatic moments accurately are a consequence of the close cancellation between the charged nuclear cores and electron distribution in space and the relatively inaccurate manner in which this distribution can at present be calculated. It is this cancellation between the effects of charges of opposite sign which determines the net electrostatic far-field and, in turn, the electrostatic moments. This is a difficulty not encountered in calculating inertial moments.

With this in mind, we had proceeded to obtain electrostatic moments from several different calculations with the objective of indicating that QSAR predictability is not an artefact of any single set of calculated moments, but a mirror of systematic variations in the electrostatics within a molecular series.

Calculations were performed on the following five molecular series: (a) 31 steroids with corticosteroid binding data [6,7,8]; (b) 37 β-carboline, pyridodiindole, and CGS compounds with affinity for the benzodiazepine receptor inverse agonist site [9,12]; (c) 15 substituted imidazoles with dissociation constant (pK_a) data [9,10]; (d) 49 substituted benzoic acids with Hammett sigma constant data [9,11]; and (e) 33 non-nucleoside reverse transcriptase HIV-1 inhibitors (NNRTI's) of the TIBO related series, with measured inhibition of cytopathic effects of HIV-1 in MT-4 cells [13].

A systematic search [14] was performed for conformer selection and the lowest energy conformer chosen for the QSAR study. A final force-field optimization was subsequently performed. Dipole and quadrupole moments were calculated by three different procedures. One method utilized the assignment of Gasteiger-Marsili charges [15] at the atomic sites. Another procedure utilized Mulliken partial charges from an AM1 MOPAC calculation [16]. The molecular dipole and quadrupolar components were then obtained by performing the appropriate sums over the atomic partial charges. Finally, Gaussian 92 [17] *ab initio* calculations were performed with either an STO-3G* or 6-31G* basis. The *ab initio* electrostatic moments are calculated from the extended electronic charge distribution associated with the molecular orbitals.

The steps of the procedure for performing the 3D QSAR calculations can then be summarized: one generates the structures and chooses the conformers to be used in the study. One then calculates the center-of-mass and determines the principal inertial components and axes for each of the conformers about their centers. Using the calculated dipolar and quadrupolar components for an arbitrary Cartesian frame of reference, the center-of-

Table 3 *Partial least squares (PLS) cross-validated q^2's*

	31 steroids (cbg)	15 imidazoles	49 benzoic acids	37 carbolines, etc.	33 TIBOS
Ab initio	0.689 (6)	0.698 (2)	0.690 (13)	0.394 (2)	0.358 (2)
MOPAC	0.675 (8)	0.722 (2)	0.567 (5)	0.378 (2)	0.374 (2)
Gasteiger	0.559 (1)	0.788 (2)	0.398 (1)	0.493 (4)	0.468 (1)

dipole is calculated for each conformer and the principal quadrupolar moments and axes obtained about this center. Dipolar, quadrupolar and displacement descriptors are then calculated with reference to the principal inertial axes translated such that its origin is superimposed on the center-of-dipole. This yielded a set of 13 descriptors used in the previous study [5]. Partial least-squares (PLS) analysis was then performed with the cross-validation 'leave-one-out' procedure. Table 3 summarizes the results obtained for the five different molecular series that were investigated with the different moment assignments; the number of optimal PLS components is listed in parentheses.

Fifteen imidazoles had been included in the training set treated previously [9,10]. For this molecular series, only 11 descriptors have been utilized, since all of the 15 molecular structures are essentially planar, the only atoms above or below the molecular plane being hydrogen atoms associated with alkane substituents. For this molecular series, the

Table 4 *Substituted imidozoles with pK_a data*

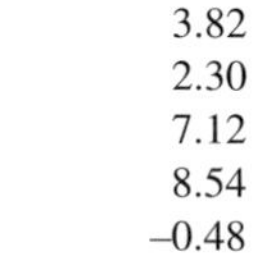

No.	R_1	R_2	pK_a	q_{xx}	q_y
1	CH_3	Br	3.82	0.837	0.025
2	CH_3	F	2.30	0.640	0.042
3	CH_3	H	7.12	0.135	1.247
4	CH_3	NH_2	8.54	0.663	0.042
5	CH_3	NO_2	–0.48	–0.244	–0.576
6	H	Br	3.79	0.898	0.153
7	H	Cl	3.55	0.772	0.161
8	H	C_2H_5	7.73	1.228	0.038
9	H	F	2.40	0.681	0.044
10	H	H	6.99	1.338	0.114
11	H	CH_3	7.86	1.344	0.017
12	H	NH_2	8.46	1.992	0.377
13	H	NO_2	–0.81	–0.408	–0.669
14	H	C_6H_5	6.48	2.794	0.000
15	H	SCH_3	5.95	2.229	0.001

inclusion of the quadrupole descriptors makes the greatest impact on the calculated q^2 for correlating with the pK_a data. With only the two components, q_{xx} and q_{yy}, the calculated q^2 is 0.69. Table 4 lists the imidazole structures, pK_a values and values of the two quadrupolar descriptors, q_{xx} and q_{yy}. When these two descriptors, as well as the principal quadrupolar moment Q, are deleted from the descriptor set of 11 values, the PLS leave-one-out calculated q^2 value is reduced to 0.24.

Comparison of cross-validated q^2's for a particular molecular series calculated with several different charge distributions is not sufficient to guarantee consistency. It is also necessary to compare the selectivity of the descriptors in correlating with the chemical/biological activity variances. In the following, *ab initio* moment calculations have been used to provide a base-line for the examination of descriptor selectivity. It should be recalled that moments obtained from these calculations are not derived from a partitioning of the charge distribution at the atomic sites, but are calculated from the distribution of electronic charge associated with the atomic basis functions.

Table 5 illustrates PLS results obtained by selecting the subset of *ab initio* CoMMA descriptors from the original 13 that optimize the q^2 for each of the five molecular series indicated. The original cross-validated leave-one-out value is given with an arrow indicating the optimization achieved by selecting the set of descriptors indicated. Results are also provided for MOPAC and Gasteiger CoMMA descriptors. The MOPAC and Gasteiger results do not, however, represent the optimization that can be achieved within each descriptor set, but indicate the q^2 value achieved by utilizing the descriptor set that optimizes the *ab initio* results, namely the set shown in Table 5. The only significant deterioration noted is associated with the Gasteiger result for the imidazoles. This indicates that the *ab initio* and Gasteiger CoMMA vectors select differently to reproduce the variances in observed activity for this molecular series.

CoMMA descriptors need not be utilized solely in 3D QSAR investigations where the number of molecules is relatively small. Such descriptors might be of value in issues related to large-scale screening or molecular diversity. For such applications, it will be necessary to utilize charge assignments that can be made rapidly. The rapid assignment of molecular charge has been a subject of continued interest [15,18,19].

5. Phosphodiesterase PDE Type III Inhibitors

An interesting example where the electrostatic moment descriptors were not found to correlate with a set of binding activity measurements is provided by the phosphodiesterase PDE type III inhibitors. This example is of interest since comparison of the electrostatic potential profiles of several of these inhibitors with the profiles of adenosine and guanosine monophosphates, the natural substrates, indicates registration of similar regions of electrostatic minima and maxima, thereby implicating electrostatic interactions as performing a fundamental role in the binding of the ligands to the receptor site [20,21]. The calculations involved comparison between protonated cyclic-amp and the neutrally charged inhibitors. Binding activity measurements [22] of the inhibitors yielding IC_{50} data were available, hence it was possible to perform a CoMMA analysis on a select set of the specific inhibitors.

Table 5 *Enhanced q^2 utilizing selected descriptors*

	31 steroids (cbg)	15 imidazoles	49 benzoic acids	37 carbolines, etc.	33 TIBOS
Ab initio	0.689 (6) → 0.753 (7)	0.698 (2) → 0.772 (2)	0.690 (13) → 0.729 (6)	0.394 (2) → 0.446 (4)	0.358 (2) → 0.489 (8)
	p, Q_{xx}, I_x, I_y, I_z, d_x, d_y	Q_{xx}, Q_{yy}, I_x, I_y, I_z	p_z, p, Q_{yy}, I_x, I_y, I_z, d_x	px, p_y, p, Q, Q_{yy}, I_x, $I_y I_z$	p_x, p, Q, Q_{xx}, I_x, d_x, d_y, d_z
MOPAC	0.675 (8) → 0.576 (3)	0.722 (2) → 0.728 (5)	0.567 (5) → 0.506 (3)	0.378 (2) → 0.413 (2)	0.374 (2) → 0.418 (2)
Gasteiger	0.559 (1) → 0.555 (7)	0.788 (2) → 0.296 (5)	0.398 (1) → 0.456 (6)	0.493 (4) → 0.542 (5)	0.468 (1) → 0.339 (1)

Table 6 *Thirty type-III-specific phosphodiesterase inhibitors*

amrinone	anagrelide	dazanone
fenoximone	milrinone	piroximone
skf92345	skf92538	skf92600
skf93403	skf93505	skf93639
skf93741	skf93880	skf94120
skf94442	skf94928	skf94934
skf95190	skf95230	skf95294
skf95295	skf95641	skf95648
skf95742	skf95764	skf95778
skf95779	skf95800	skf95807

Thirty type-III specific phosphodiesterase inhibitors [20] were chosen for investigation (Table 6). The choice involved a selection that spanned the limited range of activity reported for the entire series [20], approximately three orders of magnitude, and neglected certain of the larger more complex structures. Three structures spanning the range of activity are shown in Fig. 2. The majority of the more complex structures were not included in the analysis due to ambiguity in the choice of conformation. Most of the structures included in the analyses had few, if any, rotatable bonds. A systematic conformational search [14] was performed on each of the structures, as well as a final force-field optimization of the lowest force-field energy structure identified by the search. QSAR analyses were performed on the 30 structures with several different sets of CoMMA descriptors, as well as with the utilization of Gasteiger [15], Charge Equilibration [23] and MOPAC charges [16]. All results indicated that the only descriptor correlating with activity was the molecular weight of the molecule. Elimination of molecular weight and inertial moments from the descriptor set yielded a leave-one-out cross-validation result no better than obtained by using the average dosage as a predictor — i.e. essentially a q^2 of zero. Using the single descriptor of molecular weight yields a leave-one-out cross-validated q^2 of 0.58.

It is somewhat surprising that the electrostatic moment descriptor variances provide no correlation with the activity variances; however, such result is not inconsistent with

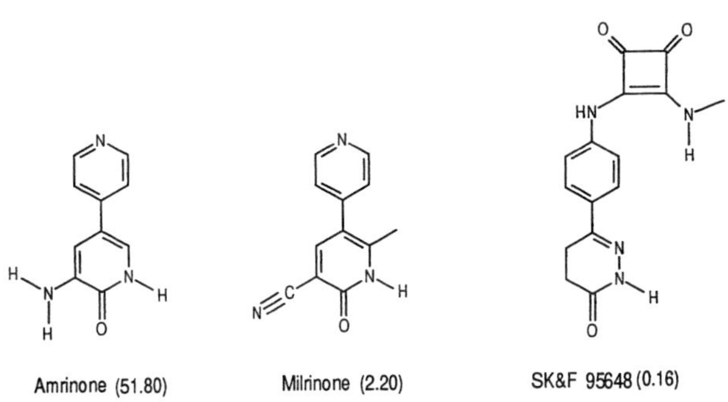

Fig. 2. Three phosphodiesterase PDE type-III structures (IC_{50} μM in parentheses).

previous findings — e.g. that 'calculations of charge, dipole moment and molecular orbital coefficients around the cyclic amide ... could not explain the relative affinities' [20]; that the difference in activity between the bipyridines, amrinone and milrinone might plausibly be associated with 'the steric interaction between the methyl substituent and the 3′,5′ hydrogen atoms of the monosubstituted pyridine ring' [24], thereby implicating steric features; and that the 'optimal interaction probably occurs through a center at a greater distance from the cyclic amide group' [20].

This result for the phosphodiesterases contrasts with results obtained for the five series treated in the previous section where the electrostatic descriptors were found to make a significant contribution to the cross-validated q^2's.

6. Summary

This chapter has reviewed certain concepts involving the identification of an expansion center that can be utilized for molecular similarity comparison between electrostatic moments of order higher than lowest non-vanishing order. It has also described how such information has been used in 3D QSAR studies and the predictive results achieved.

It should be emphasized that the inertial and electrostatic moments of a molecule are fundamental molecular characterizations that relate directly to how molecules respond to both mechanical and electrical forces. Such moments describe global molecular three-dimensional information at a most elemental level. On the other hand, the utility of such information with respect to drug discovery is in a preliminary stage of evaluation.

Several of the issues that remain to be addressed are:

1. Can the electrostatic moments be calculated with sufficient accuracy to be reliably used in general 3D QSAR investigations? In addressing massive molecular databases, can the moments be assigned rapidly and accurately? what is a lower bound on dipole moment magnitude to provide computational accuracy?
2. Will the CoMMA descriptors provide useful information with respect to molecules that consist of a greater number of rotatable bonds than those presently investigated? For large molecular databases, does the small number of CoMMA descriptors enable one to treat the conformer degrees of freedom by calculations on the fly?
3. What is the best set of descriptors to predict the activity of drugs; will higher-order moment information provide an enhancement of predictability — e.g. sensing the principal axes? How might the CoMMA descriptor set be amplified to enhance pharmacological predictability?

These as well as other issues remain to be addressed. On the other hand, having the ability to compare the higher-order electrostatic moments of different molecules, we believe, provides an enhanced perspective with respect to 3D QSAR in drug discovery.

Acknowledgement

One of the authors (B.D.S.) would like to thank Professor S.L. Price for suggesting the phosphodiesterases as an interesting molecular series for investigation.

References

1. Goldstein, H., Classical mechanics, 2nd Ed., Addison Wesley, New York, 1980.
2. Jackson, J.D., Classical electrodynamics, 2nd Ed., John Wiley, New York, 1975.
3. Platt, D.E. and Silverman, B.D., *Registration, orientation and similarity of molecular electrostatic potentials through multipole matching*, J. Comp. Chem., 17 (1996) 358–366.
4. Buckingham, A.D., *Permanent and induced molecular moments and long-range intermolecular forces*, In Hirschfelder, J.O. (Ed.) Advances in chemical physics, Vol. 12, Interscience Publishers, a division of John Wiley & Sons, New York-London-Sydney, 1967, p. 107.
5. Silverman, B.D. and Platt, D.E., *Comparative molecular moment analysis (CoMMA): 3D QSAR without molecular superposition*, J. Med. Chem., 39 (1996) 2129–2140.
6. Cramer III, R.D., Patterson, D.E. and Bunce, J.D., *Comparative molecular field analysis (CoMFA): Effect of shape on binding of steroids to carrier proteins*, J. Am. Chem. Soc., 110 (1988) 5959–5967.
7. Good, A.C., Sung-Sau, S. and Richards, W.G., *Structure–activity relationships from molecular similarity matrices*, J. Med. Chem., 36 (1993) 433–438.
8. Jain, A.N., Koile, K. and Chapman, D., *Compass: Predicting biological activities from molecular surface properties–performance comparisons on a steroid benchmark*, J. Med. Chem., 37 (1994) 2315–2327.
9. Good, A.C., Peterson, S.J. and Richards, W.G., *QSARs from similarity matrices: Technique validation and application in the comparison of different similarity evaluation methods*, J. Med. Chem., 36 (1993) 2929–2937.
10. Kim, K.H. and Martin, Y., *Direct prediction of dissociation constants (pKa's) of clonidine-like imidazoles, 2-substituted imidazoles, and 1-methyl-2-substituted-imidazoles from 3D structures using a comparative molecular field analysis (CoMFA) approach*, J. Med. Chem., 34 (1991) 2056–2060.
11. Kim, K.H. and Martin, C.M., *Direct prediction of linear free energy substituent effects from 3D structures using comparative molecular field analysis: 1. Electronic effects of substituted benzoic acids*, J. Org. Chem., 56 (1991) 2723–2729.
12. Allen, M.S., Tan, Y. and Trudell, M. Ml., Narayanan, K., Schindler, L.R., Martin, M.J., Schultz, C., Hagen, T.J., Koehler, K.F., Codding, P.W., Skolnick, P. and Cook, J.M., *Synthetic and computer-assisted analyses of the pharmacophore for the benzodiazepine receptor inverse agonist site*, J. Med. Chem., 33 (1990) 2343–2357.
13. Breslin, H.J., Kukla, M.J., Ludovici, D.W., Mohrbacher, R., Ho, W., Miranda, M., Rodgers, J.D., Hitchens, T.K., Leo, G., Gauthier, D.A., Ho, C.Y., Scott, M.K., De Clercq, E., Pauwels, R., Andries, K., Janssen, M.A.C. and Janssen, P.A.J., *Synthesis and anti-HIV-1 activity of 4,5,6,7-tetrahydro-5-methylimidazo- [4,5,1-jk][1,4]benzodiazepin-2(1H)-one (TIBO) derivatives: 3*, J. Med. Chem., 38 (1995) 771–793.
14. 'Systematic search' option under SYBYL 6.01, available from TRIPOS Associates Inc., 1699 S. Hanley Road, St. Louis, MO 63144, U.S.A. All molecular modeling was performed using SYBYL.
15. Gasteiger, J. and Marsili, M., *Iterative partial equalization of orbital electronegativity — a rapid access to atomic charges*, Tetrahedron, 36 (1980) 3219–3288.
16. Stewart, J.J.P., *MOPAC: A semiempirical program*, J. Comput.-Aided Mol. Design, 4 (1990) 1–105.
17. Frisch, M.J., Trucks, G.W., Head-Gordon, M., Gill, P.M.W., Wong, M.W., Foresman, J.B., Johnson B.G., Schlegel, H.B., Robb, M.A., Replogle, E.S., Gomperts, R., Andres, J.L., Raghavachari, K., Binkley, J.S., Gonzalez, C., Martin, R.L., Fox, D.J., Defrees, D.J., Baker, J., Stewart, J.J.P. and Pople, J.A., Gaussian 92; Revision C, Gaussian Inc., 4415 Fifth Avenue, Pittsburgh, PA 15213, U.S.A.

18. Abraham, R.J. and Grant, G.H., *Charge calculations in molecular mechanics: 10. A general parameterisation of the scheme for saturated halogen, oxygen, and nitrogen compounds*, J. Comput.-Aided Mol. Design, 6 (1992) 273–286.
19. Rappe, A.K. and Goddard III, W.A., *Charge equilibration for molecular dynamics simulations*, J. Phys. Chem., 95 (1991) 3358–3363.
20. Davis, A., Warrington, B.H. and Vinter, J.G., *Strategic approaches to drug design: 2. Modeling studies on phosphodiesterase substrates and inhibitors*, J. Comput.-Aided Mol. Design, 1 (1987) 97–120.
21. Apaya, R.P., Lucchese, B., Price, S.L. and Vinter, J.G., *The matching of electrostatic extrema: A useful method in drug design? A Study of phosphodiesterase III inhibitors*, J. Comput.-Aided Mol. Design, 9 (1995) 33–43.
22. Reeves, M.L., Leigh, B.K. and England, P.J., *The identification of a new cyclic nucleotide phosphodiesterase activity in human and guinea-pig cardiac ventricle*, Biochem. J., 241 (1987) 535–541.
23. The Rappe-Goddard charge equilibration procedure is available with Cerius2, distributed by Molecular Simulations, Inc., 9685 Scranton Road, San Diego, CA 92121, U.S.A.
24. Robertson, D.W. and Boyd, D.B., *Structural requirements for potent and selective inhibition of low-K_m, cyclic-AMP-specific phosphodiesterases*, Adv. in Second Messenger and Phosphoprotein Res., 25 (1992) 321–340.

Part III

3D QSAR Applications

The CoMFA Steroids as a Benchmark Dataset for Development of 3D QSAR Methods

Eugene A. Coats
Amylin Pharmaceuticals, Inc., 9373 Towne Centre Drive, San Diego, CA 92121, U.S.A.

1. Introduction

The publication of Comparative Molecular Field Analysis (CoMFA) in 1988 by Cramer et al. [1] ushered in a new era in quantitative structure–activity methodology by offering the possibility of dealing effectively with ligand–receptor interactions in three dimensions. The success of CoMFA and the acceptance of this methodology is attested to by the hundreds of investigations using the procedure to describe three-dimensional structure–activity relationships and to predict structural modifications for optimizing activities. CoMFA has become a very effective tool among the methods available for computer-assisted drug design, as the reader will note in chapters elsewhere in this volume. The CoMFA method itself will not be dwelt upon here, but rather it is the intent of this discussion to focus on analyses of the steroid dataset used for the initial description of the method. A search of the literature reveals a number of papers which make use of the original CoMFA steroid dataset as a means to compare modifications of the CoMFA method, as well as completely different approaches to the development of 3D QSAR. Thus, this set of steroids has become somewhat of a 'benchmark' against which investigators have attempted to measure the success (or failure) of alternative procedures.

2. The Steroid Dataset

The original data on the steroids were taken from two papers. In the first by Dunn et al. [2], the binding affinities of 21 steroids for testosterone-binding globulin (TeBG) and for corticosteroid-binding globulin (CBG) were determined. The binding data in the form of affinity constants and the steroid names are reproduced in Table 1, along with compound numbers to be used throughout these discussions. As these data are affinity constants, the larger numbers reflect higher affinity for the binding protein. Thus, following QSAR convention, one would use log K as the form of the biological activity to be employed in any QSAR analysis. These values are also given in Table 1. The structures of the 21 steroids are shown in Fig. 1 with all asymmetric centers defined. The steroids listed in Table 1 served as the training set in the original CoMFA publication, as well as in many of the subsequent papers to be discussed.

In the second report, Mickelson et al. [3] determined the binding affinities and computed the free energies of binding of 47 steroids with human corticosteroid-binding globulin. Of these, 11 steroids were in common with the first paper (those with associated $-\Delta G°$ values in Table 1) and used to derive an equation relating the two studies. This equation was used to place the binding data from the two papers on the same scale to allow the selection of an additional set of 10 steroids as a test set for predictions. The

H. Kubinyi et al. (eds.), 3D QSAR in Drug Design, Volume 3. 199–213.

Table 1 *CoMFA steroid training set: measured binding activities*

No.	Steroid[a]	CBG[b] $K \times 10^6$	CBG Log K	CBG[c] $-\Delta G°$	TeBG[b] $K \times 10^6$	TeBG Log K
1	Aldosterone	1.9	6.279	8.77	0.21	5.322
2	Androstanediol	< 0.1	5		1300	9.114
3	Androstenediol	< 0.1	5		1500	9.176
4	Androstenedione	0.58	5.763		29	7.462
5	Androsterone	0.41	5.613		14	7.146
6	Corticosterone	76	7.881	11.41	2.2	6.342
7	Cortisol	76	7.881	11.24	1.6	6.204
8	Cortisone	7.8	6.892	9.78	2.7	6.431
9	Dehydroepiandrosterone	< 0.1	5		66	7.819
10	Deoxycorticosterone	45	7.653	11.22	24	7.380
11	Deoxycortisol	76	7.881	11.18	16	7.204
12	Dihydrotestosterone	0.83	5.919		5500	9.740
13	Estradiol	< 0.1	5	6.23	680	8.832
14	Estriol	< 0.1	5		4.3	6.633
15	Estrone	< 0.1	5		150	8.176
16	Etiocholanolone	0.18	5.255		1.4	6.146
17	Pregnenolone	0.18	5.255	7.24	14	7.146
18	17-hydroxypregnenolone	< 0.1	5		2.3	6.362
19	Progesterone	24	7.380	11.14	8.8	6.944
20	17-hydroxyprogesterone	55	7.740	11.17	9.9	6.996
21	Testosterone	5.3	6.724	9.78	1600	9.204

[a] See Fig. 1 for structures.
[b] Binding affinities in M^{-1} taken from reference [2].
[c] Free energies of binding, in kcal, taken from reference [3].

equation, re-derived here (Eq. 1) using JMP [4], is very similar to that first reported [1], although there is a slight difference in

$$\text{Log } K = 1.222 + 0.578\,(-\Delta G°) \tag{1}$$

$$n = 11 \qquad r^2 = 0.979$$

the intercept. Neither the original nor the re-derived equation gives the exact log K values used in the CoMFA paper [1]. The differences are insignificant with the exception of steroids **29** and **30**. The test set steroids are listed in Table 2, together with the three sets of log K values. The compound numbers for the test set have been assigned as **22–31**, as used in most subsequent reports, while those used in the original CoMFA report are also given, in parentheses, in an attempt to avoid confusion. The structures of the 10-steroid test set are shown in Fig. 2 with all asymmetric centers defined.

CoMFA was carried out [1] on the 21-steroid test set using what have become 'standard' CoMFA conditions. Deoxycortisol, **11**, was used as a template for alignment based on carbon atoms 3, 5, 6, 13, 14 and 17. Both steric and electrostatic fields at 2.0 Å

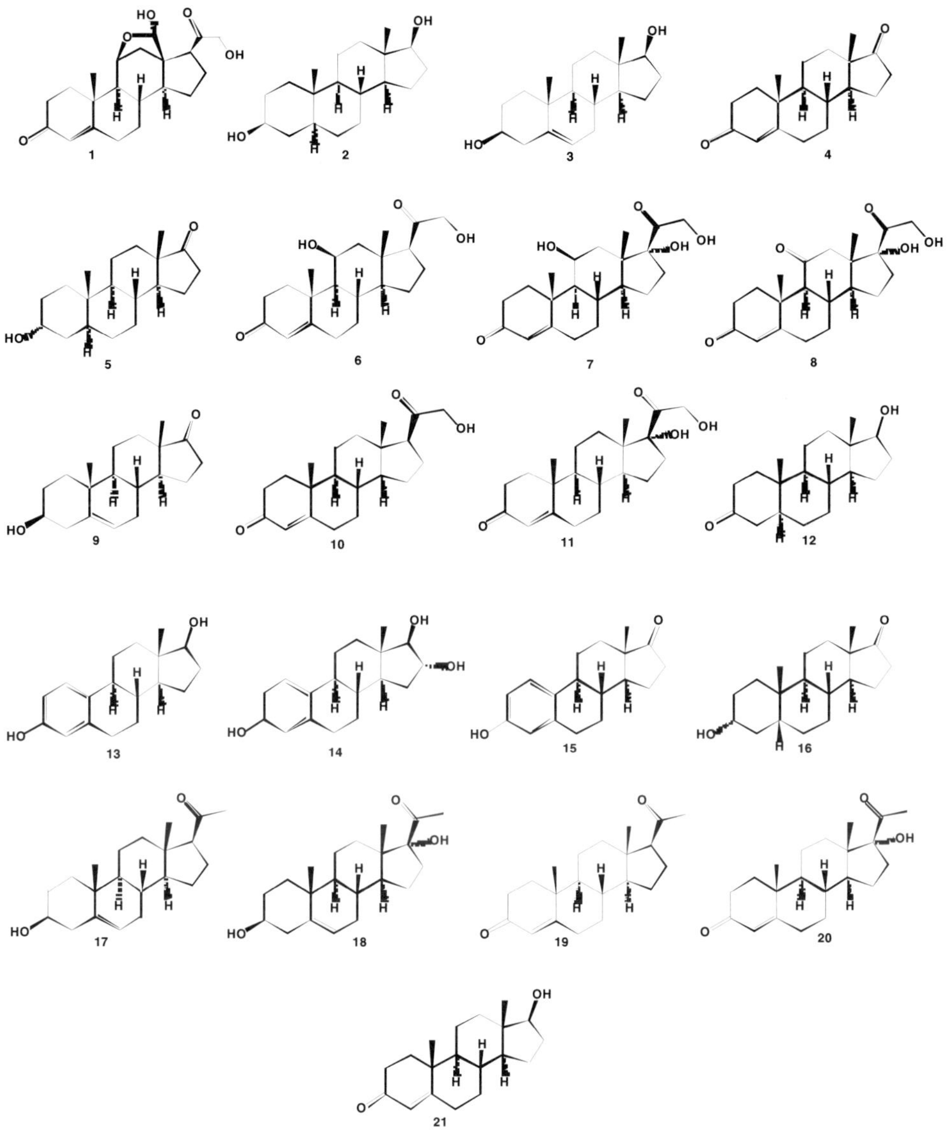

Fig. 1. Training set steroids.

resolution were employed. Four cross-validation groups were used instead of the easily reproducible leave-one-out procedure. For CBG binding data, the q^2 (cross-validated r^2) and PRESS at the two-component level were reported as 0.662 and 0.719, respectively. The q^2 and PRESS at the two-component level for TeBG binding were 0.555 and 0.849, respectively. The predicted CBG binding values for the 10-steroid test set using CoMFA derived under standard conditions, as well as those with different atom probes,

Table 2 *CoMFA steroid test set: calculated and predicted binding activities*

No.[a]	Steroid[b]	CBG[c] $-\Delta G°$	Calcd.[d] Log *K*	Calcd.[e] Log *K*	Calcd.[f] Log *K*	Pred.[g] Log *K*
22(1)	Prednisolone	10.88	7.512	7.517	7.514	7.336
23(2)	Cortisol-21-acetate	10.95	7.553	7.557	7.554	7.335
24(3)	4-pregnene-3,11,20-trione	9.61	6.779	6.783	6.779	6.961
25(4)	Epicorticosterone	10.34	7.200	7.205	7.201	7.704
26(5)	19-nortestosterone	8.51	6.144	6.147	6.143	6.838
27(6)	16α, 17-dihydroxy-4-pregnene-3,20-dione	8.69	6.247	6.251	6.247	6.788
28(7)	16α-methyl-4-pregnene-3,20-dione	10.21	7.120	7.129	7.126	7.214
29(8)	19-norprogesterone	9.78	6.817	6.881	6.877	7.269
30(9)	11β, 17, 21- trihydroxy-2α-methyl-4-pregnene-3,20-dione	11.15	7.688	7.673	7.670	7.629
31(10)	11β, 17, 21- trihydroxy-2α-methyl-9α-fluoro-4-pregnene-3,20-dione	7.91	5.797	5.800	5.796	7.676

[a] Numbers in parentheses are the steroid test set numbers used in reference [1].
[b] See Fig. 2 for structures.
[c] Free energy of binding, in kcal, from table V of reference [3].
[d] Log affinity constant from table IVB of reference [1].
[e] Log affinity constant computed from $-\Delta G°$ using: Log $K = 1.228 + 0.578\ (-\Delta G°)$, reference [1].
[f] Log affinity constant computed from $-\Delta G°$, using Eq. 1 in this work.
[g] Predicted by CoMFA from corrected training set steroids.

offset lattice definitions and variations in lattice spacing were reported. The use of this initial application of CoMFA and the steroid data as a benchmark for comparison has, unfortunately, been frustrated by a number of problems. First, as indicated above, the partial least-squares (PLS) analyses were conducted using four cross-validation groups. Since the algorithm selects these groups at random, it is virtually impossible to reproduce the cross-validated statistics, as opposed to the use of leave-one-out cross-validation where one achieves the same results each time.

A second, and far more serious difficulty was uncovered by Gasteiger and co-workers [5]. There were a large number of erroneous steroid structures included in the analyses – steroids **2**, **5**, **13**, **14**, **15**, **16**, **21** and **28** depicted in the figures of the paper [1]. Upon contacting the authors, it has been determined that the actual coordinates used for the 21-steroid training set are those currently available in the SYBYL modelling package [6] as a CoMFA tutorial, while the original coordinates of the 10 test set steroids are no longer available [7]. While this cannot be confirmed by cross-validated PLS, it is possible to recompute the results without cross-validation using the original CoMFA conditions found in the SYBYL file: 'comfa.demo'. For the 21 steroids, using PLS within SYBYL 6.3 gives r^2 (standard error) values of 0.878 (0.445) for the CBG binding data and 0.895 (0.400) for the TeBG binding data. These are essentially identical values of r^2 (standard error) as those of 0.873 (0.453) and 0.897 (0.397) for the CBG and TeBG data, respectively, found in reference [1]. It should be noted here that this SYBYL steroid dataset still contains one incorrect structure, that of androstanediol, **2**,

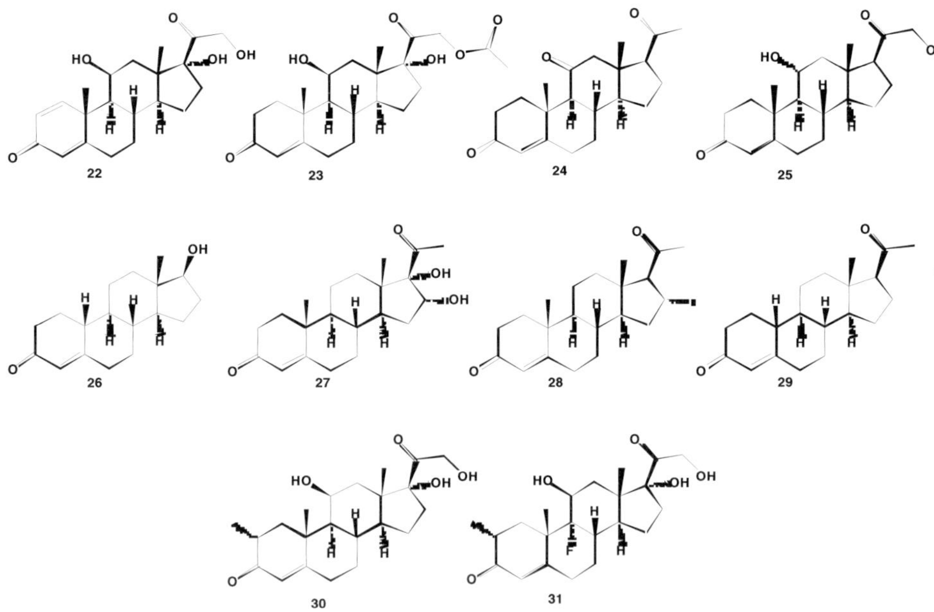

Fig. 2. Test set steroids.

where the 3-OH should be β and not α. Finally, it should be noted that the form of the biological activities in the paper is given as log 1/*K* (–log *K*), which while not erroneous, can be misleading when interpreting results. As indicated previously, the form log *K* is more appropriate here, since *K* increases with increasing affinity (activity).

Before turning to a discussion comparing analyses of the steroids, it was thought useful to recompute the CoMFA using the correct steroid structures given in Fig. 1. The androstanediol correction was made and a standard CoMFA computed without further modification to structures, or alignments. Steroid partial charges were those of Gasteiger and Marsili [8]. Combined steric and electrostatic fields at 2.0 Å resolution with a 30 kcal steric cutoff were used along with standard CoMFA scaling. For the PLS analysis, a ± 2.0 kcal filter was applied along with leave-one-out cross-validation. This afforded a q^2 (standard error of predictions) for the CBG data of 0.708 (0.668) and for the TeBG data of 0.601 (0.805), both at the two-component level. If one uses q^2 to select the optimal number of components, three is optimal for the CBG data giving 0.734 (0.657), while eight is optimal for the TeBG data giving 0.764 (0.758). Use of the CBG 21-steroid training set CoMFA for prediction of the 10-steroid test set (Fig. 2) gave results shown in Table 2.

3. Methods Applied to the Steroids

As indicated in the introduction, a number of investigators have examined modifications to the CoMFA procedures and fields, while others have devised quite different

3D QSAR methods applied to the steroid data. Many of these are described by the original authors elsewhere in this volume, so the details of each procedure will not be repeated here. Rather, the methods will be briefly summarized, with emphasis upon the statistical comparison with CoMFA, advantages or disadvantages in qualitative interpretation and indications of any errors in the steroid dataset employed.

Cross-validated R2-guided Region Selection (q^2-GRS), devised by Cho and Tropsha [9], is suggested as an alternative to GOLPE [10]. The method involves dividing the original CoMFA region into 125 small boxes from which are selected only those with q^2 above a specified cutoff level. These are then combined giving an altered region which should involve only those grid-points which are strongly related to the observed changes in biological activity. The method was applied to the TeBG and CBG binding data for the 21-steroid training set. The steroid structures and biological response data were reportedly taken directly from the SYBYL 6.0 tutorial without modification. Thus, one structural error, in androstanediol (**2**), was present in the analyses. The 'best' results as characterized by q^2 values were 0.658 for TeBG binding and 0.790 for CBG binding, both at the two-component level. Clearly, some improvement is offered by this procedure upon comparison with CoMFA results from the same coordinate set. Because the procedure is encoded in SYBYL programming language (SPL) [6], it can be readily investigated further by those using this modelling software. This publication did not include assessments of the predictive capabilities on the 10-steroid test set.

Norinder [11] has also examined possible ways to improve variable selection in CoMFA. In this study, both single mode (GOLPE) [10] and domain mode were evaluated. In single mode single grid-points were selected, while in domain mode boxes containing 3 or 4 grid-points were chosen. Variable selection was based upon the magnitude of the corresponding PLS regression coefficients. The 21-steroid dataset with CBG binding data was employed as a training set, while the ability of the process to make true predictions was checked using the 10-steroid test set. Both selection processes afforded high q^2 values but performed poorly in prediction of the test set. Direct comparison with standard CoMFA analyses of the steroid test set data is not possible here, because the tabular listing of data and steroid structural details in the paper reveal several errors. The structure for 16-α-methyl-4-pregnene-3,20-dione (**28**) is incorrect and there are errors in the experimental binding activities for compounds **16**, **17** and **26**.

Alternatives to the standard steric and electrostatic CoMFA fields were the subject of an investigation by Kellogg et al. [12]. In this work, electrotopological state (E-state) and hydrogen electrotopological state (HE-state) fields were developed and compared with steric, electrostatic and hydropathic (HINT) [13] fields for utility in CoMFA applied to the 21-steroid training set. CBG binding data and steroid structures were obtained from SYBYL 6.2, thus the previously mentioned structural error in androstanediol (**2**) was included in the analyses. Comparison is facilitated here, since the authors included all five types of fields — singly and in combination — in their evaluations. Additionally, both 1 Å and 2 Å field resolutions were considered. The quality of the correlations as measured in terms of q^2 values suggest that the new fields perform quite well: 0.803 at 1 Å resolution and three components for the combined E-state/

HE-state CoMFA as compared to 0.736 at 1 Å resolution and three components for the combined steric/electrostatic field CoMFA. Contour plots of the E-state/HE-state field CoMFA showed that changes in regions near the 3 and the 17 positions of the steroid nucleus were important in explaining the observed changes in CBG binding activities. It is important to note, however, that no prediction of the 10-steroid test set was attempted.

A series of reports have appeared in which the three-dimensional properties of a molecule are described by various procedures for mapping features or potential intermolecular interactions onto the surface of the molecule. While it is an oversimplification to suggest that these methods are similar, they do all differ from CoMFA, in that no box-like grid of interaction points is employed. In the first of these rather unique methods, Jain et al. describe Compass [14], a procedure which involves iterative selection of molecular poses, extraction of physico-chemical features computed near the van der Waals surface and construction of a statistical model, which explains the observed biological activity and can be used to predict the activities and bioactive poses of new molecules. The term 'pose' here refers to both the conformation and the alignment of a given molecule. The method employs a neural network to extract relevant features, as well as to improve pose selection and, thus, is capable of handling and developing nonlinear relationships. When Compass was applied to the 21-steroid training set, q^2 values of 0.89 for CBG binding activity and 0.88 for TeBG activity were obtained using combined steric and polar features. The resulting model was then applied to prediction of the CBG binding activities of the 10-steroid test set. The predictions were not good for the entire test set, primarily because of the quite poor prediction of steroid **31** which is the only one having a fluorine in the 9-position. Other investigators have also noted this. When the remaining nine steroids (**22–30**) were used as a test set, the predictions were quite good as assessed by a Kendall's τ value of 0.84. It must be noted at this point, however, that structure **28** of the test set contains an error, so that the predictions described are also not completely correct. There are also two errors in the biological activities given in the paper, namely the CBG binding activities for steroids **16** and **17** should be 5.255 in each instance. With the exception of the structural error, these are minor and do not detract from the intriguing results described by these authors.

In a study by Wagener et al. [5], molecular surface properties for the combined training and test set steroids were transformed into spatial autocorrelation descriptors as an alternative means of characterizing electrostatic potential. The utility of the autocorrelation vectors for the 31 steroids was investigated by principal component analysis, as well as through the use of Kohonen neural network maps. Both types of analyses afforded reasonably good classification of the CBG binding data into high, intermediate and low binding groups. Having demonstrated an apparent relationship between the spatial autocorrelation vectors and CBG binding, the new descriptors were then used as input for a multilayer back-propagation neural network. A leave-one-out cross-validation procedure was applied to the neural network analyses by running 31 separate experiments to gain an estimate of the quality of prediction. A q^2 of 0.63 was obtained with all 31 steroids, and a value of 0.84 with steroid **31** omitted. It should be noted that the CBG affinities for steroids **16** and **17**, respectively, were listed as 5.225 for each compound instead of the correct 5.255 value. This would have a slight but probably insignificant effect on these analyses, because the rank order of the steroid activities is not changed. Beyond the investigation of new methods, what is most intriguing about these results is the observation that electrostatic properties account for all of the changes in steroid binding in contrast to the CoMFA results where both electrostatic and steric effects influence activity. This apparent qualitative difference may simply suggest that the autocorrelation vectors include steric information from the molecular electrostatic potential mapped onto the van der Waals surface of the steroids.

In a more recent work, Gasteiger and co-workers [15] have investigated more fully the ability of Kohonen neural networks to be useful in mapping molecular surface properties into two dimensions and in facilitating a variety of comparisons. Arrangement of the two-dimensional Kohonen maps according to steroid binding affinity (CBG) provided a visual assessment of the ability of the method to classify compounds. Projection of the Kohonen maps back onto the van der Waals surface of the steroid helped to identify the steroid regions affecting binding.

Comparisons of shape and also a method of template comparison to generate a type of similarity analyses were presented. These offer a variety of qualitative methods to visualize the relationships between steroid structure and binding affinity offering an alternative to quantitative methods.

Hahn and Rogers [16] have also devised a method based upon molecular surfaces. This study involved the construction of a receptor surface model (RSM) from individual structures. The method was applied to the steroids where a subset of the most active molecules, **6**, **7**, **10**, **11**, **19** and **20** from Fig. 1, was used to create the receptor surface model. This afforded an aggregate molecular shape similar to a union volume surface generated in the active analog approach. Points on the surface may be parameterized with steric, electrostatic and hydrophobic properties to facilitate computation of various types of interaction between training or test set molecules and the receptor surface model. Four types of energies between molecules and the model were computed and assessed for their abilities to account for changes in CBG binding affinities of the steroids which were divided into the 21-steroid training set and the 10-steroid test set. These energies were: *E*(interact), nonbonded van der Waals and electrostatic interaction energy; *E*(inside), intramolecular strain energy of the ligand inside the receptor surface model; *E*(relaxed), from minimization of the ligand in the absence of the receptor surface model; and *E*(strain), the difference between *E*(inside) and *E*(relaxed). Two types of receptor surface model were examined: a closed and an open model. The closed surface completely encompasses the training set, while the open model contains undefined regions. These models and the corresponding energies for the steroids were evaluated using a genetic function approximation (GFA) to identify those variables, energies, which could most effectively account for the CBG binding energies. The open model, which includes an undefined region for the test steroid acetate, **23**, afforded the best results. The models can be visually examined by depicting the steroids aligned within the rendered receptor surface. The statistical results of this study may not be directly compared to those of others, because there are two errors in the steroid structures. Steroids **5** and **28** are incorrect as drawn in the paper. There are also three errors in the CBG binding affinities (steroids **16**, **17** and **26**) used.

Good et al. [17] examined the CoMFA steroids in a study of the potential applicability of molecular similarity using $N \times N$ similarity matrices where each molecule is compared to every other. Relationships between similarity and CBG binding affinities for all 31 steroids, as well as for TeBG binding affinity for 21 steroids, were developed qualitatively through the use of neural network analyses in an attempt to classify the molecules into high, intermediate and low affinity. Essentially correct classifications were achieved using electrostatic similarity matrices, while classifications based upon shape similarity were less successful. The similarity matrices were then subjected to quantitative analyses via partial least squares and the results compared with corresponding CoMFAs computed using separate and combined electrostatic and shape fields. In a second report [18], 10 similarity measures were investigated using the CoMFA steroids and 7 additional sets of molecules. Since this work employed integral similarity indices of the entire molecules, graphical depiction was not possible, thereby complicating interpretation of the results. Unfortunately, these extensive studies on similarity are marred by the apparent incorporation of numerous errors in steroid structure, as well as clerical errors in the CBG binding affinities. There are at least seven errors in structural drawings in the first paper and six in the second paper. As the dataset is available as a part of the ASP tutorial from Oxford Molecular Group [19], a check of these revealed errors in steroid structure **2**, **5**, **14**, **16**, **21** and **28** [20]. The CBG binding activities of steroid **16** and **17** are reported as 5.225 when the correct value is 5.255.

In another study of potential applications of similarity analyses, Klebe et al. [21] proposed Comparative Molecular Similarity Indices Analysis (CoMSIA) as an alternative to CoMFA. In these investigations using the CoMFA steroids as well as several other datasets, molecular alignments were achieved using mutual similarity indices (modified SEAL [22] procedure) pairwise calculated between all atoms of the molecules under study. To achieve a spatial comparison between steroids, similarity indices were enumerated for each of the aligned molecules in the dataset at regularly spaced grid-points using a common probe atom. The steroids were analyzed by CoMFA and by CoMSIA in this work which allows a direct comparison of the results. For alignments based upon the steroid nucleus as outlined in the original CoMFA publication, q^2 (PRESS) for CoMFA and CoMSIA were very comparable: 0.662 (0.719; 2 components) and 0.662 (0.763; 4 components), respectively. Using the modified SEAL alignment procedure gave similar statistical results affording q^2 (PRESS) values of 0.598 (0.832; 4 components) for CoMFA and 0.665 (0.759; 4 components) for CoMSIA. Both methods yielded comparable predictions of the additional 10-steroid test set where steroid **31** was notably an outlier as indicated in other studies. It is worthy of note here that while CoMFA was computed from combined steric and electrostatic fields, CoMSIA, in contrast, employed similarity indices derived from steric, electrostatic and hydrophobic properties. The CoMFA results were evenly weighted between steric and electrostatic properties, while CoMSIA suggests that steric properties may be insignificant while electrostatic and hydrophobic properties are of similar importance. Because of the nature of the similarity indices utilized here, it was possible to plot contours allowing visual examination of the portions of the steroid structures that were related to binding. The set of 21 training set steroids was taken from SYBYL 6.2 and, thus, the structure of androstanediol, **2**, is in error. In addition, steroid **28** of the test set is incorrect [23].

In a report detailing Comparative Molecular Moment Analysis (CoMMA), Silverman and Platt [24] have examined the potential of the moments of molecular mass and charge distribution to serve as molecular descriptors. The three principal moments of inertia, I_x, I_y and I_z relate to molecular shape while the magnitude of the dipole moment, p, and the magnitude of the principal quadrupole moment, Q, account solely for molecular charge. Descriptors that relate both shape and charge were also developed by computing the magnitudes of the dipolar components and the magnitudes of the components of displacement between the center-of-mass and the center-of-dipole with respect to the principal inertial axes, giving six additional descriptors: p_x, p_y, p_z, d_x, d_y and d_z. Finally, quadrupolar components were calculated with respect to a translated inertial reference frame whose origin coincides with the center-of-dipole, giving Q_{xx} and Q_{yy}. These 13 descriptors provided a set of three-dimensional internal molecular moment parameters which were independent of the orientation and location of the molecules in three-dimensional space. Thus, these authors have devised a set of parameters for use as independent variables which are based upon three-dimensional distribution of mass and charge. These 13 parameters were computed for the CoMFA 31 steroid training and test sets and correlations derived using PLS. Gasteiger-Marsili, AM1-Mulliken and *ab initio* charges (6-31G*) were evaluated as a basis for parameter development. The best correlations were seen using the *ab initio* charges giving q^2 values of 0.828 (3 components) for the CBG binding affinity and 0.693 (4 components) for the TeBG binding affinity of the 21-steroid training set. The test set of 10 steroids was not examined predictively, but included in a 31 molecule correlation. No attempts to interpret the correlations qualitatively were offered. Here, again, it must be noted that there are eight errors in the published set of 31 steroid structures; however, since the 21-steroid training set coordinates were taken from SYBYL 6.01, it may be assumed that the known structural error in androstanediol, **2**, was the only incorrect structure actually incorporated in the examination of the training set. Structure **28** of the test set is also incorrect.

While CoMMA is based upon deriving the parameters from atomic positions and properties, MS-WHIM (Molecular Surface-weighted Holistic Invariant Molecular), reported recently by Bravi and co-workers [25], uses the coordinates of points on the molecular surface to derive descriptors. A set of 12 MS-WHIM indices were computed from x, y and z coordinates of molecular surface points using various physico-chemical properties associated with the surface points. The MS-WHIM descriptors were computed for the 21-steroid training set, PLS analyses conducted and the results compared with those obtained from atom-based WHIM descriptors and also from CoMFA fields. While the q^2 achieved with the MS-WHIM was lower than that from CoMFA, the ability of MS-WHIM derived correlations to predict the activities of the 10-steroid test set was slightly better. As with the CoMMA procedure described previously, one difficulty with the use of MS-WHIM is that qualitative interpretation in terms of receptor ligand interactions is not possible. It should be noted that the coordinates of the 21-steroid training set were taken from SYBYL and, thus, the structure of androstanediol, **2**, is in error. Furthermore, the structure of test set steroid **28** is incorrect.

A recent paper by Schnitker et al. [26] reports the application of EGSITE (Energy and Geometry of SITE models) to the steroid datasets. In this method, binding site models are chosen in terms of a number of convex regions, such that every atom of a given molecule in a particular binding mode falls into one of the regions. The regions include solvent as well as receptor. The molecules under study are characterized by conformation, and by physico-chemical parameterization. In the current study, the steroids were characterized by molar refractivity, hydrophobicity and partial charge. In order to minimize the computations required, each molecule was divided into 7 to 10 superatoms. No alignment assumptions were made. Rather, the method proceeded by mapping superatoms into binding site regions so as to achieve the least amount of error in computed binding energies. For the 21-steroid training set, two- and three-region binding site models were obtained for CBG and for TeBG binding with q^2 values of 0.23 and 0.35 for the two-region model, slightly better than that for the three-region model. While all three physico-chemical properties were included in the models, a study of parameter importance identified molar refractivity as the most relevant parameter. Studies on the ability of the models to predict the 10-steroid test set afforded results that were, in general, comparable to other reported methods as characterized by Kendall's τ. It was not clear how one would present the results graphically in order to facilitate evaluation of the model in terms of actual binding interactions. However, studies on the importance of various superatom definitions, as well as the parameterization options, were presented. It should be noted that the structure of steroid **28** in the test set was incorrect. In addition, the CBG binding activities for steroids **16**, **17** and **26** were in error with respect to those of the original CoMFA paper.

One of the older methods proposed to account for steric effects in QSAR is that of Minimal Steric Difference (MTD) devised by Simon and co-workers. More recently, in a study by Oprea et al. [27], the MTD method was applied to both the training and the test set steroids. A hypermolecule based upon maximal superposition of the steroid structures upon 4-androstene-3-one was constructed and the MTD optimization procedure carried out. Cross-validation was conducted by dividing the 21-steroid training set into two subsets and using the model for each to predict the activities of the other. Four steroids were excluded as unique, thus leading to q^2 values of 0.704 for TeBG binding and 0.720 for CBG binding for the remaining 17 steroids. The SYBYL tutorial set of 21 steroids, which included the structural error in androstanediol, **2**, was used for the training set [28], so the numerous structural errors in the paper do not reflect the molecules actually used in the investigation. There were also two clerical errors in the binding activities of the training set. The analysis of the test set cannot be compared to other studies, because the authors chose to estimate the experimental binding activities for steroids **22–31** graphically. Structures given for test set steroids **22**, **23** and **28** were incorrect.

Vorpagel [29] has investigated the utility of Apex-3D [30] in developing an analysis of the steroids. As applied to 3D QSAR models for the steroids, the procedure involved automated pharmacophore identification, automated alignment on the pharmacophore, parameter pool specification, stepwise multiple linear regression with cross-validation (leave-one-out) and estimates of chances for fortuitous correlation. The parameter pool included pharmacophoric site indices (continuous atomic properties), global molecular properties (log P, molar refractivity) and secondary site indices (indicator variables).

Parameters were evaluated singly against both CBG and TeBG binding. Molar refractivity as well as a term called π-population-of-heteroatoms at C-3 (accounts for effect of 3-oxo) each gave significant correlations with CBG binding, while the presence of an H-bond donor at 17β was most significant for TeBG binding. The q^2 (s_{PRESS}) for the best CBG binding model was 0.897 (0.421). The ability of the model to predict the binding affinities of the test set steroids was conducted; however, the structures of steroids **27** and **28** were incorrect [31]. Apex-3D does provide an excellent graphic depiction of the pharmacophore models devised.

4. Discussion and Conclusion

Table 3 offers a summary of the methods and datasets used, as well as the results achieved in the investigations that have been described. To assist comparison, test set observed versus predicted r^2 values have been computed for all cases where true predicted log K (CBG) values are available. In considering the CoMFA steroids as a benchmark dataset for 3D QSAR methods development and comparison, a number of problems arise, as has been indicated. Most perplexing is the number of structural errors incorporated into many of the reports. The nature of the errors, the diligence of a few investigators and the availability of the 21-steroid training set coordinates have, fortunately, made some comparison possible. A further disturbing observation is the apparent lack of understanding of the biological data itself. As pointed out in the introductory paragraphs, the measured binding affinities increase with increasing activity. The description of the biological response parameter as log $1/K$ would lead to an inversion of the rank order of the activities and, thus, ultimately to a complete reversal in qualitative interpretation with respect to those structural modifications which may increase or decrease activity. This would not, of course, affect the correlation statistics; and, in fact, most investigators have used the correct log K form of the binding affinity, even while describing it erroneously as log $1/K$!

An equally serious problem comes from the choice of the 21-steroid training set and the 10-steroid test set. Kubinyi [32] pointed out that the test set contains several structural features not covered by the training set and that a better training set selection should lead to superior results. He demonstrated this in a simple one-parameter Free-Wilson analysis of the steroids. For the 21-steroid training set, a q^2 (s_{PRESS}) of 0.726 (0.630) is obtained with the presence and absence of the cycloaliphatic 4,5-double bond being used as the Free-Wilson independent variable. This equation affords an r^2_{pred} of 0.477 and s_{PRESS} of 0.733 for the 10 test set steroids. If steroids **1–12** and **23–31** (see Figs. 1 and 2) are used as a training set instead, a q^2 (s_{PRESS}) of 0.454 (0.754) is obtained. While this is clearly poorer than that afforded by analysis of the original training set, the predictivity becomes markedly better. Prediction of the 'new' test set, steroids **13–22**, gives r^2_{pred} of 0.909 and s_{PRESS} of 0.406! This serves as a further demonstration that proper consideration in the design and/or selection of any training set such that a broad variety of structural features are included is vital.

Finally, it would seem appropriate that the data making up any training set be as reliable and complete as possible. In Table 1, the original CBG affinities for the 21

Table 3 *Summary of methods and results*

Method	q^2 (comp.)[a] CBG binding	Observed vs predicted r^2 [b,c]	Errors: steroid no.[d]	Graphic interpretation[e]	Reference
CoMFA	0.665(2)	0.05, 0.12	2, 28	Yes	–[f]
CoMFA	0.734(3)	0.13, 0.71	None	Yes	–[g]
CoMFA: g^2-GRS	0.790(2)	–[h]	2	Yes	9
CoMFA: variable select	0.903(2)	–	28	Yes	11
CoMFA: E-state fields	0.803(3)	–	2	Yes	12
COMPASS	0.89	0.16, 0.70	28	Yes	14
Spatial autocorrelation	0.63[i]	–	None	No	5
Kohonen neural network	–	–	None	Yes	15
Receptor surface models	0.664	0.00, 0.68[j]	1, 5	Yes	16
Molecular similarity	0.633(5)[i]	–	2, 5, 14, 16, 21, 28	No	17, 18
CoMSIA	0.665(4)	0.18, 0.64	2, 28	Yes	21
CoMMA	0.828(3)	–	2, 28	No	23
MS-WHIM	0.631(2)	0.27, 0.62	2, 28	No	25
EGSITE	0.23	0.14, 0.46[k]	28	Yes	26
MTD	–[l]	0.06, 0.38	2, 22, 23, 28	No	27
Apex-3D	0.897	0.18, 0.76	27, 28	Yes	29

[a] Cross-validated r^2 via the leave-one-out procedure. Values are for the 21-steroid training set unless otherwise indicated. Where available and relevant to the method, the number of components is given in parentheses.
[b] Observed log *K* CBG taken from Table 2.
[c] Two r^2 values are for $n = 10$ and $n = 9$ (**31** omitted except as noted).
[d] Steroids with incorrect structures used in the analyses.
[e] Does the method allow or facilitate 3D-graphic depiction of results in terms of ligand–receptor interaction or new compound design?
[f] Computed for this review using original CoMFA settings, reference [1].
[g] Computed for this review using corrected structures and SYBYL 6.3 settings.
[h] Predictions not computed, or predicted log *K* values not reported.
[i] All 31 steroids used in the analyses.
[j] Steroid **31** included but steroid **23** omitted.
[k] Computed for EGSITE model 2.1.
[l] Leave-one-out cross-validation not reported.

steroids are given as reported by the authors of the study. The measured *K* values for steroids **2**, **3**, **9**, **13**, **14**, **15** and **18** are all listed as < 0.1. *No* binding affinities for these steroids could be determined. Thus, a third of the 21-steroid training set should be listed as 'inactive'! Given this fact, it is quite amazing that any meaningful correlation could be computed other than a classification of the steroids into broadly defined activity groups.

There may, indeed, be valid reasons for the apparent success in analyses of the steroids. The structures are attractive for 3D QSAR because of a large rigid nucleus which places potential interacting functional groups at opposite ends of the structure and which avoids any ambiguity in superposition. Thus, structural changes that influence binding affinity should be significant ones, both electrostatically and spatially. Even with the inability to measure CBG binding for seven steroids, the CBG affinities cover almost a 100-fold range, and TeBG binding affinities were measured for *all* 21 steroids. The robustness of the analytical tools employed by investigators have certainly facilitated the achievement of potentially meaningful results. And finally, in many cases, the development of new tools for 3D QSAR has not depended upon the analysis of the steroid set alone, but rather researchers have gone on to evaluate their methods against additional, varied datasets.

References

1. Cramer, R.D., III, Patterson, D.E. and Bunce, J.D., *Comparative molecular field analysis (CoMFA): 1. Effect of shape on binding of steroids to carrier proteins*, J. Am. Chem. Soc., 110 (1988) 5959–5967.
2. Dunn, J.F., Nisula, B.C. and Rodbard, D., *Transport of steroid hormones: Binding of 21 endogenous steroids to both testosterone-binding globulin and corticosteroid-binding globulin in human plasma*, J. Clin. Endocrin. Metab., 2 (1981) 58–68.
3. Mickelson, K.E., Forsthoefel, J. and Westphal, U., *Steroid-protein interactions: Human corticosteroid binding globulin–some physicochemical properties and binding specificity*, Biochemistry, 20 (1981) 6211–6218.
4. JMP Statistical Discovery Software, Version 3.1, SAS Institute Inc., Cary, NC, U.S.A.
5. Wagener, M., Sadowski, J. and Gasteiger, J., *Autocorrelation of molecular surface properties for modeling corticosteroid binding globulin and cytosolic Ah receptor activity by neural networks*, J. Am Chem., Soc., 117 (1995) 7769–7775.
6. Tripos Inc., 1699 S. Hanley Road, St. Louis, MO 63144, U.S.A.
7. Patterson, D.E., personal communication.
8. Gasteiger, J. and Marsili, M., *Iterative partial equalization of orbital electronegativity: A rapid access to atomic charges*, Tetrahedron, 36 (1980) 3219–3288.
9. Cho, S.J. and Tropsha, A., *Cross-validated R2-guided region selection for comparative molecular field analysis: A simple method to achieve consistent results*, J. Med. Chem., 38 (1995) 1060–1066.
10. Baroni, M., Costantino, G., Cruciani, G., Riganelli, D., Valigi, R. and Clementi, S., *Generating optimal linear PLS estimations (GOLPE): An advanced chemometric tool for handling 3D QSAR problems*, Quant. Struct.-Act. Rel., 12 (1993) 9–20.
11. Norinder, U., *Single and domain mode variable selection in 3D QSAR applications*, J. Chemometrics, 10 (1996) 95–105.
12. Kellogg, G.E., Kier, L.B., Gaillard, P. and Hall, L.H., *E-state fields: Applications to 3D QSAR*, J. Comput.-Aided Mol. Design, 10 (1996) 513–520.
13. Abraham, D.J. and Kellogg, G.E., *The effect of physical organic properties on hydrophobic fields*, J. Comput.-Aided Mol. Design, 8 (1994) 41–49.
14. Jain, A.N., Koile, K. and Chapman, D., *Compass: Predicting biological activities from molecular surface properties–performance comparisons on a steroid benchmark*, J. Med. Chem., 37 (1994) 2315–2327.
15. Anzali, S., Barnickel, G., Krug, M., Sadowski, J., Wagener, M., Gasteiger, J. and Polanski, J., *The comparison of geometric and electronic properties of molecular surfaces by neural networks: Application to the analysis of corticosteroid-binding globulin activity of steroids*, J. Comput.-Aided Mol. Design, 10 (1996) 521–534.
16. Hahn, M. and Rogers, D., *Receptor surface models: 2. Application to quantitative structure–activity relationships studies*, J. Med. Chem., 38 (1995) 2091–2102.
17. Good, A.C., So, S. and Richards, W.G., *Structure–activity relationships from molecular similarity matrices*, J. Med. Chem., 36 (1993) 433–438.

18. Good, A.C., Peterson, S.J. and Richards, W.G., *QSARs from similarity matrices: Technique validation and application in the comparison of different similarity evaluation methods*, J. Med. Chem., 36 (1993) 2929–2937.
19. Automated Similarity Package, Oxford Molecular Group, Oxford, U.K.
20. Sadowski, J., personal communication.
21. Klebe, G., Abraham, U. and Mietzner, T., *Molecular similarity indices in a comparative analysis (CoMSIA) of drug molecules to correlate and predict their biological activity*, J. Med. Chem., 37 (1994) 4130–4146.
22. Kearsley, S.K. and Smith, G.M., *An alternative method for the alignment of molecular structures: Maximizing electrostatic and steric overlap*, Tetrahedron Comput. Methodol., 3 (1990) 615–633.
23. Abraham, U. and Kubinyi, H., personal communication.
24. Silverman, B.D. and Platt, D.E., *Comparative molecular moment analysis (CoMMA): 3D QSAR without molecular superposition*, J. Med. Chem., 39 (1996) 2129–2140.
25. Bravi, G., Gancia, E., Mascagni, P., Pegna, M., Todeschini, R. and Zaliani, A., *MS-WHIM, new 3D theoretical descriptors derived from molecular surface properties: A comparative 3D QSAR study in a series of steroids*, J. Comput.-Aided Mol. Design, 11 (1997) 79–92.
26. Schnitker, J., Gopalaswamy, R. and Crippen, G.M., *Objective models for steroid binding sites of human globulins*, J. Comput.-Aided Mol. Design, 11 (1997) 93–110.
27. Oprea, T.I., Ciubotariu, D., Sulea, T.I. and Simon, Z., *Comparison of the minimal steric difference (MTD) and comparative molecular field analysis (CoMFA) methods for analysis of binding of steroids to carrier proteins*, Quant. Struct-Act. Relat., 12 (1993) 21–26.
28. Oprea, T.I., personal communication.
29. Vorpagel, E.R., *Analysis of steroid binding using apex-3D and 3D QSAR models*. 210th American Chemical Society Meeting, Chicago, 1995, COMP-0125.
30. Golender, V.E. and Vorpagel, E.R., *Computer-assisted pharmacophore identification*, In Kubinyi, H. (Ed.) 3D-QSAR in drug design: Theory, methods, and applications, ESCOM, Leiden, The Netherlands, 1993, pp. 137–149.
31. Vorpagel, E.R., personal communication.
32. Kubinyi, H., *A general view on similarity and QSAR studies*, In van de Waterbeemd, H., Testa, B. and Folkers, G. (Eds.) Computer-assisted lead finding and optimization, Proceedings of the 11th European Symposium on Quantitative Structure–Activity Relationships, Lausanne, Switzerland, Verlag Helvetica Chimica Acta and VCH: Basel, Weinheim, 1997, pp. 7–28.

Molecular Similarity Characterization Using CoMFA

Thierry Langer
Institut für Pharmazeutische Chemie, Leopold-Franzens-Universität Innsbruck, Innrain 52a, A-6020 Innsbruck, Austria

1. Introduction

Similarity is an instantly recognizable and universally experienced abstraction capability of humankind that is ubiquitous in scope, interdisciplinary in nature and boundless in its ramifications. It is, therefore, not surprising that in recent years similarity studies have become the focus of interest within various disciplines of the biological, medical, physical and social sciences [1]. A highly notable feature is that similarity is never absolute and, thus, no absolute measure of similarity exists. Therefore, similarity always has to be defined using subjective terms. Efforts to quantify similarity are, in all cases, associated with some degree of arbitrariness: what appears to be similar to one mind may not necessarily be so to another. Within the drug-development context, the concept of molecular similarity has proven to be one of the most important tools that can be used to provide new design ideas [2]. Molecular similarity, however, is also a highly complex notion that can only be described with reference to the immediate use for which it is intended and, therefore, different measures of similarity have to be formulated for each eventual use [3]. In drug design, different notions of molecular similarity are used based on molecular formulae, molecular graphs, molecular skeletons, their atom types and positions, their conformations, their van der Waals surfaces or their molecular fields. Determination of molecular similarity based on the latter will be the goal of this chapter.

2. Molecular Similarity: A Basic Concept in Drug Design

All notions of similarity are based on recognition at patterns followed by attempts of pattern classification. The reverse of molecular similarity is complementarity; in between lies molecular dissimilarity, which often is needed as crucial information by molecular designers, who wish to generate sets of dissimilar molecular structures that share common (similar) features. The search for pattern and for classification rules are fundamental problems in molecular similarity research. If two molecules have to be considered, their shapes, electron densities, etc. can be compared by using similarity indices such as those of Carbó [4] or Hodgkin [5]. If the similarity between more than two molecules is to be defined and the search for features can be done stepwise, the problem gets even bigger since the question arises how to weight the different features. Another layer of complexity is added by conformational flexibility of molecular structures [6]. Therefore, it is not surprising that there is still no generally agreed algebraic expression of similarity — or even what is meant by molecular similarity. However, the general concept is well established in the basic drug-design context, and the number of papers dealing with molecular similarity studies is still increasing. Some recent

H. Kubinyi et al. (eds.), 3D QSAR in Drug Design, Volume 3. 215–231.

examples covering diverse areas of molecular similarity research are given in references [7]–[12].

3. The Use of Molecular Fields for Similarity Description

Basically, molecular similarity can be expressed in terms of shape, electrostatic potential, surface hydrophobicity and hydrogen-bonding capacity. As molecules interact with their binding sites through their molecular fields, it appears also justified to define molecular similarity by field comparison, if certain conditions are fulfilled. In general, fields originating from molecular properties, such as the electrostatic potential, are continuous. The term 'field' usually refers to a potential or other scalar property; in fact, molecular fields are derivatives of a potential and, therefore, are vector quantities. For instance, the molecular electrostatic potentials of molecules may be easily calculated at any position in the surrounding space, resulting in continuous scalar quantities. The derivatives of this potential give the vector field, which is far more complicated to use for similarity assessment, since at each point there are three values (one for each main axis of the Cartesian space) of the field to be considered. In the molecular modelling context, so-called 'interaction energy fields' have been shown to be useful for establishing quantitative structure–activity relationships — e.g. using the CoMFA approach [13]. Fields used for these studies represent the discrete type of fields since they consist of a three-dimensional matrix of scalar values obtained by calculating interaction energies at all grid-points of a defined lattice between a probe and the molecule.

A major problem in 3D QSAR studies which is still far from being solved is the alignment definition — i.e. the correct and self-consistent superposition of all molecular structures under investigation. This remains also the main issue if such fields are used for similarity assessment. Therefore, the application of molecular field analysis for similarity determination is limited to those cases where an unambiguous alignment definition is provided.

The crucial step then becomes the question how to analyze the interaction energy matrices. A suitable method has been proposed by Martin and co-workers within the framework of 3D QSAR [14]: they applied multivariate statistical methods, namely principal components analysis (PCA) and cluster analysis, based on steric potential interaction energy matrices for a comparative molecular field analysis of shape properties. The latent variables obtained after PCA of a huge data matrix as statistical scores are often called principal properties (PPs) and represent in an appropriate way each multidimensional system by a few descriptors. Since PPs are orthogonal to each other, they are particularly suitable as design variables [16]: applying criteria of experimental design using PPs as descriptors, one is able to select the most informative combinations of substituents or molecules of a series. Moreover, PPs can also be used in pairs or triplets to describe substituents linked to each substitution site in a given series of molecules sharing a common skeleton, instead of traditional QSAR descriptors that are mimicked in the best possible way.

However, as has been pointed out [15,17], the direct derivation of 3D PPs from interaction energy matrices obtained by CoMFA is not obvious, since additionally to the alignment and conformational flexibility problem, doubts exist on the congruency of the descriptor matrix. Clementi et al. [18] have proposed to overcome the latter by auto- and cross-correlation and covariance (ACC) transforms that have been developed, together with Fourier transforms, to account for the dependencies between consecutive observations: it has been found that PCA on the ACC matrix of a CoMFA field gave results which limit to a certain extent the dependency upon the way of orientation of substituents. However, also utilizing this technique, the field descriptor derived PPs of each molecule still depend heavily upon many subjective choices in their derivation, such as selection of the appropriate geometry, alignment of orientation, type of force field, type of charge calculation, etc. Thus, much care has to be taken if such scales should be used in retrieval of information.

4. The Use of CoMFA for Similarity Determination: Case Studies

4.1. Characterization of amino acids

Since the quantitative description of amino acids is crucial for deriving quantitative structure–activity relationships of peptides, much effort has been spent on the derivation of appropriate descriptors of amino acid properties. A large body of both experimental and theoretical data has been produced over the last 50 years, and recently, the PPs approach has been successfully used in peptides QSAR [19]. Also 3D QSAR methods have been implicated to derive novel parameters: Norinder [20] has characterized amino

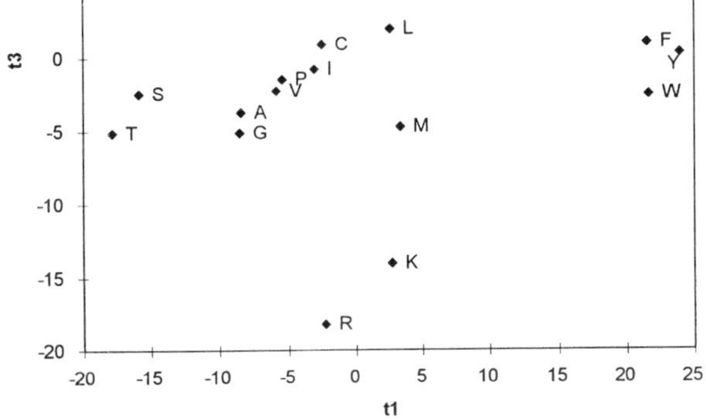

Fig. 1. Plot of the object scores. First (t1) versus third (t3) principal component. The 20 coded amino acids are denoted by their standard one-letter code (taken from reference [21]).

acids using interaction molecular descriptors calculated from three types of fields (the nonbonded and charge–charge interactions and the molecular lipophilic potential) and the PPs were then used as independent variables in the PLS analysis of a set of bradykinin potentiating peptides. It has to be noted that the QSAR models obtained were satisfactory; however, in this study, little attention was paid by the author to the amino acids classification according to design criteria.

In another recent study, Cocchi et al. [21] have characterized the 20 coded amino acids by their interaction energies calculated by the program GRID [22] and multivariate data analysis; the aim of this paper was to extend further the amino acids characterization in the context of the principal properties approach. They used six different probes mimicking various functional groups which can be involved in peptide–peptide interactions (H_2O, $-COO^-$, $-CO$, $-NH_2$, $-NH_3^+$, $-CH_3$). PCA of the interaction energies data matrix has been done to derive amino acid PPs and compare the obtained classification with the previously published z-scales [23] calculated by a multiproperties matrix containing both experimental data and empirical constants of amino acids. As already stated, the *a priori* problem of such studies is the specification of an alignment rule for superpositioning and the consideration of conformational flexibility. In this context, weight was put rather on a consistent overlapping of the side chains than to do a systematic search of all energetically accessible conformations, which was achieved by strictly superimposing the functional carboxy and amino groups and the C_β atoms. The residues were aligned by flexible fitting to the atoms of the side chain of the reference molecule arginine having the longest side chain. By GRID calculations a data matrix of 20 objects and 1050 variables was obtained. After scaling the data in order to

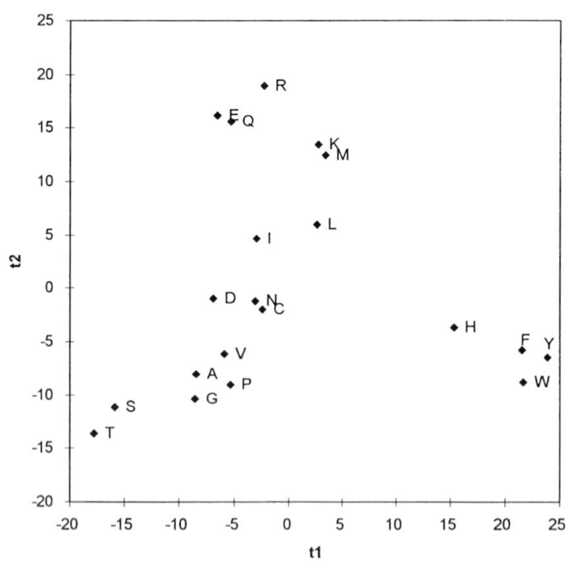

Fig. 2. Plot of the object scores. First (t1) versus second (t2) principal component. The 20 coded amino acids are denoted by their standard one-letter code (taken from reference [21]).

Table 1 *Grouping of the 20 coded amino acids according to the eight octants in the three-dimensional factorial design cube (from reference [21])*

Amino acid	Design coordinates		
T, S, A, G, P, V	–	–	–
D, N, C	–	–	+
R, K, I	–	+	–
E, Q	–	+	+
W	+	–	–
F, H, Y	+	–	+
M	+	+	–
L	+	+	+

let all the probes equally contribute to the models, a PCA was done to calculate new principal properties and to classify the amino acids. According to the authors, seven components are significant and explain about 72% of the total data variance. The first PC is interpreted to contain a blending of size and polarizability effects; whereas PC_2 is less interpretable, PC_3 is shown to distinguish between plus and minus charged amino acids, thus representing mainly electrostatic effects. The object scores for each amino acid are reproduced in Figs. 1 and 2. In both plots the amino acids are grouped, according to the features of the side chains, into aromatic, small nonpolar and charged, whereas Ser and Thr are two extremes, what is explained by their small side chain bearing an hydroxy group capable of H-bond interactions on the C_β atom. However, the dimensionality is still seven; a lot of information is lost about the amino acid grouping when looking at two dimensions at a time.

In Table 1 the amino acids are divided into eight groups representing the octant subspaces according to the signs of their coded t-scales. This subdivision can be used in the design of test series for peptide QSAR. In the present study, the PPs have finally been used to model the activity of a set of 58 dipeptides acting as inhibitors of angiotensin converting enzyme (ACE). PLS analyses have been done independently on the first six GRID derived PPs, as well as on the whole interaction energy data matrix. Moreover, inhibitory activity values have been predicted starting from a model generated with a subset of eight dipeptides spanning approximately a fractional factorial design in t_1 and t_2. The results of all models are satisfactory. As far as peptide–QSAR modelling is concerned, the direct use of the calculated probe interaction energies as amino acid descriptors gave slightly better results (a three-component PLS model of the 1050 original descriptors explains 89% of the total Y variance) than the use of GRID PPs (a one-component PLS model of the GRID derived t_1–t_6 scales explains 74% of the total Y variance).

The authors claim that their new amino acid descriptors are advantageous to the previously derived z-scales [23]: (i) they permit discrimination between plus and minus charged amino acids, (ii) Gly and Trp are not found to be outliers and (iii) His lies closer to the other aromatic amino acids. However, it has later been pointed out [15] that the different lengths of the side chains give interactions with the probes at different grid nodes and, therefore, may simply result in a ranking of amino acid scores, which classify them with little further information with respect to previously defined, traditional PPs.

4.2. *Characterization of heteroaromatic residues*

We have recently reported [24,25] on the results of our studies aimed at the multivariate characterization of heteroaromatic moieties using the CoMFA approach, together with the Tripos [26] or the GRID force field, respectively. The driving force for these studies was the fact that in medicinal chemistry one of the major problems when dealing with isosteric or bioisosteric replacement [27] in heterocyclic systems is the selection of the *a priori* most promising candidates among several dozens of possible rings. A large number of descriptors has been available for such fragments, and recently PPs for heteroaromatic systems based on both empirical and theoretical data have also been derived in view of their relevance as building blocks to a large number of compounds of pharmaceutical interest [28]. Until that time, descriptors of heteroaromatics, or therefrom derived principal properties, respectively, have been measured or calculated only for entire systems, taking no account of differences in the anchoring positions of such fragments in a given molecule. It is well known, however, that properties of heteroaromatic moieties may drastically vary upon variation of the substitution position, thus the need of descriptors appropriate for describing such effects.

In a first step [24], we examined 16 different aromatic ring systems appearing in a total of 37 isomers (Fig. 3), in order to check the principal usefulness of molecular similarity characterization using molecular interaction energy fields. All molecules were aligned as shown in Fig. 3, using a connection bond to a dummy atom located in the origin of a Cartesian coordinate system, the aromatic rings being placed in the XY plane. All statistical calculations were performed within the QSAR module of the SYBYL molecular modelling software [29]: interaction energies between the heteroaromatic moieties and the probe atoms were calculated at a total of 4158 grid-points with 1 Å spacing in a lattice of $18 \times 21 \times 11$ ($X = -5$ to 12, $Y = -10$ to 10, $Z = -5$ to 5) using the default Lennard-Jones and Coulomb potential functions and the standard Tripos CoMFA probes (the $C.sp^3$ probe was used for calculation of steric interactions and the H^+ probe for calculation of electrostatic interactions, respectively). A PCA (factor analysis without axes rotation) was done on the descriptor matrix and a classification of the heteroaromatic substituents into families was performed using the SYBYL hierarchical clustering procedure of the obtained PCs. The thereby obtained clustering dendrogram is reproduced in Fig. 4; in this type of diagram, the most similar compounds cluster together at the lowest levels.

It has been argued [15,17] that 3D PPs may suffer from major drawbacks when not properly derived. In our special case, the conformational flexibility problem does not exist and the alignment definition assuming a hypothetical binding pocket in which the heteroaromatic moieties would all align in a plane according to the dipole moment vector is straightforward: a possible 180° rotation would just lead to PPs with inverted signs. The possible influence of the substituent parts of the heteroaromatic rings is mini-

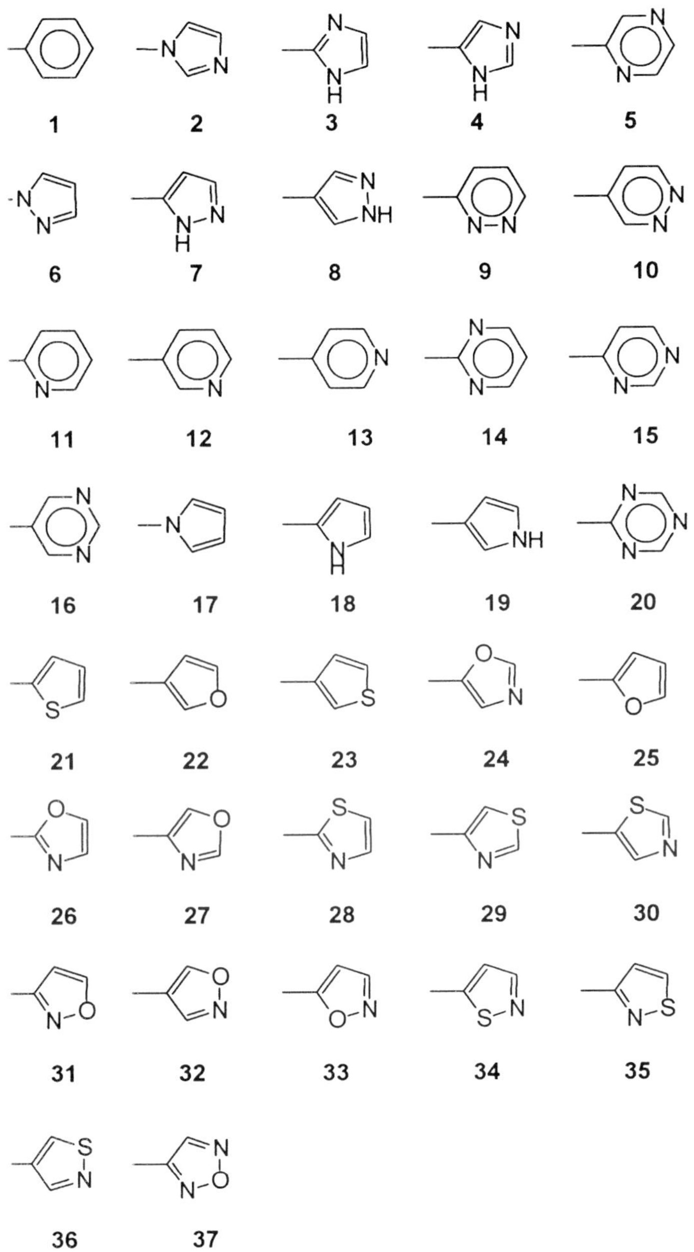

Fig. 3. Heteroaromatic systems investigated in reference [24].

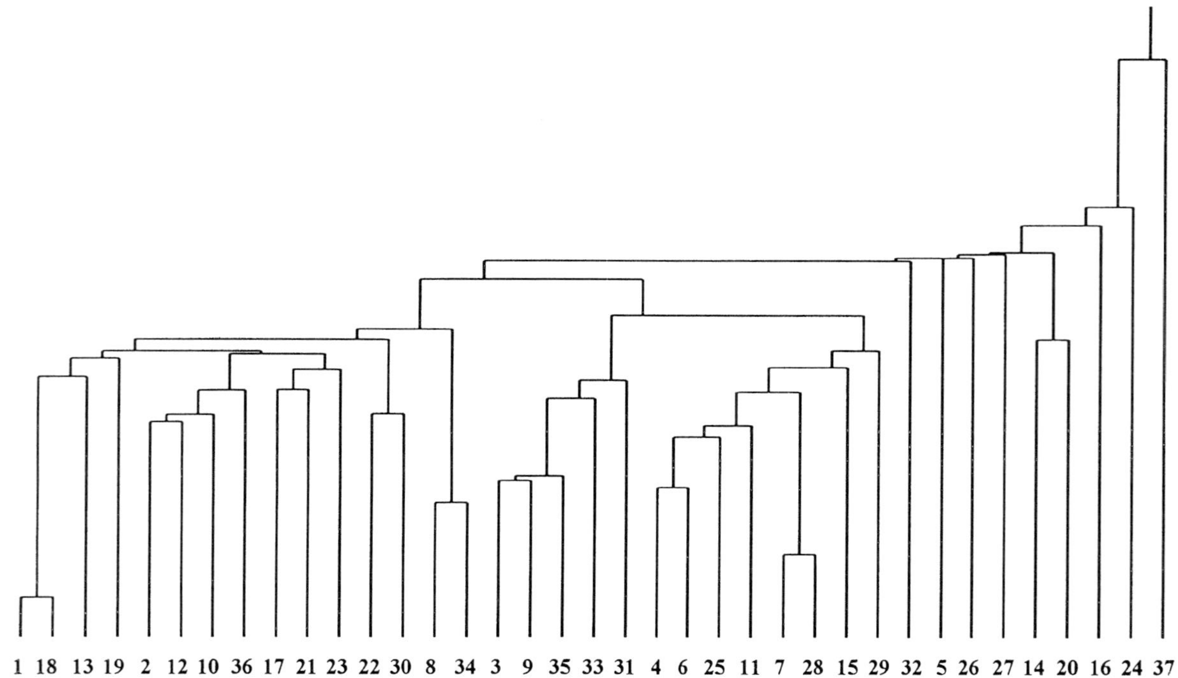

Fig. 4. *Dendrogram obtained after hierarchical clustering of principal components 1–3 calculated from the CoMFA descriptor matrix; compounds numbering as in Fig. 3 (taken from reference [24]).*

mized by the connecting dummy atom. However, a problem still may be seen in the parameters of the force field used: parameterization of sulfur atoms might render heteroaromatic ring systems containing sulfur atoms different from other systems — giving rise to different clusters and, therefore, different possible representative systems.

We, therefore, extended the previously described study also to other bicyclic systems [25], using this time the GRID force-field atom parameters: a total of 72 aromatic moieties (five- and six-membered monocyclic and benzo-fused bicyclic heteroaromatics containing one or two heteroatoms, as listed in Table 2) were analyzed using a total of six GRID multiatom probes (H_2O, Alkyl–OH, Carbonyl–O, Csp^3, Aromatic C, NH_4^+), considered as a representative selection among the variety of the main interaction modes with amino acids, in order to mimic possible interactions of the molecule with a putative receptor. The alignment was chosen in a consistent way, the aromatic rings being placed in the *XY* plane in such a way that the dipole moment vectors of all compounds were pointing into the same subspace. Interaction energies between the heteroaromatic moieties and the probes were calculated at a total of 3553 grid-points with 1 Å spacing in a lattice of 19 × 17 × 11 Å ($X = -6$ to 12, $Y = -8$ to 8, $Z = -5$ to 5). The first three principal components explaining 78% of the total variance (PC_1: 38%; PC_2: 31%; PC_3: 9%) were extracted and used for further calculations. A classification of the heteroaromatic substituents into families was again performed, using a complete linkage hierarchical clustering procedure of the obtained PCs. The obtained clustering dendrogram is reproduced in Fig. 5. In fact, the results gained in this case are in better agreement with common chemical knowledge — e.g. phenyl is located in the same cluster as 2- and 3-thienyl; the π electron deficient heteroaromatic moieties 3- and 4-pyridyl are found in the same cluster as 4-pyridazinyl; and five-membered π electron-rich heteroaromatics are located in one cluster, like 1-pyrrolyl, 3-pyrrolyl and 5-thiazolyl.

The PPs were finally used also to model the activity of a set of 16 3-[(arylmethyl)-amino]-5-ethyl-6-methylpyridin-2(1*H*)-one derivatives acting as specific inhibitors of HIV-I reverse transcriptase [30]. As shown, a satisfactory QSAR equation (Eq. 1) could be calculated using the first two principal components suggesting that a significant correlation exists between the GRID-derived PPs and differences in biological activities related to bioisosteric heteroaromatic modifications in the test compounds:

$$pIC_{50} = 0.467(\pm 0.207)PC_1 - 0.712(\pm 0.131)PC_2 - 1.145(\pm 0.185)$$
$$n = 16 \quad r = 0.85 \quad s = 0.57 \quad F = 16.62 \tag{1}$$

In an independent study, Clementi et al. [31] have characterized a set of 44 different heteroaromatic systems by 13 descriptors derived by GRID. The main difference to the previously described studies is the fact that the PPs calculated here refer to the whole heteroaromatic moiety and not to a specific substitution position. The data matrix comprised the best interaction energy (maximum negative value) obtained for each ring system using nine GRID probes (six single- and three multiatom probes), together with four descriptor variables representing both hydrated volumes and surfaces. The best attractive energies for each probe are independent of their grid location, thus bypassing

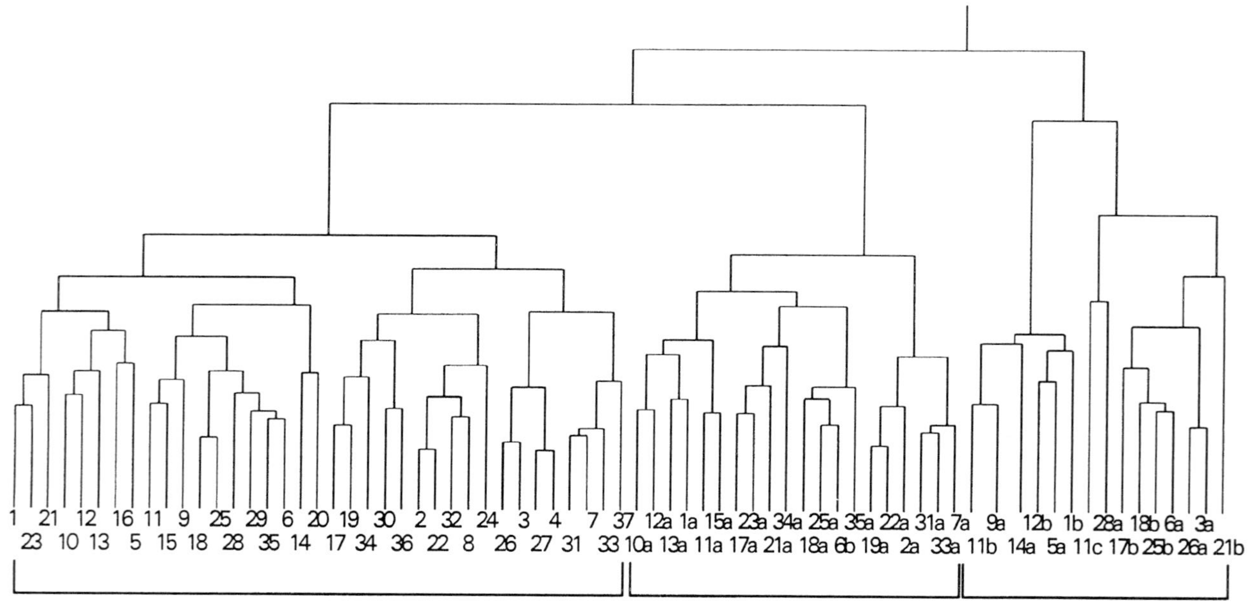

Fig. 5. *Dendrogram obtained after hierarchical clustering of principal components 1–3 calculated from the GRID descriptor matrix; compounds numbering as in Table 2 (taken from reference [25]).*

Table 2 *Aromatic systems investigated in reference [25] and principal properties derived from GRID interaction energy fields*

Compound no.	Fragment	PC_1 scores	PC_2 scores	PC_3 scores
1	Phenyl	–0.497	–0.039	–0.161
1a	1-Naphthyl	1.545	0.424	–0.644
1b	2-Naphthyl	0.721	–1.667	1.884
2	1-Imidazolyl	–0.852	0.834	1.016
2a	1-Benzimidazolyl	1.363	0.960	0.510
3	2-Imidazolyl	–1.112	0.184	–0.586
3a	2- Benzimidazolyl	–0.250	–2.072	–0.986
4	4-Imidazolyl	–1.104	0.434	–0.627
5	Pyrazinyl	–0.893	0.185	1.109
5a	2-Quinoxalyl	0.390	–1.591	2.986
6	1-Pyrazolyl	–0.916	0.296	–0.650
6a	2-Indazolyl	–0.059	–1.978	–1.162
6b	1-Indazolyl	1.378	0.425	–0.780
7	3-Pyrazolyl	–0.877	0.533	–0.332
7a	3-(1H)-Indazolyl	1.387	0.531	–0.286
8	4-Pyrazolyl	–1.002	0.785	0.754
9	3-Pyridazinyl	–0.611	0.239	–0.137
9a	3-Phthalazyl	0.606	–1.556	1.823
10	4-Pyridazinyl	–0.707	0.799	0.648
10a	4-Cinnolyl	1.464	1.186	0.162
11	2-Pyridyl	–0.567	–0.111	–0.997
11a	1-Isoquinolyl	1.494	0.343	–1.467
11b	3-Isoquinolyl	0.654	–1.780	1.022
11c	2-Quinolyl	–0.334	–1.573	–1.957
12	3-Pyridyl	–0.501	0.339	0.847
12a	4-Isoquinolyl	1.479	0.778	0.079
12b	3-Quinolyl	0.601	–1.458	2.485
13	4-Pyridyl	–0.606	0.534	0.043
13a	4-Quinolyl	1.455	0.943	–0.732
14	2-Pyrimidinyl	–0.898	–0.257	–0.555
14a	2-Quinazolyl	0.330	–1.908	1.150
15	4-Pyrimidinyl	–0.738	0.478	–0.924
15a	4-Quinazolyl	1.436	0.885	–1.385
16	5-Pyrimidinyl	–1.073	0.590	0.778
17	1-Pyrrolyl	–0.847	0.396	0.203
17a	1-Indolyl	1.443	0.578	–0.028
17b	2-Isoindolyl	0.035	–1.847	–0.134
18	2-Pyrrolyl	–0.773	0.275	–0.339
18a	1-Isoindolyl	1.383	0.273	–0.469
18b	2-Indolyl	–0.102	–1.954	–1.176
19	3-Pyrrolyl	–0.821	0.615	0.558
19a	3-Indolyl	1.383	0.576	0.473
20	2-(1,3,5-Triazinyl)	–1.164	0.331	–0.687
21	2-Thienyl	–0.650	0.221	0.514
21a	1-Benzo[c]thienyl	1.499	0.473	–0.071
21b	2-Benzo[b]thienyl	0.175	–1.842	0.824

Table 2 *Cont'd.*

Compound no.	Fragment	PC_1 scores	PC_2 scores	PC_3 scores
22	3-Furyl	–0.789	0.693	0.831
22a	3-Benzo[b]furyl	1.342	0.605	0.659
23	3-Thienyl	–0.685	0.114	–0.064
23a	3-Benzo[b]thienyl	1.573	0.382	–0.289
24	5-Oxazolyl	–1.114	0.777	1.258
25	2-Furyl	–0.798	0.334	–0.398
25a	1-Benzo[c]furyl	1.360	0.475	–0.887
25b	2-Benzo[c]furyl	–0.104	–1.954	–1.424
26	2-Oxazolyl	–1.162	0.123	–0.652
26a	2-Benzoxazolyl	–0.380	–2.149	–1.254
27	4-Oxazolyl	–1.155	0.446	–0.691
28	2-Thiazolyl	–0.657	0.165	–0.878
28a	2-Benzothiazolyl	–0.240	–1.864	–1.969
29	4-Thiazolyl	–0.740	0.118	–0.849
30	5-Thiazolyl	–0.566	0.737	1.049
31	3-Isoxazolyl	–0.884	0.643	–0.138
31a	3-Benzo[d]isoxazolyl	1.396	0.724	–0.152
32	4-Isoxazolyl	–1.086	0.958	0.969
33	5-Isoxazolyl	–0.937	0.743	0.002
33a	3-Benzo[d]isoxazolyl	1.336	0.795	–0.081
34	5-Isothiazolyl	–0.847	0.855	0.929
34a	3-Benzo[d]isothiazolyl	1.463	1.026	0.763
35	3-Isothiazolyl	–0.779	0.120	–0.911
35a	3-Benzisothiazolyl	1.504	0.329	–1.111
36	4-Isothiazolyl	–0.688	0.697	1.077
37	Oxadiazolyl	–1.237	0.857	–0.206

the problems of developing 3D PPs. A PCA was carried out on the block-weighted matrix and a four-components model was obtained. From examination of the score and loading plots for all the principal components, the following interpretation is given by the authors: the first PP (explaining 40% of the total variance) describes the change from hydrophobicity to hydrophilicity of the heteroaromatic moiety since it is related to the negative volumes and surfaces and to the best interaction energies of all probes. Consequently, it separates the systems investigated into three groups: the hydrophobic 5-membered moieties and their benzo derivatives, the hydrophilic nitrogen bases, and azines and azoles. The second component (explaining 16% of the total variance) illustrates the H-bonding capacity of the systems since it separates the H-bonding acceptors from the H-bonding donors: on the one hand, azoles and azines, and on the other hand, diazoles and pyridones. The third component (again, explaining 16% of the total variance) measures shape and hydrophobicity; it is mainly determined by positive surfaces and volumes leading to a rough separation between monocyclic and bicyclic systems. The fourth PC (explaining 10% of the total variance) indicates the capability of multiple interaction modes of the molecules with the positively charged probe amidine, which leads to a slight separation of the systems containing oxygen or sulfur from those

Table 3 *Separation of heteroaromatic systems according to 16 subspaces calculated from PCA of GRID-derived descriptor set (taken from reference [31])*

Compound	Design coordinates			
Pyridine, pyrazine, imidazole, thiazole, 4-pyridone	–	–	–	–
Furan, oxazole, isoxazole	–	–	–	+
Quinoline, isoquinoline, benzoxazole, acridine, quinoxaline, 1,5-naphthyridine, quinazoline	–	–	+	–
Benzofuran	–	–	+	+
Pyrazole	–	+	–	–
Thiophen, pyrrole	+	–	+	–
Pyrimidine, pyridazine, triazine, purine, tetrazole	+	–	–	+
Indole, carbazole, isoindole	–	+	+	–
Benzothiophen	–	+	+	+
Benzimidazole, benzothiazole	+	–	+	–
Pteridine, cinnoline, 1,8-naphthyridine	+	–	+	+
2-Pyridone, thymine, uracil, xanthine	+	+	–	–
Cytosine	+	+	–	+
Indazole	+	+	+	–
Adenine	+	+	+	+

containing nitrogen. The main separation trends are reproduced in Scheme 1, a compounds listing according to their belonging to 16 factorial subspaces is given in Table 3.

In summary, it may be concluded that this study leads to the definition of groups of heteroaromatic systems that are in good agreement with chemical sense, except for some of the acidity/basicity categorization. However, since the systems under investigation required four PPs for a thorough description, the straightforward application of a factorial design criterion, selecting one representative for each of the 16 (2^4) subspaces listed in Table 3, is far too demanding since it requires the synthesis of at least 16 molecules to control a single site substituted by a heteroaromatic system. Therefore, the authors propose that a better approach would be to consider the clustering of the heteroaromatics in the PP space, which can be achieved using a cluster analysis procedure. The number of significant clusters defines the number of significant components extracted by PCA as being equal to the number of clusters minus one; therefore, in this case, five different clusters were found, and according to the authors, it might be sufficient to take into account only five systems to span at best the heteroaromatic space. Another possibility for solving this problem is usage of D-optimal design, which would also select a minimum of five systems in the four PP space. A larger number would better cover the domain of the possible structural variations. Therefore, from comparison of results obtained by cluster analysis and PCA it is proposed to select the following 10 heteroaromatics: pyrrole, thiophen, indole, benzothiophen, pyridine, imidazole, quinoline, benzimidazole, uracil and purine. However, the problem of the substitution position of the heteroaromatic systems still remains unsolved using the

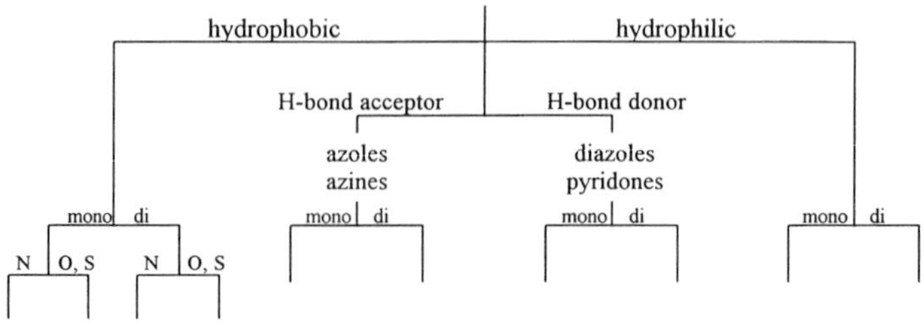

Scheme 1. Separation of heteroaromatics according to the PCA results (taken from reference [31]).

results presented in this study. For medicinal chemists, this may be of little help since, as already mentioned, it is well known that the properties of a heterocyclic ring heavily depend on its substitution position. In a study recently published by McGuire et al. [32], this question has been raised; they characterized a total of 59 different aromatic ring systems appearing in a total of 100 isomers using a total of 10 classical QSAR parameters, together with multivariate data analysis. The limited number and also the nature of the parameters used in this study, however, may cast doubt on the general applicability of the PCs obtained.

4.3. Characterization of aromatic and aliphatic substituents

Van de Waterbeemd et al. [17,33] have investigated the utility of CoMFA-derived descriptors for structure–property correlations of a total of 59 common substituents linked to aromatic and aliphatic skeletons. From the interaction energy matrices calculated using the default Tripos probes (C.sp^3, charge +1), sets of PPs have been each extracted for steric and electrostatic fields, both separately and joined together. It has been demonstrated that the CoMFA-derived 3D QSAR parameters are highly correlated with the traditional ones. In a projection of the PCs of the 3D CoMFA field descriptors into the loadings plot of 86 commonly used descriptors, the authors show that only the first PC of the steric field correlates with traditional steric descriptors and the first PC of the electrostatic field correlates with well-known Hammett constants. The first two PCs of the mixed steric-electrostatic field appear to be related to steric and electrostatic properties, respectively. The other PCs have been shown to be not significant. The advantage of using the CoMFA approach for calculating steric, electrostatic or lipophilic descriptors is that it can be applied to any substituent and does not rely on the availability of published compilations containing the desired substituent values.

However, problems are encountered when deriving 3D PPs for large and conformationally flexible substituents. The authors have used different alignment procedures of the substituents linked to an aromatic ring and a methyl group, respectively: 'random', 'rule-based' and 'sphere-filling'. In the 'rule-based' alignment, polar and nonpolar portions have been overlapped in the best possible way. In the 'sphere-filling' mode, the substituents have been oriented in such a way that taken all together they fill a sphere at the point of attachment. All calculations have been done using a 1 Å grid spacing and the effect of different box orientation has been studied indicating that a significant influence exists upon both alignment and grid position. Use of ACC transforms has been proposed to overcome some of the problems with generation of 3D PPs. In this study, it has been shown that the 3D ACC transforms used take into account neighbor effects, thus leading to more or less continuous molecular interaction fields, and that they are congruent and, therefore, independent of alignment within the grid lattice. After the transform procedure, PCA gives a model in which the first two principal components already explain 85% of the total variance, which is far more than extracted by the corresponding fields matrix (55–65%, depending upon the superposition model). The first PC is easily recognized as steric, and the second as electrostatic PC.

5. Conclusion

In this chapter, a brief review of different studies aimed at the characterization of molecular similarity using comparative molecular field analysis, together with multivariate data analysis, is given. The results obtained so far suggest that, using principal properties derived from a descriptor matrix calculated from fields within a CoMFA approach, a characterization of molecules according to similarity criteria is feasible. It has to be pointed out, that the application of this procedure still suffers from some major drawbacks (alignment problem, field congruency, etc.) in deriving 3D PPs and, therefore, the descriptors obtained for the series under investigation should not be considered as general-purpose 3D descriptors. When carefully used in series close to those whence they have been generated, however, they can serve as variables valuable both in experimental design and classical QSAR.

References

1. Rouvray, D.H., *The evolution of the concept of molecular similarity*, In Johnson, M.A. and Maggiora, G.M. (Eds.) Concepts and applications of molecular similarity, John Wiley, Inc. New York, 1990, pp. 15–42.
2. Dean, P.M., *Defining molecular similarity and complementary for drug design*, In Dean, P.M. (Ed.) Molecular similarity in drug design, Blackie Academic and Professional, London, U.K., 1995, pp. 1–23.
3. Dean, P.M., *Molecular similarity*, In Kubinyi, H. (Ed.) 3D QSAR in Drug design: Theory, Methods and Applications, ESCOM, Leiden, The Netherlands, 1993, pp. 150–172.
4. Carbó, R., Leyda, L. and Arnau, M., *An electron density measure of the similarity between two compounds*, Int. J. Quantum Chem., 17 (1980) 1185–1189.
5. Hodgkin, E.E. and Richards, W.G., *Molecular similarity based on electrostatic potential and electric field*, Int. J. Quantum Chem. Quantum Biol. Symp., 14 (1987) 105–110.
6. Leach, A.R., *The treatment of conformationally flexible molecules in similarity and complementarity searching*, In Dean, P.M. (Ed.) Molecular similarity in drug design, Blackie Academic & Professional, London, U.K., 1995, pp. 57–88.
7. Rozas, I., Du, Q. and Arteca, G.A., *Interrelation between electrostatic and lipophilicity potentials on molecular surfaces*, J. Mol. Graph., 13 (1995) 98–108.
8. Burgess, E.M., Ruell, J.A., Zalkow, L.H. and Haugwitz, R.D., *Molecular similarity from atomic electrostatic multipole comparisons: Application to anti-HIV drugs*, J. Med. Chem., 38 (1995) 1635–1640.
9. Benigni, R., Cotta-Ramusino, M., Giorgi, F. and Gallo, G., *Molecular similarity matrices and quantitative structure–activity relationships: A case study with methodological implications*, J. Med. Chem., 38 (1995) 629–635.

10. Briem, H. and Kuntz, I.D., *Molecular similarity based on DOCK-generated fingerprints*, J. Med. Chem., 39 (1996) 3401–3408.
11. Montanari, C.A., Tute, M.S., Beezer, A.E. and Mitchell, J.C., *Determination of receptor-bound drug conformations by QSAR using flexible fitting to derive a molecular similarity index*, J. Comput.-Aided Mol. Design, 10 (1996) 67–73.
12. Jain, A.N., Koile, K. and Chapman, D., *Compass: Predicting biological activities from molecular surface properties — performance comparisons on a steroid benchmark*, J. Med. Chem., 37 (1994) 2325–2327.
13. Cramer III, R.D., Patterson, D.E. and Bunce, J.D., *Comparative molecular field analysis (CoMFA): 1. Effect of shape on binding of steroids to carrier proteins*, J. Am. Chem. Soc., 110 (1988) 5959–5967.
14. Lin, T.C., Pavlik, P.A. and Martin, Y.C., *Use of molecular fields to compare series of potentially bioactive molecules designed by scientists or by computer*, Tetrahedron Comput. Methodol., 3 (1990) 723–738.
15. Clementi, S., Cruciani, G., Baroni, M. and Costantino, G., *Series design*, In Kubinyi, H. (Ed.) 3D QSAR in drug design: Theory, methods and applications, ESCOM, Leiden, The Netherlands, 1993, pp. 567–582.
16. Wold, S., Sjöström, M., Carlson, R., Lundstedt, T., Hellberg, S., Skagerberg, B., Wirkström, C. and Öhman, J., *Multivariate design*, Anal. Chim. Acta., 191 (1986) 17–32.
17. Van de Waterbeemd, H., Clementi, S., Costantino, G., Carrupt, P.-A. and Testa, B., *CoMFA derived substituent descriptors for structure–property correlations*, In Kubinyi, H. (Ed.) 3D QSAR in drug design: Theory, methods, and applications, ESCOM, Leiden, The Netherlands, 1993, pp. 697–707.
18. Clementi, S., Cruciani, G., Riganelli, D., Valigi, R., Costantino, G., Baroni, M. and Wold, S., *Autocorrelation as a tool for a congruent description of molecules in 3D QSAR studies*, Pharm. Pharmacol. Lett., 3 (1993) 433–438.
19. Hellberg, S., Sjöström, M., Skagerberg, B. and Wold, S., *Peptide quantitative structure–activity relationships: A multivariate approach*, J. Med. Chem., 30 (1987) 1127–1135.
20. Norinder, U., *Theoretical amino acid descriptors: Application to bradykinin potentiating peptides*, Peptides, 12 (1991) 1223–1227.
21. Cocchi, M. and Johansson, E., *Amino acids characterization by GRID and multivariate data analysis*, Quant. Struct.-Act. Relat., 12 (1993) 1–8.
22. Goodford, P., *A computational procedure for determining energetically favourable binding sites an biologically important macromolecules*, J. Med. Chem., 28 (1985) 849–857.
23. Hellberg, S., Sjöström, M., Skagerberg, B. and Wold, S., *On the use of multipositionally varied test series for quantitative structure–activity relationships*, Acta Pharm. Jugosl., 37 (1987) 53–65.
24. Langer, T., *Molecular similarity determination of heteroaromatics using CoMFA and multivariate data analysis*, Quant. Struct.-Act. Relat., 13 (1994) 402–405.
25. Langer, T., *Molecular similarity determination of heteroaromatic ring fragments using GRID and multivariate data analysis*, Quant. Struct.-Act. Relat., 15 (1996) 469–474.
26. Clark, M., Cramer III, R.D. and Van Opdenbosch, N., *Validation of the general purpose Tripos 5.2 force field*, J. Comput. Chem., 10 (1989) 982–1012.
27. Wermuth, C.G., *Molecular variations based on isosteric replacements*, In Wermuth, C.G. (Ed.) The practice of medicinal chemistry, Academic Press, London, U.K. 1996, pp. 203–237.
28. Caruso, L., Katritzky, A.R. and Musumarra, G., *Classical and magnetic aromaticities as new descriptors for heteroaromatics in QSAR: 3. Principal properties for heteroaromatics*, Quant. Struct.-Act. Relat., 12 (1993) 146–151.
29. SYBYL, Versions 6.01, 6.03, 6.2, Tripos Associates, St. Louis, MO, U.S.A.
30. Saari, W.S., Wai, J.S., Fisher, T.E., Thomas, C.M., Hoffman, J.M., Rooney, C.S., Smith, A.M., Jones, J.H., Bamberger, D.L., Goldman, M.E., O'Brien, J.A., Nunberg, J.H., Quintero, J.C., Schleif, W.A., Emini, E.A. and Anderson, P.S., *Synthesis and evaluation of 2-pyridinone derivatives as HIV-1-specific reverse transcriptase inhibitors*, J. Med. Chem., 35 (1992) 3792–3802.
31. Clementi, S., Cruciani, G., Fifi, P., Riganelli, D., Valigi, R. and Musumarra, G., *A new set of principal properties for heteroaromatics obtained by GRID*, Quant. Struc.-Act. Relat., 15 (1996) 108–120.

32. Gibson, S., McGuire, R. and Rees, D.C., *Principal components describing biological activities and molecular diversity of heterocyclic aromatic ring fragments*, J. Med. Chem., 39 (1996) 4065–4072.
33. Van de Waterbeemd, H., Carrupt, P.-A., Testa, B. and Kier, L.B., *Multivariate data modeling of new steric, topological and CoMFA-derived substituent parameters*, In Wermuth, C.G. (Ed.) Trends in QSAR and Molecular Modelling 92, ESCOM, Leiden, The Netherlands, 1993, pp. 69–75.

Building a Bridge between G-Protein-Coupled Receptor Modelling, Protein Crystallography and 3D QSAR Studies for Ligand Design

Ki Hwan Kim

Department of Structural Biology, D46Y AP10-2, Pharmaceutical Products Division, Abbott Laboratories, 100 Abbott Park Road, Abbott Park, IL 60064-3500, U.S.A.

1. Introduction

The technique of comparative molecular modelling of protein structures has been known for some time, and there are a large number of guanine nucleotide-binding protein coupled receptor (GPCR) model structures obtained utilizing this technique. Likewise, a growing number of three-dimensional quantitative structure–activity relationship (3D QSAR) studies have been described on various GPCR ligands using the Comparative Molecular Field Analysis (CoMFA) methodology (see the chapter by Ki Hwan Kim in this volume for a listing). Nonetheless, there are only a few studies that have utilized both techniques for ligand design. Several explanations are possible for this. The most probable reason might be that there are still many uncertainties in the current GPCR models, even though these GPCR models would be refined as the technique improves and additional experimental data become available. A similar statement can be made for the CoMFA methodology, which was invented for the situations where the 3D structure of macromolecule is not known, and this is where it is most frequently used. However, a growing number of CoMFA studies take advantage of the known 3D structure of macromolecule. A third reason for the small number of studies utilizing both techniques might be that many scientists may be an expert on one methodology but not both.

As both the GPCR modelling and CoMFA studies progress, examples of the use of both techniques in a study will certainly grow. In some cases, the experts in the field of protein modelling and three-dimensional quantitative structure–activity (3D QSAR) studies may cooperate to bring the two together. Certainly, more and more scientists will become familiar with both techniques.

The objective of this report is to build a bridge between the two techniques: 3D protein modelling and the 3D QSAR approach of CoMFA, toward the common goal of ligand design. Toward this goal, three examples are described below where both CoMFA and a GPCR model were used in a study. Seven more examples are summarized to examine how the protein structures and CoMFA results were used together in other than GPCRs.

2. G-protein Coupled Receptors

GPCRs, also known as seven transmembrane (7TM) receptors or heptahelix receptors, form a large family of membrane proteins that have seven hydrophobic regions corresponding to 7TM α-helices (7TMHs). GPCRs are found in a wide range of

H. Kubinyi et al. (eds.), 3D QSAR in Drug Design, Volume 3. 233–255.

organisms and are functionally diverse. Receptors in this family are believed to be involved in the transmission of signals across membranes to the interior of the cell. When a signaling molecule, an agonist, binds to the GPCR on the extracellular side of the cell membrane, the GPCR is activated and interacts with a heterotrimeric guanine nucleotide-binding protein (G protein) on the intracellular side. The activated G protein then initiates a second messenger system of intracellular signaling.

GPCRs bind a variety of ligands ranging from small biogenic amines to peptides, small proteins and large glycoproteins. All members of the GPCRs are thought to have the same basic structure in the transmembrane domain. This is mainly due to sequence similarities and their common ability to activate G proteins to initiate signal transduction. The hydrophobic 7TMHs regions of the receptors are located within the cell membrane and span the phospholipid bilayer seven times. These highly conserved hydrophobic transmembrane helices are connected by highly diverse hydrophilic loops. The N-terminus of the receptors is located on the extracellular side and the C-terminus on the intracellular side.

2.1. *Receptor structure*

The overall structural features of the GPCR family are characterized by seven 20–25 amino acid sequences in length that are believed to represent the transmembrane-spanning hydrophobic regions of the proteins. Each receptor is believed to have an extracellular N-terminal region that varies in length from less than 10 amino acids (adenosine receptors) to several hundred (metabotropic glutamate receptors) and an intracellular C-terminal region. The majority of intracellular and extracellular loops are thought to be 10–40 amino acids long, although the third intracellular loop and the C-terminal sequence may have more than 150 residues. The overall size of these receptors varies significantly from less than 300 amino acids of adrenocorticotrophin hormone receptor to more than 1100 amino acids for the metabotropic glutamate receptors [1].

The structure of the 7TM segments has not been characterized by X-ray crystallography or magnetic resonance spectroscopy. Based on structural similarities with bacteriorhodopsin [2], these regions are predicted to be α-helices that form a ligand binding pocket. The orientation of the helices (clockwise or anti-clockwise) remains unclear, although anti-clockwise orientation (seen from outside) seems to be more plausible [1]. Among the GPCRs, only rhodopsin has been structurally characterized by cryoelectron microscopy and confirmed to have transmembrane seven-helix bundles [3] (see section 3 for more information).

2.2. *Subfamilies of GPCRs*

The GPCRs are often divided into different families by sequence homology [1,4]. Three most distinct families of GPCRs are the (1) opsin type, (2) peptide hormone receptor type and (3) metabolic glutamate receptor type. Members of the opsin family constitute the majority of GPCRs [1].

All of the opsin-type receptors show a high degree of amino acid conservation within their seven transmembrane α-helices, while those of the hormone receptor type show homology within the class but not with the opsin-type receptors. The metabolic glutamate receptors show no homology with the GPCRs of the opsin or hormone receptor types.

The majority of the residues in the hydrophobic transmembrane domain are conserved, whereas the residues in the hydrophilic loop regions are more divergent. The primary sequence identity in the 7TM domain ranges from 85–95% for species homologs of a given receptor to 60–80% for related subtypes of the same receptor, to 35–45% for other members of the same family, down to 20–25% for unrelated GPCRs [5,6].

Although the primary sequences among GPCRs are quite diverse, the overall structural features of the GPCRs are highly conserved, reflecting their common mechanism of action. Various criteria can be used to classify the over 300 currently known GPCRs. While only low-sequence homology is found in the loop regions, the 7TM regions contain a number of residues that are conserved for several or all receptor types; for example, the disulfide bridge between a cysteine residue at the top of TM3 and another cysteine residue in the second extracellular loop is common in all GPCRs [1]. Most of the receptors identified so far belong to the opsin-like subfamily characterized by a small N-terminal segment that is highly glycosylated. They have highly conserved residues in the transmembrane segments: Asn-18 on TM1, Asp-10 on TM2, Arg-26 on TM3 and Asn-16 on TM7. Closely related receptors have a number of additional conserved residues [1].

2.3. *GPCR sequences*

Today, there are over 770 GPCR sequences from all species listed in the SWISS-PROT Protein Sequence Databank (Table 1); this number changes very rapidly. The most represented species are as follows: human, 186; rat, 139; mouse, 96; bovine, 33; chicken, 24; pig, 21; xenopus, 17; guineapig, 16; dog, 14; drosophila, 14; *C. elegans*, 13; rabbit, 11; and goldfish, 9.

2.4. *Ligand binding mode*

There are two main hypotheses regarding the interaction of a ligand and its receptor [1]. In the first and older hypothesis, agonists and antagonists are believed to bind in a similar manner to the receptor. An agonist binds to the receptor and induces a conformational change that causes signal transduction, whereas an antagonist binds without a conformational change. However, in the second hypothesis [7], GPCRs are assumed to exist in at least two conformations. The active conformation interacts with G proteins, but the inactive (resting or uncoupled) conformation cannot bind G proteins. The inactive form usually predominates in the resting state. If a ligand binds to the active conformation with high affinity, the active conformation becomes the dominant species present, and the ligand is called an agonist. If a ligand binds to the active conformation with moderate affinity and the resulting concentration of the active conformation is low

Table 1 *Number of reported GPCR sequences (as of 18 March 1997)*

Receptor	No.	Receptor	No.
Family 1 (Opsin type)			
Acetylcholine (muscarinic) receptors	23	Neurotensin receptors	4
Adenosine and adenine nucleotide receptors	28	Odorant/olfactory and gustatory receptors	53
Adrenergic receptors	45	Opioid peptide receptors	15
Angiotensin receptors	16	Opsins	78
Bombesin receptors	8	Orphan receptors	69
Bradykinin receptors	5	Other receptors	5
Cannabinoids receptors	7	Platelet activating factor receptors	4
Chemokines and chemotactic factor receptors	34	Prostanoid receptors	29
Cholecystokinin/gastrin receptors	8	Releasing hormone receptors	10
Dopamine receptors	32	Serotonin receptors	49
Endothelin receptors	10	Somatostatin receptors	16
Glycoprotein hormone receptors	21	Tachykinin receptors	15
Histamine receptors	9	Proteinase-activated receptors	7
Melanocortins receptors	17	Vasopressin/oxytocin receptors	13
Melanotonin receptors	20	Viral receptors	11
Neuropeptide Y receptors	10		
Other families			
Calcitonin, corticotropin releasing factor, glucagon, etc.	53	Metabotropic glutamate and calcium receptors	19
Fungal receptors	10	Slime mold receptors	3
Archebacterial receptors	14		

but displays detectable efficacy, the ligand is called a partial agonist. A ligand that binds to both conformations and does not change their ratio is called a competitive antagonist. If a ligand binds to the inactive conformation and reduces the amount of the active conformation, it is called an inverse agonist.

2.5. *Ligand binding site*

The location of ligand binding site differs depending on the type of GPCRs. Mutagenesis and biophysical studies of several GPCRs indicate that small molecule agonists and antagonists bind to a hydrophilic pocket buried in the transmembrane core of the receptor [4]. On the other hand, peptide ligands bind to both the extracellular and transmembrane domains [8]. The binding sites of agonists and antagonists of small peptides are different, whereas the binding sites of larger peptide hormones and endothelin are larger and overlapping for both agonists and antagonists [1,9–14].

A detailed discussion on the binding sites of various ligands are presented in recent review papers [1,5,8].

3. Molecular modelling of GPCRs

Quantitative structure–activity relationships, the three-dimensional structures of receptors, and the biochemical mechanism of the drugs all provide important information for ligand design. However, due to the lack of three-dimensional structures of these membrane protein receptors, the structural insights have been inferred with the aid of three-dimensional computer models.

As noted above, a major feature in the amino acid sequence of GPCRs is the occurrence of seven hydrophobic helical regions. This feature provided a rationale for modelling GPCRs based on the bacteriorhodopsin structure.

The first three-dimensional model of rhodopsin was prepared in 1986 [2], based on the high-resolution electron cryo-microscopy structure of bacteriorhodopsin (3.5 Å in X and Y directions and 10 Å in Z direction), determined by Henderson and co-workers [3]. In 1993, 9 Å resolution electron density projection map of GPCR bovine rhodopsin was reported [15]. The projection maps of bacteriorhodopsin and rhodopsin clearly showed the 7TMHs. However, the spatial organization of the TMHs in rhodopsin appeared to be different from that of bacteriorhodopsin [3].

The structures of both bacteriorhodopsin and rhodopsin provided significant information toward the three-dimensional structure modelling of GPCRs [3]. All three-dimensional models of GPCRs are essentially constructed after one of these two structures. Some people used the coordinates of the structures in a homology modelling, whereas others used the structures only as a guide to the helical packing.

The use of the bacteriorhodopsin structure was questioned because bacteriorhodopsin is not a GPCR and does not have high amino acid sequence homology with GPCRs, despite the fact that it has seven transmembrane helices (7TMHs) similar to the GPCR 7TM helix regions [16,17]. However, bacteriorhodopsin has a functional resemblance to mammalian opsin and is functionally related to rhodopsin which is a GPCR. Therefore, bacteriorhodopsin was assumed to be structurally homologous to the GPCRs. Unlike bacteriorhodopsin, bovine rhodopsin is a GPCR, and some people preferred to use the rhodopsin structure as a template over the bacteriorhodopsin structure.

Since the reported electron diffraction projection map of bovine rhodopsin is quite different from that of bacteriorhodopsin, comparison of bacteriorhodopsin and rhodopsin structures has been instructive in assessing the 3D structure of the GPCRs. Considering the experimental evidence of rhodopsin and the results of 204 GPCR sequence analysis, Baldwin [18] proposed a probable arrangement of the seven helices which differs considerably from the previously constructed models based on the bacteriorhodopsin structure. On the other hand, Hoflack et al. [19] compared the electron diffraction maps of both proteins and suggested that bacteriorhodopsin and bovine rhodopsin have the same, or a very similar, transmembrane helix packing. They claimed that the differences in the projection of the backbone structures became strikingly similar after the structure was rotated by 15° around an axis perpendicular to the seven helices.

3.1. General procedures of GPCR modelling

The extra- and intracellular loop regions are conformationally flexible, and their modelling structures are much less reliable than the 7TM regions [20]. Thus, the modelling of only the 7TMHs is usually attempted.

The following six-step procedure is usually employed for the homology-based modelling of the 7TMs.

1. *Sequence alignment*: although considerable sequence homology between 7TMs exists between various GPCRs, it can be very low with certain receptors. A strict alignment with that of bacteriorhodopsin or rhodopsin determines the start and end of each TMH, as well as the rotation of each TMH in relation to the six other helices. Various properties are considered in the sequence alignment such as hydropathy, hydrophobic and hydrophilic nature of the TM bundle and the existence and function of conserved residues in a particular receptor sequence, as well as site-directed mutagenesis information.
2. *Backbone construction*: the seven helices corresponding to TM 1–7 are constructed with fixed ϕ and ψ values. Most conserved amino acids are distributed on the same face of the α-helices. Proline-containing helices are kinked due to the lack of hydrogen-bonding donor capacity of proline. Since the positions of the prolines in the GPCRs and bacteriorhodopsin are not conserved, the kinked helices in bacteriorhodopsin cannot be used directly as templates for the proline-containing TM of GPCRs. In such cases, these helices are constructed with a kink typical of a proline-containing α-helix [21]. 7TMHs may also be built based on the standard helix builder [22].
3. *Modelling TM bundle*: in each of the seven helices corresponding to TM 1–7, side chains are rotated to avoid van der Waals overlap and subsequently geometry optimized. The resulting helices are positioned to form the TM bundle using the backbone of bacteriorhodopsin or rhodopsin as a template.
4. *Helix orientations*: most hydrophobic residues of the sequence are considered to constitute TMHs. The TMHs are amphiphilic and should have the hydrophobic face located on the outside toward the lipid layer. On the other hand, the polar face of the TMHs is located at the relatively hydrophilic interior of the TM bundles. The conserved residues are considered to be important for the function or structure of the receptor, and they is on the inside of the TMHs or in an area that is facing other helices.
5. *The intra- and extracellular loops* are added if desired, based on a loop-searching procedure.
6. *The geometry of the whole protein structure* is optimized by energy minimization, using molecular mechanics or molecular dynamics calculations and using certain constraints to fix the positions of the helices relative to each other.

3.2. Three-dimensional molecular models

Most of the earlier models were based on the structure of bacteriorhodopsin. Analysis of the sequence alignment of the GPCR superfamily was reviewed by Probst et al. [6] and Baldwin [18]. The earlier 3D GPCR models were reviewed by Strader et al. [5,8] and the structural characterization and binding sites of GPCRs were recently reviewed by

Beck-Sickinger [1], who also listed some of the most important ligands that bind to over 100 different GPCRs. A large number of GPCR models are described in the literature [11,18,19,21–57]. The 3D coordinates of some of these models are available from various web sites (see the web site information below).

Although these models will undoubtedly be modified as additional experimental data (such as those from receptor mutagenesis) become available, they still provide a visual model that can help one to formulate hypotheses and design new ligand molecules.

3.3. Web sites of GPCR and protein engineering

There are a number of World Wide Web (WWW) sites [58], relevant to GPCRs and protein engineering. Some of the selected sites are listed below. The GPCR web sites offer many GPCR models, and their 3D coordinates can be downloaded. Swiss-Model provides a WWW server for an automated protein modelling of user-defined transmembrane helices [59]:

Secondary structure prediction:	
nnpredict	http://www.cmpharm.ucsf.edu:80/~nomi/nnpredict.html
PredictProtein	http://embl-heidelberg.de/predictprotein/
Structure database and visualization:	
Protein Data Bank	http://www.pdb.bnl.gov/
RasMol	http://www.umass.edu/microbio/rasmol/
3D-structure prediction and G-protein coupled receptors:	
GPCR Database	http://receptor.mgh.harvard.edu/GCRDBHOME.html
Swiss-Model	http://expasy.heuge.ch/swissmod/SWISS-MODEL.html
NCBI GenBank	http://www.ncbi.nlm.hin.gov/
SWISS-PROT Sequence Data Bank	http://receptor.mgh.harvard.edu/GCRDBHOME.html
GPCRDB:GPCR 3D models	http://swift.embl-heidelberg.de/7tm/models/models.html
	http://mgddk1.niddk.nih.gov:8000/GPCR.html

3.4. Limitation of GPCR models

The limitations of the 3D structures of GPCRs based on the bacteriorhodopsin were discussed with respect to the structural information of rhodopsin, as well as the principles of homology modelling [4,60]. The main problem in modelling GPCRs is the low sequence homology of the receptors to that of bacteriorhodopsin or rhodopsin. It makes the sequence alignment difficult using bacteriorhodopsin or rhodopsin as a template. In addition, the resolution of the bacteriorhodopsin or rhodopsin structure is low, and neither of the structures may be an ideal template structure. Likewise, the relative positioning of the transmembrane domain is approximate, and the conformation of some loops is not explicitly taken into account within the model. The hydropathy analyses and primary sequence alignments of GPCR do not allow one to define precisely the 7TMHs, which leads to uncertainties about exactly where the helices start and end and their relative position in the membrane. Interpretation of mutagenesis data and the use of the results can be quite subjective, and the 3D models are static representations and do not represent the dynamic structure.

Many pitfalls in protein sequence alignments and predictions of 3D structure were also discussed by Rost and Valencia [61].

4. CoMFA Studies on GPCRs in Conjunction with Models of the Receptors

Despite the limitations of the current 3D models, a few authors attempted to use information from both a relevant protein model and 3D QSAR. These studies are summarized below.

4.1. Melatonin receptor

Based on the helical structure of bacteriorhodopsin, Sugden et al. [51], proposed a model for melatonin binding. Recently, Navajas et al. [62], also proposed a melatonin-binding mode in the G-protein-coupled melatonin model. Sugden et al. used the melatonin receptor sequence from *Xenopus laevis* melanophores, whereas Navajas et al. used the sequences of several vertebrate melatonin receptors. The binding mode proposed by these two groups differ considerably.

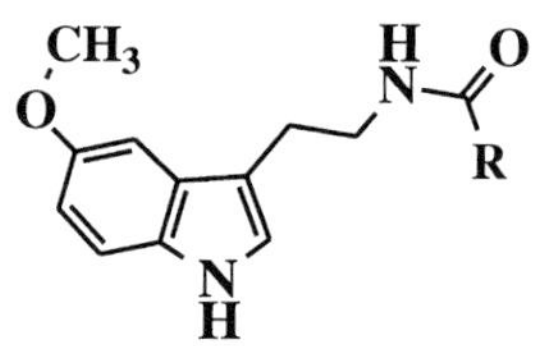

Melatonin analogs

In a 3D QSAR study, Navajas et al. [62] first developed a CoMFA model from 28 melatonin analogs. The AM1-minimized lowest energy conformations of melatonin analogs were superimposed over the melatonin molecule as the reference, and the inverse logarithm of the relative binding affinity (K_i) was used as the dependent variable in CoMFA. The probes used were an sp^3 carbon with a +1 charge, an sp^3 oxygen and a hydrogen; the grid spacing used was 2 Å; for other CoMFA conditions, default settings were used.

From different CoMFA models, Navajas et al. chose the 5-component model from the oxygen probe as the best one due to the favorable statistics of the model. The final CoMFA model has the following statistics (L = number of PLS latent variables):

$$N = 28 \quad L = 5 \quad R^2 = 0.97, \quad RMSE = 0.37, \quad Q^2 = 0.76, \quad RMSE_{cv} = 0.98$$

The activities of three other compounds were predicted from the model with reasonable accuracy for two: predicted (measured) K_i = 1.2 (1.0), 44 (45) and 3.4 (562). A large deviation between the predicted and observed K_i values for the third compound (5-benzyloxy-N-acetyltryptamine) was likely to be due to the fact that the original set of compounds did not include any with such a large substituent at position 5.

The G-protein-coupled melatonin model was then examined along with the CoMFA model to locate and dock melatonin analogs into the binding site. The following four SAR criteria were used for the docking of melatonin analogs: (1) The 5-methoxy group of melatonin is specifically recognized and selectively differentiated from the corresponding 5-hydroxy group; a bulky hydrophobic substituent at the 5-position is not tolerated; and the oxygen at 5-position is selectively recognized, together with the methyl group attached. (2) The oxygen of the N-acetyl group of melatonin is specifically recognized, and this recognition site is about 10.8 Å away from the 5-methoxy group. (3) The docking of melatonin at its binding site is stabilized by an aromatic interaction between the receptor and the indole moiety of melatonin. (4) The methoxy and N-acetyl groups are recognized in a plane which is outside the plane of the aromatic interaction.

Based on these criteria, Navajas et al. proposed a binding mode in which melatonin fits into the hydrophilic binding cleft formed by the extracellular ends of helices V and VII and the middle part of helix VI of the G-protein-coupled melatonin model. The recognition of the functional moieties of the indole occurred through interaction with fully conserved amino acid residues present in the 15 different melatonin receptors but not in other members of the G-protein-coupled receptor family.

Sugden et al. [15] proposed that melatonin binds into the binding cleft formed by isoleucine I-25 in helix II, serine S-10 in helix III, asparagine N-21 and valine V-24 in helix IV and tryptophan W-16 in helix VI. This contrasts with Navajas et al.'s proposal which suggested that the binding cleft of melatonin was formed by valine V-7 and histidine H-10 in helix V, serine S-6 and alanine A-10 in helix VI, and phenylalanine F-9 in helix VII. Navajas et al. claimed that, when placed in the rhodopsin-based model, many of the specific amino acid residues proposed by Sugdon et al. pointed toward the lipid bilayer and other helices rather than toward the hydrophilic pocket. Therefore, Navajas et al. claimed that these residues must not be able to interact with the functional groups of the melatonin molecule. However, the reverse may also be true if the specific amino acid residues proposed by Navajas et al. are placed in the bacteriorhodopsin-based model of Sugdon et al.

Because of these conflicting proposals, Navajas et al. suggested that site-directed mutagenesis may provide the answers regarding the contribution of each suggested amino acid residue to the recognition of melatonin in the G-protein-coupled melatonin receptor.

Thus, Navajas et al. utilized both a GPCR structure and CoMFA in their study to orient the ligands into the binding site and to generate a new hypothesis to be tested in a later study.

4.2. Serotonin receptor (5-HT$_{1A}$ receptor)

Gaillard et al. [63] developed a CoMFA model from 185 5-HT$_{1A}$ receptor ligands including 101 arylpiperazines, 30 aryloxypropanolamines and 54 tetrahydropyridylindoles. In the CoMFA study, the energy-minimized conformations of these compounds were superimposed by manual geometrical fitting over (1-(2-methoxyphenyl)-4-[4-(2-phthalimido)butyl]piperazine as the reference. The inverse logarithm of the relative binding affinity (K_i) was used as the dependent variable in CoMFA. The probe used was an sp^3 carbon with a +1 charge, and the grid spacing used was 1.5 Å. In addition, lipophilic field was used.

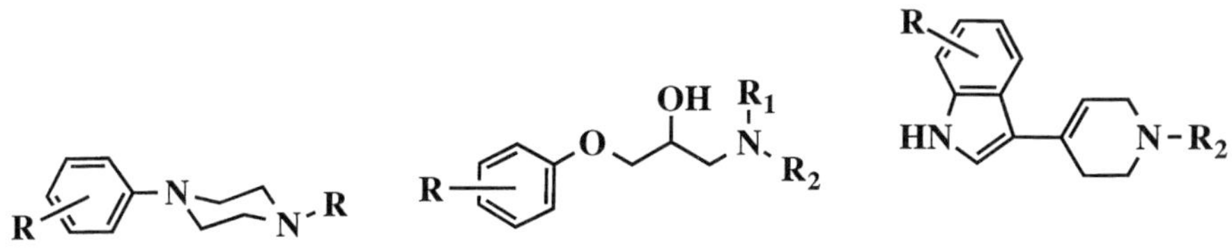

Arylpiperazines Aryloxypropanolamines Tetrahydropyridylindoles

The final CoMFA model was derived from the steric, electrostatic and lipophilic fields and had the following statistics:

$$N = 185 \quad L = 3 \quad R^2 = 0.85, \quad RMSE = 0.55, \quad Q^2 = 0.66$$

In order to validate the CoMFA model, Gaillard et al. compared the model with the binding site of the 5-HT_{1A} receptor model proposed by Kuipers et al. [14]. The 5-HT_{1A} receptor model was constructed using bacteriorhodopsin as the structural template.

Gaillard et al. claimed that their CoMFA model gave remarkable analogies with the receptor model. The receptor model showed an electron-rich region (Thr-200) close to the 5-substituent of the indole ring, a polar region (Asn-386) near the hydroxy group of aryloxypropanolamines, a forbidden steric region (Asp-116) near the basic nitrogen and an electron-rich region (Ser-199) close to nitrogen of the indole ring of tetrahydropyridylindoles. The receptor model also indicated that a large region was allowed for the nitrogen substituent between helices III, VI and VII. This observation was also compatible with the CoMFA model. In addition, the CoMFA model suggested additional interactions around the aromatic moiety of aryloxypropanolamines and around the nitrogen substituent.

4.3. Histamine receptor

Dove et al. [64] used 34 2-phenyl and 2-heteroarylhistamine derivatives to investigate QSAR and pharmacophoric elements necessary for H_1 agonism. The energy-minimized conformations of these compounds were superimposed by aligning the histamine moieties. In the CoMFA study, the pD_2 values obtained from isolated organs were used as the dependent variable. The grid spacing used was 1.5 Å, and lipophilic field f and π of a m-substituent were also included:

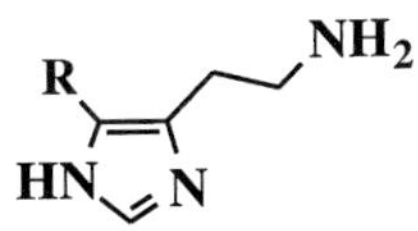

2-phenyl and 2-heteroarylhistamines

Two CoMFA models obtained with and without the lipophilic fields were as follows. The contribution from the steric and electrostatic fields were almost equal, and the lipophilic contribution was 7% when it was included.

Lipophilic fields	L	R^2	$RMSE$	Q^2	$RMSE_{cv}$
without	3	0.78	0.38	0.49	0.58
with	5	0.88	0.29	0.54	0.57

Dove et al. [64] constructed models of the rat H_1 receptor helices assuming that helix *V* contained the agonist-specific binding site: one based on Trumpp-Kallmeyer et al.'s alignment [65], and the other based on Yamashita et al.'s alignment [66]. Between the two models, the authors preferred the second model, based on the crystal structure of bacteriorhodopsin. The helices were then minimized with 2-(m-MeO-phenyl)-histamine bound at the active site. According to the authors, the ligand fit vertically between the helices and possibly interacted with Asp-107, Asn-198 and Thr-194. They suspected that Trp-165 and His-166 might be responsible for the sterical constraints in para and (somewhat weaker) in the meta position of 2-phenylhistamines and also for favored positive charges. They suggested that both models more or less correspond to the CoMFA results, even though the second model was more probable.

As in the case of Sugden et al. [51] on the melatonin receptor discussed above, Dove et al. used their CoMFA results to dock the ligands into the histamine receptor and to choose a more probable GPCR model.

5. Bridges between Other Protein Structures and CoMFA

The structures of macromolecule can be obtained from X-ray crystallography or NMR spectroscopy as well as from protein homology modelling and used for ligand design in various ways in 3D QSAR studies: they are used for alignment of the ligand molecules, ligand docking, interpretation and comparison of CoMFA models. It would be instructive to examine how different studies bridged the protein structures and CoMFA. A few selected examples are presented below.

5.1. Papain structure and its substrates

In a CoMFA study of papain catalyzed hydrolysis of phenyl N-benzoyl glycinates (HIP) and phenyl N-methanesulfonyl glycinates (MSG), Carriere et al. [67] used the X-ray structure of papain for ligand docking. In this case, they took the protein structure to support the hypothesis used in the original QSAR by comparing the results of classical QSAR, CoMFA and the enzyme structure.

The initial QSAR reported by Smith et al. [68] was as follows:

$$\log 1/K_m = 0.57\sigma + 1.03\pi_3' + 0.61\,MR_4 + 3.80$$
$$n = 25, \quad R^2 = 0.823, \quad s = 0.208, \quad Q^2 = 0.740$$

In this equation, K_m is the Michaelis-Menten binding constant, and σ and MR_4 are the Hammett electronic substituent constant and the molar refractivity of the para substituent, respectively. Special attention was given to the parameter π_3', the hydrophobic substituent constant referring to only the more hydrophobic of the two meta groups. The initial working hypothesis involved in this parameter was that only meta hydrophobic substituents could contact an enzymic hydrophobic counterpart, whereas the hydrophilic groups could be placed into a polar environment (aqueous solvent surrounding the enzyme surface).

In their CoMFA study, Carriere et al. selected the papain active site from the X-ray crystallographic structure of N-Cbz-PheAlaCH_2–papain complex (ZPACK) [69]. This was done by choosing all the amino acid residues with 12 Å radius from the sulfur atom of Cys-25. After constructing the models of HIP and MSG using standard bond lengths and angles from SYBYL fragment library, they were docked into the binding site. All of the starting conformations of HIP, MSG and the enzyme–substrate complexes of the active site were then fully optimized by MNDO, AM1 and AMBER force fields, respectively, in SYBYL.

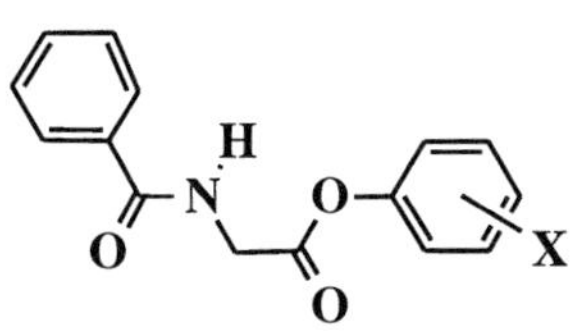

Phenyl N-benzoyl glycinates (HIP)

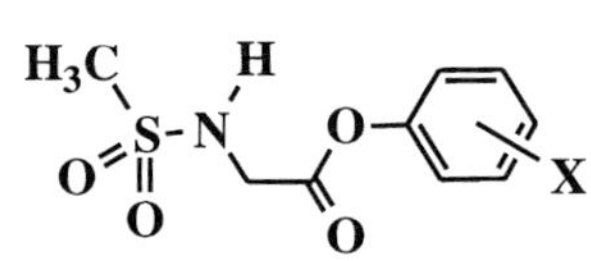

Phenyl N-methansulfonyl glycinates (MSG)

Two alignments (S and T orientations) were used in the CoMFA and molecular docking study. In the T orientation, the meta substituents were oriented in the active site in such a way that they occupied a large hydrophobic region defined by the side chains of Trp-26, Val-133, Leu-134, Val-157, Tyr-67 and Pro-68. In the S orientation, the meta hydrophobic substituents were oriented as above, whereas the meta hydrophilic substituents were placed in hydrophilic regions mainly composed by the Gln-19 and Ser-176. Both orientations maintained the hydrogen-bonding network in a same manner. Then CoMFA was performed using AM1 charges in 2 Å spacing grids using an sp^3 carbon probe with a +1 charge.

An inferior CoMFA model was obtained from the T alignment ($Q^2 = 0.24$) than the S alignment ($Q^2 = 0.73$). Similar results were obtained from the MSG series: T ($Q^2 = -0.10$) and S ($Q^2 = 0.52$). Therefore, the authors concluded that the results supported the initial hypothesis formulated in the classical QSAR model on the basis of hydrophobicity of π'_3.

Alignment	*L*	R^2	*SD*	Q^2
		phenyl N-benzoyl glycinates (HIP)		
S	7	0.980	0.076	0.726
T	6	0.931	0.134	0.238
		phenyl N-methanesulfonyl glycinates (MSG)		
S	1	0.822	0.155	0.525
T	1	0.586	0.237	–0.104

5.2. *Glycogen phosphorylase structure and its inhibitors*

One of the key steps in CoMFA is selection of the bioactive conformation for each ligand and its alignments. The binding modes of ligands can be unpredictable, even in the presence of several X-ray structures of similar compounds.

In a CoMFA study for the glycogen phosphorylase inhibition, Watson et al. [70] used the experimentally determined ligand–macromolecule three-dimensional structures as a most reliable source for the alignment and bound conformations of each of the ligands. In this way, they could avoid the problems and potential errors in selecting the bioactive conformation and their alignments. In this study, the three-dimensional enzyme structure and CoMFA were used to gain insight about the binding modes of individual molecule and to design a tighter binding inhibitor.

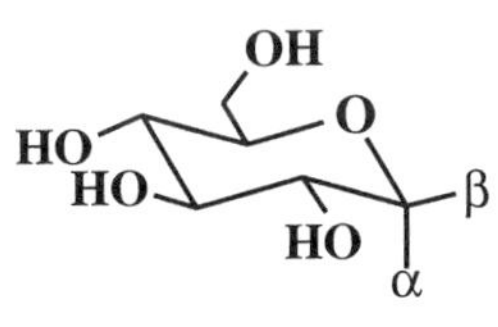

Glucose derivatives

However, even when the bioactive conformation and alignment are not an issue, there are still a number of other practical problems in CoMFA model development. They include selection of appropriate probes and eliminating irrelevant variables from the initial interaction energy matrix. Including irrelevant variables can lead to overfitting and chance correlation and have detrimental effects on the model selection and the model's predictive ability. (See the chapter by K.H. Kim et al. in this volume.)

Cruciani and Watson [71] used three-dimensional structures not only for determining the bioactive conformations and alignment, but also for selecting the most appropriate pretreatment procedures in CoMFA. The CoMFA was performed with 36 glucose analogs in 1 Å spacing grids using the GRID phenolic OH probe. From a number of possible data pretreatment and variable selection procedures in a CoMFA study for the inhibition of glycogen phosphorylase, they chose the method of autoscaling on a subset of variables. The subset of variables were preselected using a D-optimal algorithm (procedure 2) as the most appropriate pretreatment procedures to eliminate a reasonable amount of noise. Their argument for the selection was as follows. Although autoscaling performed on the entire dataset (procedure 1) gave better Q^2 and $RMSE_{cv}$, the CoMFA model from the data produced chance correlations; the chance correlations were reflected in the overestimation of regions where it was known from the three-dimensional structure that there were no possibilities of such interactions. There were several such regions between Asn-284, Asp-283 and Leu-136 that were predicted to be important but were known from the binding study to play no significant role. On the other hand, the predicted CoMFA coefficient contour map from procedure 2 for ligand–enzyme interactions and the experimental regions identified by the X-ray crystallographic binding studies showed good agreement: the interactions at the catalytic site residues Gly-675, Ser-674, His-377, Tyr-573 and Asn-484 were well predicted. For this reason, they selected one with slightly inferior Q^2 and $RMSE_{cv}$ as the final model:

Data pretreatment	L	R^2	$RMSE$	Q^2	$RMSE_{cv}$
Procedure 1	4	0.93	0.44	0.81	0.74
Procedure 2	3	0.92	0.48	0.80	0.78

Procedure 1: autoscaling performed on the entire dataset.
Procedure 2: autoscaling on a subset of variables that were preselected using a D-optimal algorithm.

They claimed that numerical comparison such as Q^2 or $RMSE_{cv}$ between models obtained from different pretreatments of the same dataset was not sufficient to select the best model unless the CoMFA coefficient contour map was compared with the enzyme X-ray structure.

5.3. *Aromatase structure and its inhibitors*

The study by Recanatini [72] on the aromatase inhibitors can be considered somewhat similar to the GPCR study. In this study, the CoMFA results were compared with the homology modeled protein structure developed by Laughton et al. [73,74]. In a study of 29 non-steroidal aromatase inhibitors related to fadrozole, Recanatini developed a CoMFA model for the *in vitro* inhibitory activity on the human placental aromatase. The CoMFA study was performed using an sp^3 carbon atom with a +1 charge as the probe and a 2 Å grid spacing. The final model was derived from the AM1 geometries and charges with an atom-by-atom alignment and had the following statistics:

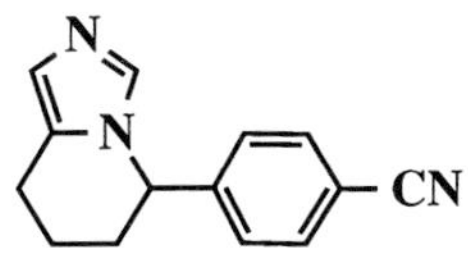

Fadrozole

$$N = 60 \quad L = 3 \quad R^2 = 0.949, \quad RMSE = 0.185, \quad Q^2 = 0.763, \quad RMSE_{cv} = 0.398$$

Laughton et al. [73,74] derived a three-dimensional model of aromatase on the basis of the cytochrome P-450$_{cam}$ X-ray structure and the sequence of the cytochrome P-450$_{arom}$. Assisted by site-directed mutagenesis, they identified some active site residues and examined their interactions with a steroid ligand.

Recanatini claimed that some of the observations reported by Laughton et al. were consistent with their CoMFA results. For example, Laughton et al. placed the phenyl rings of Phe-234 and Phe-235 near the α region of the steroid. CoMFA results indicated that the p-cyanophenyl group in fadrozole occupied the same region and interacted with the phenylalanine phenyl rings of Phe-234 and Phe-235. Laughton et al.'s model revealed the presence of His-475 in the area close to the C4 position of the steroid. This area appeared to represent the steric limitation of the hydrophobic site revealed by the CoMFA model. The positive steric coefficient contours in CoMFA corresponding to the meta positions of the p-cyanophenyl ring of fadrozole might correspond to the Tyr-244 on the β face and Ile-305 on the α face of the steroid severely restricting the space available to the D ring.

Thus, in this study, the modeled three-dimensional protein structure was used to compare and show agreements between the active site of the modeled structure and CoMFA results.

5.4. *Rhinovirus structure and its non-steroidal inhibitors*

In a study for the antipicornavirus activity associated with disoxaril analogs, Diana et al. [75] used the X-ray structure of human rhinovirus-14 for the orientation and conformation of ligand molecules in their CoMFA study. Compounds whose X-ray structures were not available were modeled from a similar compound whose bound comformation was known.

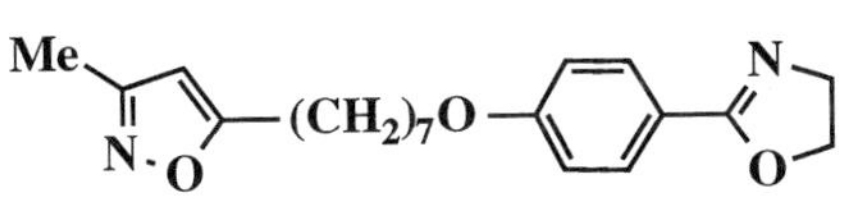

Disoxaril

Artico et al. [76] extensively modified the disoxaril structure to find a new class of potent and selective human rhinovirus-14 inhibitors. Due to the lack of X-ray crystallographic data of the studied compounds and structural similarity to disoxaril and its analogs, they used the X-ray structures of disoxaril and related analogs to model some of their compounds. The crystal structure of an analog was also used for superimposing these compounds for CoMFA study. They also used a protein crystal structure for docking a disoxaril analog to study its binding mode. From 17 compounds, they obtained the following CoMFA model using an sp^3 carbon atom with a +1 charge as the probe and a 2 Å grid spacing:

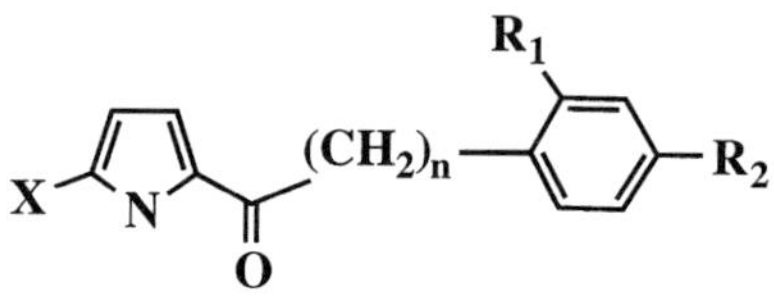

Modified disoxaril compounds

$$N = 17 \quad L = 5 \quad R^2 = 0.97 \quad RMSE = 0.11, \quad Q^2 = 0.72, \quad RMSE_{cv} = 0.38$$

This work provides an example where the protein structure was used to model and superimpose a series of extensively modified structures for a CoMFA study.

5.5. *Acetylcholinesterase (AChE) structures and its inhibitors*

Cho et al. [77] used the three available enzyme–inhibitor complex structures to align a series of 60 chemically diverse acetylcholinesterase inhibitors, shown below:

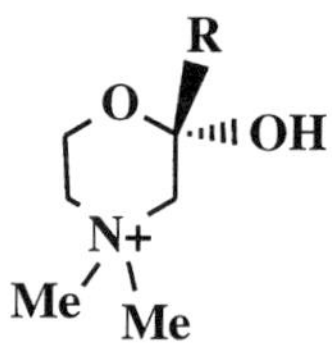

R - NMe_3+

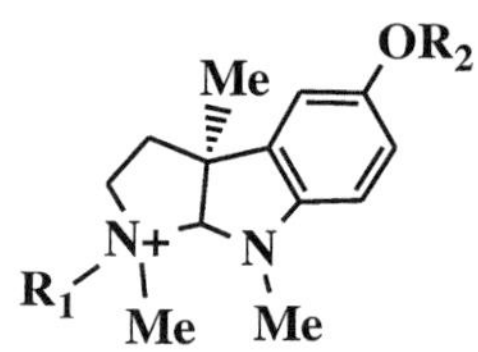

Various AChE inhibitor structures used

They extracted the structures of enzyme-bound ligands, and optimized their geometries. The structures of three inhibitors were then used as templates to determine a plausible bioactive conformation and orientation of their close analogs. The superposition was accomplished by rms fitting of selected atoms, as well as the field fitting and manual rotation of selected torsion angles.

The CoMFA was performed using Q^2-guided region selection procedures in 1 Å spacing grids using an sp^3 carbon atom with a +1 charge. The following CoMFA model was obtained:

$$N = 60 \quad L = 7 \quad R^2 = 0.932, \quad RMSE = 0.451, \quad Q^2 = 0.734, \quad RMSE_{cv} = 0.894$$

Then they used the enzyme crystal structure to compare the CoMFA results. Normally, CoMFA contour maps are not considered to be comparable to the active site, and such comparisons should be exercised with extreme care. However, when the alignment is based on the target protein structure, as in this study, there may be certain correlations. Cho et al. [77] claimed that the location of the contour coefficient maps was consistent with what was known about the active site of AChE; the sterically favorable regions occupied cavities in the AChE active site, whereas the sterically unfavorable regions overlapped with enzyme atoms.

Although such a correlation was less obvious with the electrostatic fields, positive-charge favorable regions were found in the vicinity of residues that could accommodate positive charges (Glu-199, Ser-200, Ser-226 and Glu-327). However, the negative-charge favorable regions were found to be near the residues of Phe-288, Phe-290, Phe-330 and Phe-331, and the interpretation was less obvious.

Tong et al. [78] conducted a CoMFA study with different AChE inhibitors, N-benzylpiperidines. They did not use any X-ray structure for alignments due to the lack of appropriate enzyme–inhibitor complex structure. After deriving a CoMFA model, however, they initiated molecular dynamics simulations of AChE inhibitor complexes of these inhibitors in order to validate and refine their alignments. These results are not yet reported:

Various AChE inhibitor structures

5.6. Human immunodeficiency Virus (I) structure and its inhibitors

Oprea et al. used inhibitor bound enzyme X-ray structures not only to align the molecules for a CoMFA study, but also to evaluate the CoMFA results by comparing the CoMFA coefficient contour maps with the binding site structure [79].

Five different alignments were examined in their CoMFA study with various HIV-1 inhibitors, as shown below. One alignment (I) was obtained using field-fit of neutral structures, and the other alignment (V) was obtained using field-fit of the active site minimized charged structures. The CoMFA was performed with 59 inhibitors in 2 Å spacing grids using an sp^3 carbon atom with a –1 charge. The results from two alignments were discussed in greater detail:

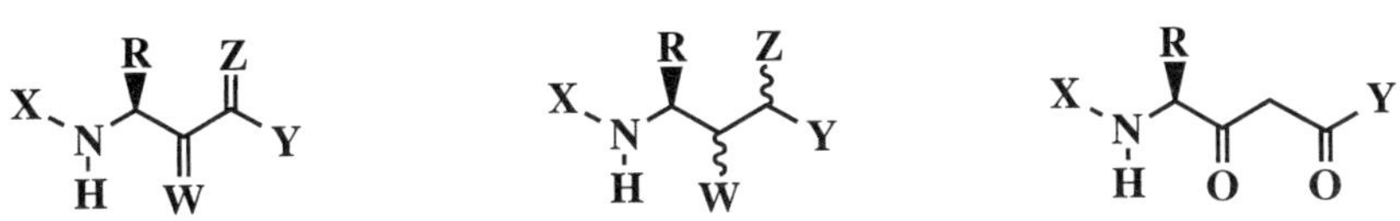

Various HIV-1 inhibitor structures

Alignments I and V yielded CoMFA models with the statistics shown below. The model from alignment I had R^2_{cv} and $RMSE_{cv}$ of 0.78 and 0.67, respectively, whereas the model from alignment V had R^2_{cv} and $RMSE_{cv}$ of 0.64 and 0.50, respectively. These models showed predictability for the test set of 34 compounds with R^2_{pred} and average error of prediction (AEP) of 0.68, 0.46 and 0.56, 0.64, respectively. Based on the statistical results, however, the authors could not draw any conclusions as to which of the two models was better:

Alignment	N	L	R^2	$RMSE$	Q^2	$RMSE_{cv}$	R^2_{pred}	AEP
I	59	6	0.98	0.15	0.78	0.67	0.68	0.46
V	59	7	0.99	0.13	0.64	0.50	0.56	0.64

Then they compared the CoMFA coeffient contour maps with the binding site structure. Significant differences in the contour maps were observed from the two alignments. Several residues that were important to ligand-binding were found to have corresponding steric and/or electrostatic CoMFA fields. For example, beneficial steric contacts could be overlapped with Arg-108 in S3, with Asp-30 in S2, Ile-50 and Gly-49 in S1, and Pro-81, Ile-150, Gly-148 and Gly-149 in S1′–S2′ pockets. Likewise, Asp-30 corresponded with the blue electrostatic (negative fields favorable) region in S2, Asp-25 was found in the vicinity of the blue contours in front of S1–S1′ and Gly-149 corresponded to the blue contour region in S1′ pocket.

Although the use of the enzyme structure was helpful in examining the CoMFA results, the comparisons also revealed limitations of the models, as some key residues were not overlapped with CoMFA fields.

5.7. *Dihydrofolate reductase structure and its inhibitors*

In a study of triazines inhibiting dihydrofolate reductase (DHFR), Greco et al. [80] used the X-ray structure information of a triazine–DHFR complex for the bioactive conformation and alignment for the ligands. Thus, all the geometry optimized structures were oriented based on two criteria: (1) the local dipole moment of the substituent had to be aligned as much as possible with that of the OCH_2 moiety in the crystal structure, and (2) the steric bulk of the substituent had to be smallest in the direction of the triazine nucleus. The molecules were superimposed by an rms fit between all the heavy atoms in common with the phenyltriazine ring:

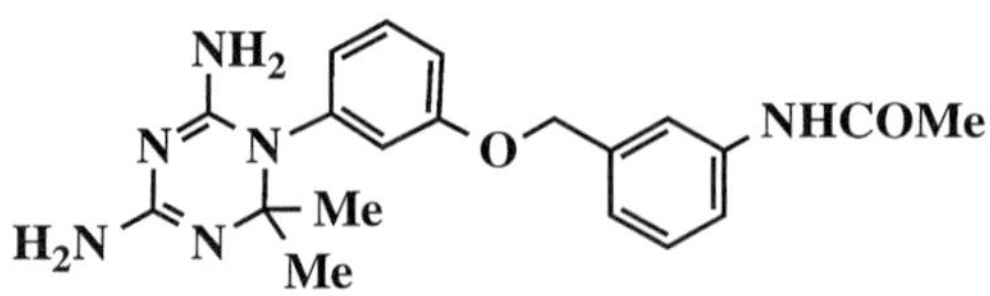

Triazine analog

After developing the CoMFA model shown below (with 35 inhibitors in 2 Å spacing grids using an sp^3 carbon atom with a +1 charge), they compared the CoMFA coefficient contour maps with the enzyme active site of the known X-ray structure. The authors indicated that the negative steric contours were near the residue Ile-60 within the active site of DHFR, and the positive and negative electrostatic contours were near the phenyl ring of Phe-34 and the guanidine moiety of Arg-70, respectively, at the active site:

$$N = 35 \quad L = 4 \quad R^2 = 0.90, \quad RMSE = 0.32, \quad Q^2 = 0.76, \quad RMSE_{cv} = 0.49$$

This is an example where the 3D structure of a ligand–enzyme complex was available, and the authors could define almost unambiguously the alignment rule and the bioactive conformation for the ligands. In addition, the authors had *a priori* knowledge of the physico-chemical factors which modulate activity from the published QSAR equations. Thus, the authors could compare the results of 2D, 3D QSAR (CoMFA) and the inhibitor–enzyme complex structure.

Unlike the work of Greco et al. described above, no consideration of the three-dimensional enzyme structures was given in the CoMFA study by Kroemer et al. [81,82], even though the X-ray structures of dihydrofolate reductase have been known and available in the PDB databank for some time.

6. Concluding Remarks

The methodologies of both homology modelling in GPCRs and the CoMFA approach of 3D QSAR are still in a stage of development; and there are still a number of limitations and weaknesses in these methods. None the less, significant advances have been made during the past several years in both fields. We have already seen that the two approaches are bridged together in many examples with other proteins.

Although there are only a few studies that have utilized both techniques for ligand design in the field of GPCRs, there is no doubt that more bridges will be built between the two approaches. It is the author's hope that this study becomes a small step toward building many bridges between the two very exciting and promising methodologies toward the common goal of ligand design.

References

1. Beck-Sickinger, A.G., *Structural characterization and binding sites of G protein-coupled receptors*, Drug Discov. Today, 1 (1996) 502–513.
2. Findlay, J.B.C. and Pappin, D.J.C., *The opsin family of proteins*, Biochem. J., 238 (1986) 625–642.
3. Henderson, R., Baldwin, J.M., Ceska, T.A., Zemlin, F., Beckmann, E. and Downing, K.H., *Model for the structure of bacteriorhodopsin based on high-resolution electron cryo-microscopy*, J. Mol. Biol., 213 (1990) 899–929.
4. Hoflack, J., Trumpp-Kallmeyer, S. and Hibert, M., *Molecular modeling of G protein-coupled receptors*, In Kubinyi, H. (Ed.) 3D QSAR in drug design: Theory, methods and applications, ESCOM, Leiden, The Netherlands, 1993, pp. 355–372.
5. Strader, C.D., Fong, T.M., Tota, M.R., Underwood, D. and Dixon, R.A.F., *Structure and function of G protein-coupled receptors*, Annu. Rev. Biochem., 63 (1994) 101–132.

6. Probst, W.C., Snyder, L.A., Schuster, D.I., Brosius, J. and Sealfon, S.C., *Sequence alignment of the G protein-coupled receptor superfamily*, DNA Cell Biol., 11 (1992) 1–20.
7. Lefkowitz, R., Cotecchia, S., Samama, P. and Costa, T., *Constitutive activity of receptors coupled to guanine nucleotide regulatory proteins*, Trends Pharmacol. Sci., 14 (1993) 303–307.
8. Strader, C.D., Fong, T.M., Graziano, M.P. and Tota, M.R., *The family of G protein-coupled receptors*, FASEB J., 9 (1995) 745–754.
9. Gether, U., Johansen, T.E., Snider, R.M., Lowe III, J.A., Nakanishi, S. and Schwartz, T.W., *Different binding epitopes on the NK1 receptor for substance P and a non-peptide antagonist*, Nature, 362 (1993) 345–348.
10. Rosenkilde, M.M., Cahir, M., Gether, U., Hjorth, S.A. and Schwartz, T.W., *Mutations along transmembrane segment II of the NK-1 receptor affect substance P competition with non-peptide antagonists but not substance P binding*, J. Biol. Chem., 269 (1994) 28160–28164.
11. Sautel, M., Rudolf, K., Wittneben, H., Herzog, H., Martinez, R., Munoz, M., Eberlein, W., Engle, W., Walker, P. and Beck-Sickinger, A.G., *Neuropeptide Y and the non-peptide antagonist BIBP 3226 share an overlapping binding site at the human Y1 receptor*, Mol. Pharmacol., 50 (1996) 285–292.
12. Schwartz, T.W. and Wells, T.N.C., *Is there a 'lock' for all agonist 'keys' in 7TM receptors?*, Trends Pharmacol. Sci., 17 (1996) 213–216.
13. Samana, P., Cotecchia, S., Costa, T. and Lefkowitz, R.J., *A Mutation-induced activated state of the b2-adrenergic receptor*, J. Biol. Chem., 268 (1993) 4625–4636.
14. Kuipers, W., van Wijngaaden, I. and Ijzerman, A.P., *A model of the serotonin 5-HT1A receptor: Agonist and antagonist binding sites*, Drug Des. Discuss., 11 (1994) 231–249.
15. Schertler, G.F.X., Villa, C. and Henderson, R., *Projection structure of rhodopsin*, Nature, 362 (1993) 770–772.
16. Soppa, J., *Two hypotheses — one answer: Sequence comparison does not support an evolutionary link between halobacterial retinal proteins including bacteriorhodopsin and eukaryotic G protin-coupled receptors*, FEBS Lett., 342 (1994) 7–11.
17. Donnelly, D., Findlay, J.B.C. and Blundell, T.L., *The evolution and structure of aminergic G protein-coupled receptors*, Receptors Channels, 2 (1994) 61–78.
18. Baldwin, J.M., *The probable arrangement of the helices in G protein-coupled receptors*, EMBO J., 12 (1993) 1693–1703.
19. Hoflack, J., Trumpp-Kallmeyer, S. and Hibert, M., *Re-evaluation of bacteriorhodopsin as a model for G protein-coupled receptors*, Trends Pharmacol. Sci., 15 (1994) 7–9.
20. Rost, B., Casadio, R., Fariselli, P. and Sander, C., *Transmembrane helices predicted at 95% accuracy*, Protein Sci., 4 (1995) 521–533.
21. Nordvall, G. and Hacksell, U., *Binding-site modeling of the muscarinic m1 receptor: A combination of homology-based and indirect approaches*, J. Med. Chem., 36 (1993) 967–976.
22. Hutchins, C., *Three-dimensional models of the D_1 and D_2 dopamine receptors*, Endocrine J., 2 (1994) 7–23.
23. Batlle, M., Campillo, M., Giraldo, J. and Pardo, L., *Computer-aided drug design of selective 5-hydroxytryptamine 1A receptor ligands using a three-dimensional model*, In Sanz, F., Giraldo, J. and Manaut, F. (Eds.) QSAR and molecular modeling: Concepts, computational tools and biological applications, J.R. Prous Science Publishers, Barcelona, Spain, 1995, pp. 541–544.
24. Bourdon, H., Trumpp-Kallmeyer, S., Hoflack, J., Hibert, M. and Wermuth, C.G., *Modeling of muscarinic M1 agonists: Study of their interaction with the M1 receptor*, In Sanz, F., Giraldo, J., and Manaut, F. (Eds.) QSAR and molecular modeling: Concepts, computational tools and biological applications, J.R. Prous Science Publishers, Barcelona, Spain, 1995, pp. 514–518.
25. Burbach, J.P.H. and Meijer, O.C., *The structure of neuropeptide receptors*, Eur. J. Pharmacol.-Mol. Pharmacol., 227 (1992) 1–18.
26. Chou, K.-C., Carlacci, L., Maggiora, G.M., Parodi, L.A. and Schulz, M.W., *An energy-based approach to packing the 7-helix bundle of bacterirhodopsin*, Protein Sci., 1 (1992) 810–827.
27. Cronet, P., Sander, C. and Vriend, G., *Modeling of transmembrane seven helix bundles*, Protein Eng., 6 (1993) 59–64.
28. Dahl, S.G., Edvardsen, I. and Sylte, I., *Molecular dynamics of dopamine at the D_2 receptor*, Proc. Natl. Acad. Sci. U.S.A., 88 (1991) 8111–8115.

29. De Benedetti, P.G., Menziani, M.C., Fanelli, F. and Cocchi, M., *The heuristic-direct approach to QSAR analysis of ligand-G-protein coupled receptor complex*, In Sanz, F., Giraldo, J., and Manaut, F. (Eds.) QSAR and molecular modeling: Concepts, computational tools and biological applications, J.R. Prous Science Publishers, Barcelona, Spain, 1995, pp. 526–527.
30. Dijkstra, G.D.H., Tulp, M.T.M., Hermkens, P.H.H., van Maarseveen, J.H., Scheeren, H.W. and Kruse, C.G., *Synthesis and receptor-affinity profile of N-hydroxytryptamine derivatives for serotonin and tryptamine receptors: A molecular-modeling study*, Recl. Trav. Chim. Pays-Bas., 112 (1993) 131–136.
31. Edvardsen, O., Sylte, I. and Dahl, S.G., *Molecular dynamics of serotonin and ritanserin interacting with the 5-HT2*, Mol. Brain Res., 14 (1992) 166–178.
32. Egner, U., Gerbling, K.P., Hoyer, G.-A., Kruger, G. and Wegner, P., *Design of inhibitors of photosystem II using a model of the D1 protein*, Pestic. Sci., 47 (1996) 145–158.
33. Fanelli, F., Menziani, M.C., Cocchi, M. and De Benedetti, P.G., *Comparative molecular dynamics study of the seven-helix bundle arrangement of G protein-coupled receptors*, J. Mol. Struct. (Theochem), 333 (1995) 49–69.
34. Findlay, J.B.C. and Donnelly, D. (Ed.), The superfamily: molecular modeling, Springer-Verlag, Berlin, Germany, 1993, pp. 17–31.
35. Grotzinger, J., Engels, M., Jacoby, E., Wollmer, A. and Strassburger, W., *A model for the C5a receptor and for its interaction with the ligand*, Protein Eng., 4 (1991) 767–771.
36. Hibert, M., Hoflack, J., Trumpp-Kallmeyer, S., Paquet, J.-L., Leppik, R., Mouillac, B., Chini, B., Barberis, C. and Jard, S. (Ed.), *Three-dimensional structure of G protein-coupled receptors: From speculations to facts*, Elsevier Science, Amsterdam, The Netherlands, 1996.
37. Humblet, C., Lunney, E.A. and Mirzadegan, T. (Ed.), *Docking ligands in the receptor cavity: What have we learned*?, ESCOM, Leiden, The Netherlands, 1993, pp. 35–43.
38. Kenakin, T., *Receptor conformational induction versus selection: All part of the same energy landscape*, Trends Pharmacol. Sci., 17 (1996) 190–191.
39. Krause, G., Kuhne, R. and Hubel, S. (Ed.), *G protein-coupled receptors, glucagon type: How to overcome the alignment/fit dilemma to the bacteriorhodopsin template*, J.R. Prous Science Publishers, Barcelona, Spain, 1995, pp. 531–533.
40. Kuipers, W., Kruse, C.G., van Wijngaarden, I., Standaar, P.J., Tulp, M.T.M., Veldman, N., Spek, A.L. and Ijzerman, A.P., *5-HT_{1A}-versus D_2-receptor selectivity of flesinoxan and analogous N4-substituted N1-arylpiperazines*, J. Med. Chem., 40 (1997) 300–312.
41. Livingstone, C.D., Strange, P.G. and Naylor, L.H., *Molecular modeling of D_2-like dopamine receptors*, Biochem. J., 287 (1992) 277–282.
42. Luo, X., Zhang, D. and Weinstein, H., *Ligand-induced domain motion in the activation mechanism of a G protein-coupled receptor*, Protein Engng., 7 (1994) 1441–1448.
43. Maloney Huss, K. and Lybrand, T.P., *Three-dimensional structure for the β_2 adrenergic receptor protein based on computer modeling studies*, J. Mol. Biol., 225 (1992) 859–871.
44. Menziani, M.C., Cocchi, M., Fanelli, F. and De Benedetti, P.G., *Theoretical QSAR analysis on three dimensional models of the complexes between peptide and non-peptide antagonists with the ET_A and ET_B receptors*, In Sanz, F., Giraldo, J., and Manaut, F. (Eds.) QSAR and molecular modeling: Concepts, computational tools and biological applications, J.R. Prous Science Publishers, Barcelona, Spain, 1995, pp. 519–525.
45. Moereels, H. and Leysen, J.E., *Novel computational model for the interaction of dopamine with the D_2 receptor*, Recept. Channels, 1 (1993) 89–97.
46. Nederkoorn, P.H.J., va Lenthe, J.H., van der Goot, H., den Kelder, G.M.D.-O. and Timmerman, H., *The agonistic binding site at the histamine H2 receptor: 1. Theoretical investigations of histamine binding to an oligopeptide mimicking a part of the fifth transmembrane α-helix*, J. Comput.-Aid. Mol. Design, 10 (1996) 461–478.
47. Nero, T.L., Iakovidis, D. and Louis, W.J., *Molecular modeling of the human β_1-adrenoceptor*, In Sanz, F., Giraldo, J., and Manaut, F. (Eds.) QSAR and molecular modeling: Concepts, computational tools and biological applications, J.R. Prous Science Publishers, Barcelona, Spain, 1995, pp. 528–530.
48. Pardo, L., Ballesteros, J.A., Osman, R. and Weinstein, H., *On the use of the transmembrane domain of the bacteriorhodopsin as a template for modeling the three-dimensional structure of guanine nucleotide-binding regulatory protein-coupled receptors*, Proc. Natl. Acad. Sci. U.S.A., 89 (1992) 4009–4012.

49. Sagara, T., Egashira, H., Okamura, M., Fujii, I., Shimohigashi, Y. and Kanematsu, K., *Ligand recognition in mu opioid receptor: Experimentally based modeling of mu opioid receptor binding sites and their testing by ligand docking*, Bioorg. Med. Chem., 4 (1996) 2151–2166.
50. Sankararamakrishnan, R. and Vishveshwara, S., *Characterization of proline-containing α-helix (helix F model of bacteriorhodopsin) by molecular dynamics studies*, Proteins: Struct. Funct. Genet., 15 (1993) 26–41.
51. Sugden, D., Chong, N.W.S. and Lewis, D.F.V., *Structural requirements at the melatonin receptor*, Br. J. Pharmacol., 114 (1995) 618–623.
52. Sylte, I., Edvardsen, O. and Dahl, S.G., *Molecular modeling of UH-301 and 5-HT$_{1a}$ receptor interactions*, Protein Eng., 9 (1996) 149–160.
53. Teeter, M.M., Froimowitz, M., Stec, B. and DuRand, C.J., *Homology modeling of the dopamine D_2 receptor and its testing by docking of agonists and tricyclic antagonists*, J. Med. Chem., 37 (1994) 2874–2888.
54. Trumpp-Kallmeyer, S., Chini, B., Mouillac, B., Barberis, C., Hoflack, J. and Hilbert, M., *Towards understanding the role of the first extracellular loop for the binding of peptide hormones to G protein-coupled receptors*, Pharm. Acta Helv., 70 (1995) 255–262.
55. Weinstein, H. and Zhang, D., *Receptor models and ligand-induced responses: New insights for structure–activity relations*, In Sanz, F., Giraldo, J., and Manaut, F. (Eds.) QSAR and molecular modeling: Concepts, computational tools and biological applications, J.R. Prous Science Publishers, Barcelona, Spain, 1995, pp. 497–507.
56. Yamamoto, Y., Kamiya, K. and Terao, S., *Modeling of human thromboxane A2 receptor and analysis of the receptor-ligand interaction*, J. Med. Chem., 36 (1993) 820–825.
57. Zhang, S. and Weinstein, H., *Signal transduction by a 5-HT$_2$ receptor: A mechanistic hypothesis from molecular dynamics simulations of the three-dimensional model of the receptor complexed to ligands*, J. Med. Chem., 36 (1993) 934–938.
58. Baxevanis, A.D., Makalowski, W., Ouellette, B.F.F. and Recipon, H., *Web alert protein engineering*, Curr. Opinion Biotech., 7 (1996) 462.
59. Peitsch, M.C., Herzyk, P., Wells, T.N.C. and Hubbard, R.E., *Automated modeling of the transmembrane region of G protein-coupled receptor by Swiss-Model*, Receptors Channels, 4 (1996) 161–164.
60. Hibert, M.F., Trumpp-Kallmeyer, S., Hoflack, J. and Bruinvels, A., *This is not a G protein-coupled receptor*, Trends Pharmacol. Sci., 14 (1993) 7–12.
61. Rost, B. and Valencia, A., *Pitfalls of protein sequence analysis*, Curr. Opinion Biotech., 7 (1996) 457–461.
62. Navajas, C., Kokkola, T., Poso, A., Honka, N., Gynther, J. and Laitinen, J.T., *A rhodopsin-based model for melatonin recognition at its G protein-coupled receptor*, Eur. J. Pharmacol., 304 (1996) 173–183.
63. Gaillard, P., Carrupt, P.-A., Testa, B. and Schambel, P., *Binding of arylpiperazines, (aryloxy) propanolamines, and tetrahydropyridylindoles to the 5-HT$_{1A}$ receptor: Contribution of the molecular lipophilicity potential to three-dimensional quantitative structure–affinity relationship models*, J. Med. Chem., 39 (1996) 126–134.
64. Dove, S., Kuhne, R. and Schunack, W., *H_1 agonistic 2-heteroaryl and 2-phenylhistamines: CoMFA and possible receptor binding sites*, In Sanz, F., Giraldo, J., Manaut, F. (Eds.) QSAR and molecular modeling: Concepts, computational tools and biological applications, Proceedings of the 10th European Symposium on Structure–Activity Relationships: QSAR and Molecular Modeling, Barcelona, Spain, September 4–9, 1994, J.R. Prous Science Publishers, Barcelona, 1995, pp. 427–432.
65. Trumpp-Kallmeyer, S., Hoflack, J., Bruinvels, A. and Hibert, M., *Modeling of G-protein-coupled receptors: Application to dopamine, adrenaline, serotonin, acetylcholine, and mammalian opsin receptors*, J. Med. Chem., 35 (1992) 3448–3462.
66. Yamashita, M., Fukui, H., Sugama, K., Yoshiyuki, H., Ito, S., Mizuguchi, H. and Wada, H., *Expression cloning of a cDNA encoding the bovine histamine H_1 receptor*, Proc. Natl. Acad. Sci. U.S.A., 88 (1991) 11515–11519.
67. Carriere, A., Altomare, C., Barreca, M.L., Contento, A., Carotti, A. and Hansch, C., *Papain catalyzed hydrolysis of aryl esters: A comparison of the Hansch, docking and CoMFA methods*, Farmaco, 49 (1994) 573–585.

68. Smith, R.N., Hansch, C., Kim, K.H., Omiya, B., Fukumura, G., Selassie, C.D., Jow, P.Y.C., Blaney, J.M. and Langridge, R., *The use of crystallography, graphics, and quantitative structure–activity relationships in the analysis of the papain hydrolysis of X-phenyl hippurates*, Arch. Biochem. Biophys., 215 (1982) 319–328.
69. Drenth, J., Kalk, K.H. and Swen, H.M., *Binding of chloromethyl ketone substrate analogues to crystalline papain*, Biochem., 15 (1976) 3731–3738.
70. Watson, K., Mitchell, E.P., Johnson, L.N., Cruciani, G., Son, J.C., Bichard, C.J.F., Fleet, G.W.J., Oikonomakos, N.G., Kontou, M. and Zographos, S.E., *Glucose analogue inhibitors of glycogen phosphorylase: From crystallographic analysis to drug prediction using GRID force-field and GOLPE variable selection*, Acta Cryst., D51 (1995) 458–472.
71. Cruciani, G. and Watson, K.A., *Comparative molecular field analysis using GRID force-field and GOLPE variable selection methods in a study of inhibitors of glycogen phosphorylase b*, J. Med. Chem., 37 (1994) 2589–2601.
72. Recanatini, M., *Comparative molecular field analysis of non-steroidal aromatase inhibitors related to fadrozole*, J. Comput.-Aid. Mol. Design, 10 (1996) 74–82.
73. Laughton, C.A., Zvelebil, M.J.J.M. and Neidle, S., *A detailed molecular model for human aromatase*, J. Steroid Biochem. Mol. Biol., 44 (1993) 399–407.
74. Zhou, D., L., C.L., Laughton, C.A., Korzekwa, K.R. and Chen, S., *Mutagenesis study at a postulated hydrophobic region near the active site of aromatase cytochrome P450*, J. Biol. Chem., 269 (1994) 19501–19508.
75. Diana, G.D., Nitz, T.J., Mallamo, J.P. and Treasurywala, A.M., *Antipicornavirus compounds: Use of rational drug design and molecular modeling*, Antivir. Chem. Chemother., 4 (1993) 1–10.
76. Artico, M., Botta, M., Corelli, F., Mai, A., Massa, S. and Ragno, R., *Investigation on QSAR and binding mode of a new class of human rhinovirus-14 inhibitors by CoMFA and docking experiments*, Bioorg. Med. Chem., 4 (1996) 1715–1724.
77. Cho, S.J., Garsia, M.L.S., Bier, J. and Tropsha, A., *Structure-based alignment and comparative molecular field analysis of acetylcholinesterase inhibitors*, J. Med. Chem., 39 (1996) 5064–5071.
78. Tong, W., Collantes, E.R., Chen, Y. and Welsh, W.J., *A comparative molecular field analysis study of N-benzylpiperidines as acetylcholinesterase inhibitors*, J. Med. Chem., 39 (1996) 380–387.
79. Oprea, T.I., Waller, C.L. and Marshall, G.R., *3D QSAR of human immunodeficiency virus (I) protease inhibitors: 3. Interpretation of CoMFA results*, Drug Des. Discovery, 12 (1994) 29–51.
80. Greco, G., Novellino, E., Pellecchia, M., Silipo, C. and Vittoria, A., *Effects of variable section on CoMFA coefficient contour maps in a set of triazines inhibiting DHFR*, J. Comput.-Aided Mol. Design, 8 (1994) 97–112.
81. Kroemer, R.T. and Hecht, P., *A new procedure for improving the predictiveness of CoMFA models and its application to a set of dihydrofolate reductase inhibitors*, J. Comput.-Aid. Mol. Design, 9 (1995) 396–406.
82. Kroemer, R.T. and Hecht, P., *Replacement of steric 6-12 potential-derived interaction energies by atom-based indicator variables in CoMFA leads to models of higher consistency*, J. Comput.-Aid. Mol. Design, 9 (1995) 205–212.

A Critical Review of Recent CoMFA Applications

Ki Hwan Kim,[a] Giovanni Greco,[b] and Ettore Novellino[c]
[a] *Department of Structural Biology, D46Y, AP10-2, Pharmaceutical Products Division, Abbott Laboratories, 100 Abbott Park Road, Abbott Park, IL 60064-3500, U.S.A.*
[b] *Dipartimento di Chimica Farmaceutica e Tossicologica, Università di Napoli 'Federico II', Via Domenico Montesano 49, 80131 Naples, Italy*
[c] *Dipartimento di Scienze Farmaceutiche, Università di Salerno, Piazza Vittorio Emanuele 9, 84048 Penta (Salerno), Italy*

1. Introduction

Comparative molecular field analysis (CoMFA) is a technique for determining three-dimensional quantitative structure–activity relationships (3D QSAR). In a standard CoMFA procedure, a bioactive conformation of each compound under study is chosen, and all the structures are superimposed in a manner defined by the supposed mode of interaction with the target macromolecule. Then, the steric and the electrostatic fields of these molecules are calculated with a probe atom, such as sp^3 carbon atom with +1 charge, at regularly spaced (1 or 2 Å) points of a three-dimensional grid. Sometimes other fields or physico-chemical parameters are also included. The calculated energy values and other descriptor values are then analyzed with the partial least-squares (PLS) statistical technique. The optimum number of components for the CoMFA model is selected based on the cross-validation test results. The final CoMFA model is derived using the optimum number of components selected. The results are usually displayed as coefficient contour maps. A good CoMFA model should show satisfactory statistical significance, explanatory capability of the variance in the activity of the compounds in the training set and predictive power of the potency of new compounds.

This work describes the CoMFA studies published since 1993. Any aspects of the standard CoMFA procedure or the works described in the previous volume [156L]* of this book or those subjects that are extensively discussed in other chapters of this volume are not discussed in any detail. For such subjects, readers are referred to the corresponding chapters in this volume.

There are many choices to be considered in a CoMFA analysis: [134L] biological data, selection of compounds and series design, generation of three-dimensional structure and charges of the ligand molecules, conformational analysis and establishment of the bioactive conformation of each molecule, alignment of the molecules, position of the lattice points, choice of force fields and calculation of the interaction energies, statistical analysis of the data and the selection of the final model, display of the results in contour maps and their interpretations, and design and forecasting the activity of unknown compounds.

Those studies reported in the last few years can be largely divided into two groups. The first group includes those that studied various aspects of CoMFA procedures to

* References in the format [xxL] are to citations in the last chapter of this volume.

H. Kubinyi et al. (eds.), 3D QSAR in Drug Design, Volume 3. 257–315.

improve the method. The second group includes those that applied the method to various research problems. Many studies focused on both issues. In the following sections, each of these main topics will be reviewed.

An introduction to the CoMFA procedures is described in recent reviews [61L,127L, 134L,173L,313L]. For various 3D QSAR approaches, readers are referred to the corresponding chapters in this volume.

2. General Aspects of CoMFA Applications

2.1. *Series design and selection of the training set*

Series design refers to the process of selecting a set of compounds to be included in a study, with the aim of gaining the maximum amount of information possible with a minimum number of compounds. Three major issues in choosing compounds are (1) minimization of collinearity between the predictor properties, (2) maximization of variance of these properties and (3) mapping of substituent space with the smallest number of compounds [134L]. The choice of the compounds for synthetic priority and testing is crucial in the early stage of a project aimed at optimizing the desired activity of a lead while reducing or eliminating undesired properties by structural modifications.

The selection of a subset of compounds that represent the total set is important not only in series design, but also in the selection of compounds for a training set in 3D QSAR analysis. A CoMFA model from a well-designed set of compounds is expected to improve the interpretability and the predictiveness of a CoMFA model. Several studies devoted to this subject were previously discussed [1,2,53L], including the use of latent variables or principal properties (PPs), factorial designs, fractional factorial designs or D-optimal designs based on PPs, auto- and cross-covariance-based 3D PPs, principal components and cluster analysis based on CoMFA energy fields.

Caliendo et al. [39L] investigated the factorial design approach as a series design method for selecting a training set for a CoMFA study. They studied the Michaelis constant (log $1/K_m$) values of 71 N-acyl-L-amino acid esters as α-chymotrypsin substrates. After calculating CoMFA steric and electrostatic fields, the first three principal components were extracted from a principal component analysis (PCA) on the CoMFA energy fields. Two different training sets (set A and set B) of 12 compounds were selected based on the factorial design. Set A was selected based on equal weight of the three components, and set B was chosen based on weighted principal components accounting for the relative sizes of the principal component eigenvalues. In addition, 50

Table 1 *CoMFA analysis of Michaelis constants (log $1/K_m$) of 71 N-acyl-L-amino acid esters as α-chymotrypsin substrates, using compounds selected by factorial design based on the first three principal components of the CoMFA fields (L = number of latent variables)*

Set	N	L	Q^2	s_{cv}	R^2_{pred}	s_{pred}
A	12	3	0.864	0.501	0.603	0.941
B	12	1	0.648	1.031	0.767	0.732

additional sets of 12 compounds were chosen by a random selection procedure. Then, CoMFA models were derived from each of the 52 sets, and the resulting models were used to predict the binding affinity of the remaining 59 compounds. Their results (Table 1) showed that the composition of the training set dramatically influenced the cross-validation results. It is interesting that, although set A gave better cross-validation results than set B, the CoMFA model from set B forecasted the binding affinity of 59 compounds more accurately. The authors concluded that set B was made of more balanced compounds than set A; 42% of the 50 randomly selected sets yielded a model that was superior to that of set B. These results suggested that although the probability of selecting informative series from a random selection may be far from zero, in the absence of a proper series design strategy there is a risk of deriving a poorly predictive CoMFA model.

Another series design procedure was investigated by Novellino et al. [40L,201L], who applied cluster analysis on the first three principal components generated from the interaction energies of CoMFA. They assessed the efficiency of their procedures by (1) deriving a CoMFA model from compounds forming a rationally designed training set, (2) predicting the biological activity of remaining compounds using the CoMFA model and (3) comparing the R^2 and s values with those from the cross-validation using all compounds. Cluster analysis on the principal component scores divided the 71 compounds into 12 clusters. From each of the 12 clusters, the most representative member was chosen. The CoMFA model from 12 compounds was then used to forecast the activity of the remaining 59 compounds. The quality of the cross-validation ($Q^2 = 0.83$ and $s_{cv} = 0.70$) was comparable to those of the CoMFA model derived from all 71 compounds ($Q^2 = 0.87$ and $s_{cv} = 0.58$). Based on the results (Table 2), Novellino et al. concluded that the training set of 12 compounds selected through cluster analysis was a representative set of the whole molecules.

Mabilia et al. [168L] used GOLPE to select 18 compounds from 28 angiotensin II antagonists. The CoMFA models derived from either 18 compounds or 28 compounds were similar in statistics. Interestingly, for the prediction of 5 external compounds, the reduced set yielded better results than the original set.

2.2. *Geometries and optimizations*

When a set of molecules is available for analysis, the first task is to build their 3D structures. Two aspects should be considered in this step: how to represent the structures accurately, and how to determine the bioactive conformation.

Table 2 *CoMFA analysis of Michaelis constants (log 1/K$_m$) of 71 N-acyl-L-amino acid esters as α-chymotrypsin substrates using compounds selected by cluster analysis of the first three principal components of the CoMFA fields*

N	L	Q^2	s_{cv}	R^2	s
12	2	0.827	0.697	0.961	0.331
71	3	0.868	0.583	0.906	0.492

Many times the X-ray structures of related compounds are a source of initial geometry, and sometimes they are also a source of bioactive conformation [19L,49L, 68L,76L,79L,117L,205L,260L,265L,266L,275L,289L]. Different levels of computational methods are used for the optimization of the initial geometries. Although molecular mechanics or semiempirical quantum mechanics are most often used, a higher level of accuracy was sometimes sought [275L].

Since the molecular fields of each aligned molecule are calculated using the positions of its atoms, the results of a CoMFA depend on the geometries of the compounds. Then, how much does the quality of molecular geometry affect CoMFA? A number of papers dealt with this important issue.

In a study with 36 aryl sulfonamides tested as antagonists of endothelin receptor subtype-A (ET_A), Krystek et al. [154L] studied the effects of crudely optimized geometries and simple charge calculations on the CoMFA results. The crude structures were based on the Tripos fragment library, which had been derived from average geometries from the Cambridge Structural Database. In some cases, this led to non-optimum conformations. These crude structures also carried simply and quickly determined atomic charges. The analysis yielded a three-component model with the Q^2 and s_{cv} values of 0.50 and 0.83 and the fitted R^2 and s values of 0.91 and 0.35, respectively. When the geometries were optimized, there was essentially no change in the CoMFA results, with $Q^2 = 0.49$, $s_{cv} = 0.84$, $R^2 = 0.91$, and $s = 0.37$.

The problem of generating realistic structures was also investigated by Horwitz et al. [117L] with a set of antitumor thioxanthenones. For one model compound, the authors compared the geometries optimized by semiempirical quantum mechanics methods (MNDO, AM1 and PM3 as implemented in MOPAC 6.0) with that optimized by *ab initio* calculations using the HF/6-31G* basis set. Based on the CoMFA results, they selected PM3 as the method of choice to optimize fully all the compounds of the training set.

Recanatini [224L] derived statistically similar models from a set of non-steroidal aromatase inhibitors using the structures minimized by the Tripos force field or by the AM1 Hamiltonian; the former structures used Gasteiger-Marsili charges, whereas the latter used AM1 charges. The results are summarized in Table 3.

The relatively low sensitivity of CoMFA on the quality of the molecular geometries receives further support from the findings of Oprea et al. [207L]. A CoMFA model forecasted the inhibitory potencies, expressed as pIC_{50}, of 36 test set molecules docked into a semi-rigid model of the HIV-1 protease. These molecules were predicted with their geometries minimized in the active site, as well as with the energy-minimized structures in vacuum using the Tripos force field. The first geometries were somewhat distorted

Table 3 *CoMFA results of 29 non-steroidal aromatase inhibitors*

Geometry optimization method	L	Q^2	s_{cv}	R^2	s
Tripos	2	0.74	0.41	0.88	0.28
AM1	3	0.76	0.40	0.95	0.18

since the active site was kept rigid about backbone atoms and water molecules. The results from the two sets of geometries showed that the differences in the predicted pIC_{50} values were all less than 0.3 log unit.

Hocart et al. [113L] also investigated the influence of geometries optimized at two different accuracy levels. Interestingly, the CoMFA models derived from the fully minimized peptide structures produced less accurate predictions than did the models derived from the less fully minimized structures. One possible cause for such paradoxical results may result from the energy minimization of highly flexible molecules in vacuum. The authors observed that many changes occurred during the final minimization, including formation of an additional hydrogen bond. Thus, full minimization might have overemphasized intramolecular interactions, whereas the bioactive conformations are influenced by intermolecular interactions with the receptor atoms. Poor alignment could be another reason. From a statistical standpoint, a 'disordered' alignment implies an increased level of noise in PLS analysis. A possible solution to this type of problem might be introducing constraints aimed at optimizing the degree of overlap among different ligands or, more simply, adopting less stringent convergence criteria.

These studies suggest that very accurate geometries are not essential to obtain a reasonable CoMFA model. No article has yet appeared reporting that crude geometries yielded a significantly worse CoMFA model from one built with high-quality geometries. However, such a diminished role of molecular geometries in CoMFA may not be totally unreasonable because the typical grid spacing employed in CoMFA studies is 2 Å, and even 1 Å grid spacing is large compared to the relatively small differences between the 'crude' and 'accurate' molecular structures.

2.3. *Charges*

2.3.1. *Partial atomic charges*

There are many methods for calculating partial atomic charges. They range from simple Gasteiger-Hückel charge calculations and semiempirical quantum mechanical approaches to a number of methods for fitting charges of the electrostatic field around a molecule. There are limits to how accurately atomic charges can reproduce molecular electrostatics.

How important is the method of atomic charge calculations? Although a number of researchers investigated this issue from the early days of CoMFA [4], there seems to be no consensus answer.

Table 4 *CoMFA results of angiotensin converting enzyme and thermolysin inhibitors*

Method of charge calculation	ACE			Thermolysin		
	L	Q^2	R^2_{pred} ($N = 20$)	L	Q^2	R^2_{pred} ($N = 11$)
Gasteiger-Hückel	8	0.66	0.53	11	0.70	0.29
PM3 (Mulliken population analysis)	7	0.66	0.46	5	0.54	0.44

For example, in a study of the 5-HT_3 receptor binding affinity of 39 piperazino-pyrrolo-thieno-pyrazines, Bureau et al. [30L] compared the CoMFA results obtained with the partial charges computed from electrostatic potential, quantum mechanically calculated charges using 6-31G* basis set, and Gasteiger-Hückel charges. The electrostatic potential charges yielded a model with the cross-validated Q^2 and *SEP* values of 0.46 and 1.48, respectively, and the fitted R^2 and *RMSE* values of 0.86 and 0.76 of a five-component model. In contrast, the Gasteiger-Hückel charges yielded an inferior model with the cross-validated Q^2 and *SEP* values of 0.32 and 1.59, respectively, and the fitted R^2 and *RMSE* values of 0.52 and 1.33 of a two-component model.

In a study of 37 benzodiazepine receptor ligands, Kroemer et al. [153L] examined 17 different methods at three different levels of theory to calculate charges and their effects on CoMFA. Gasteiger-Marsili, semiempirical (MNDO, AM1 and PM3) and *ab initio* (HF/STO-3G, HF/3-21G* and HF/6-31G*) charges were included. Semiempirical and *ab initio* electron populations were derived both from the Mulliken population analysis and from fitting the charges to the molecular electrostatic potential (ESPFIT charges). In addition, the molecular electrostatic potentials from *ab initio* calculations were mapped directly onto the CoMFA grid-points. The ESPFIT-derived charges yielded higher Q^2 values than those based on charges calculated for Mulliken population analysis. However, the simple Gasteiger-Marsili charges did not give the worst model. The Q^2 and SE_{cv} values of various electrostatic CoMFA models ranged 0.39–0.53 and 1.04–1.16 respectively, whereas those of various CoMFA models with both fields ranged 0.61–0.77 and 0.76–0.94, respectively.

Waller et al. [260L] compared the effect on CoMFA of using charges calculated using the Gasteiger-Hückel and PM3 methods for angiotensin converting enzyme (ACE) and thermolysin inhibitors (Table 4). In the ACE inhibitor series, the two methods gave nearly identical Q^2 values. PM3 charges performed slightly less well in forecasting the potencies of 20 chemically diverse ligands. External predictions of additional analogs belonging to three different chemical classes yielded very similar R^2_{pred} values. For the thermolysin inhibitors, a higher Q^2 was achieved using the Gasteiger-Hückel charges, but the PM3 method provided more accurate external predictions for 11 test compounds.

In a study of non-steroidal aromatase inhibitors related to fadrozole, Recanatini [224L] reported similar models from the geometries and charges obtained with AM1 and those with the MAXIMIN2 molecular mechanics optimized geometries and Gasteiger-Marsili charges: Q^2 from Gasteiger-Marsili charges was 0.74 for two-component model, whereas Q^2 from the AM1 charges was 0.76 for three-component model.

Table 5 *CoMFA results of non-peptidic angiotensin II antagonists*

	Charges calculated by Mulliken population analysis (MNDO)		Charges calculated by the empirical method of Gasteiger-Marsili	
Alignment	*L*	Q^2	*L*	Q^2
g	5	0.48	3	0.46
x	3	0.52	6	0.52

Table 6 *CoMFA results on the mutagenic activity of a set of 5H-furan-2-one derivatives*

LUMO field	N	L	Q^2	s_{cv}	R^2	s
MNDO	16	3	0.330	3.285	0.733	2.077
AM1	16	3	0.733	2.125	0.925	1.101
PM3	16	3	0.742	2.041	0.913	1.183

Belvisi et al. [19L] also compared Gasteiger-Marsili and MNDO charges calculated for a series of non-peptidic angiotensin II antagonists (modelled in two alternative alignments called *g* and *x*) and obtained similar cross-validated statistics from both alignments (Table 5).

The mutagenic activity of 16 5H-furan-2-one derivatives was correlated with the LUMO field by Navajas et al. [194L]. The MNDO, AM1 and PM3 Hamiltonians were employed to optimize fully each molecule, as well as to generate its LUMO field according to the SYBYL implementation. Only the AM1 and PM3 methods gave satisfactory CoMFA models (Table 6).

Different results were reported by Folkers et al. [91L]. Gasteiger-Marsili and semiempirical charges yielded similar statistical results, and the semiempirical ESPFIT and *ab initio* ESPFIT charges yielded similar results but better than the Gasteiger-Marsili and semiempirical charges. The MEPs mapped directly onto the CoMFA grid-points did not yield superior results to the ESPFIT-derived potentials. Their study showed that electrostatic fields resulting from different calculation methods influenced the CoMFA results greatly.

Krystek et al. [154L] also studied the relative influence of the geometries and charges. They studied the effects of simple charge calculations on the CoMFA models for 36 aryl sulfonamide antagonists of the endothelin receptor subtype-A (ET_A) receptor. As noted above, crude structures and simply determined atomic charges yielded a three-component CoMFA model with Q^2 and s_{cv} values of 0.50 and 0.83, respectively, and fitted R^2 and SE values of 0.91 and 0.35. However, when the charges were refined, the results improved substantially, even though the crude geometry for molecules was used: a four-component model with the Q^2 and s_{cv} values of 0.65 and 0.71, respectively. Similar results were obtained from the refined charges (PM3) and optimized geometries: a six-component model with the cross-validated Q^2 and s_{cv} values of 0.70 and 0.69, respectively, and the fitted R^2 and s values of 0.94 and 0.30. The results suggest that it is more important to have refined charge sets than refined molecular geometries.

Judging from the studies where different charge calculation methods have been compared, the overall impression is that semiempirical quantum mechanics approaches (MNDO, AM1, PM3) often produced charges which were adequate for CoMFA. However, simpler methods, such as Gasteiger-Marsili and Gasteiger-Hückel, quite often yielded results of comparable or only slightly worse quality. On the other hand, many successful CoMFA studies have been reported using relatively crude charges as a valid surrogate of semiempirical or *ab initio* wavefunctions. Thus, when dealing with a large training set, one might confidently employ a simple technique to check rapidly whether the electrostatic field is a relevant descriptor. Alternatively, to save computation time, several methods might be employed on a smaller group of compounds to select the most efficient one.

2.3.2. Charged molecules

When ionizable compounds are involved, one must decide which protonation state of the molecule to use in the calculation. Li et al. [329L] studied the inhibition of spermidine transport into L1210 cells by 46 polyamine analogs. The compounds contained from one to four cationic groups and were primary, secondary and tertiary alkylamines. All are positively charged at physiological pH. None-the-less, in order to get the best CoMFA model, they used different protonation states of ionization in the calculation. For the compounds with amino groups with pK_a values above 8, the positively charged group was used in the calculation, and when the pK_a of the functional group was less than 5, the neutral species were used. In the cases where the pK_a fell between 5–7, both charged and uncharged structures were included separately in the calculations. No compounds had the pK_a between 7–8. For the aziridine analogs, the protonated form was used if the pK_a value was above 6, and both if the pK_a was below 6.

Tong et al. [249L] also studied the effect of ionization in a study with two classes of acetylcholinesterase inhibitors, N-benzylpiperidine benzisoxazoles (NBPBs) and 1-benzyl-4-[2-(N-benzoylamino)ethyl]-piperidines (NBEPs). They investigated the influences of charged species on CoMFA using both neutral and protonated species, although the compounds involved were thought to be protonated at physiological pH. A better CoMFA model was obtained from the protonated species and two different alignments.

In a study on 93 chemically diverse inhibitors of HIV-1 protease, Marshall and his co-workers [206L,266L] also examined the effects of molecular charges. From five different alignments of 59 molecules in a test set, the two best results were obtained from alignment I and V. In alignment I, the molecules in their neutral form were aligned by field fit to the enzyme-bound X-ray structure of the most closely related compound followed by local energy minimization. In alignment V, the molecules in the protonated forms were put into the enzyme active site and energy minimized with the protein backbone and essential water molecules treated as rigid aggregates. The CoMFA models obtained from the two alignments have the statistics shown in Table 7. Interestingly, the electrostatic contribution in both models were similar: alignment I indicated 64% electrostatic and 36% steric, whereas alignment V indicated 68% electrostatic and 32% steric.

The robustness of each CoMFA model was evaluated by predicting the inhibitory potencies of 34 test set compounds belonging to three different chemical classes. Although the model from the charged species yielded a slightly lower R^2_{pred}, the low predictivity of the model (alignment V) was partially due to the negatively charged molecules in the test set. None of the training set compounds was an anion. Based on the statistical results, the authors could not conclude which of the two models was better.

Table 7 *CoMFA results of chemically diverse inhibitors of HIV-1 protease*

Alignment	N	L	Q^2	s_{cv}	R^2	s	R^2_{pred}
I	59	6	0.78	0.55	0.98	0.15	0.68
V	59	7	0.64	0.68	0.99	0.13	0.56

2.4. *Bioactive conformations and their alignment*

In CoMFA, selection of bioactive conformations and their alignments are the two most crucial steps. Not only do they often significantly influence the results, but they are also critical in the design of new molecules.

2.4.1. *Bioactive conformations*

When experimental structures of the ligand–macromolecule complex are available for all compounds, selecting the bioactive conformation is not an issue [64L,270L]; but this is not usually the case. More typically, if the bound structure of only one or a few compounds are known, they are used as a basis for constructing the bioactive conformation of related compounds [49L,76L,275L].

When no structural information is available, various computational approaches have been used for determining the bioactive conformation. A conformationally restricted compound is very helpful for determining the bioactive conformation, as in the study of angiotensin II AT_1 and AT_2 receptor antagonists by de Laszlo et al. [72L] When the molecules under study are conformationally flexible and no rigid molecules are available, the determination of bioactive conformations is more complicated. Many approaches that can be used in such cases were reviewed in the previous volume of this book [5–7].

Some authors used the global minimum energy conformation as the bioactive conformation, [302L], while others used higher-energy conformation (by up to 12 kcal/mol above the global minimum conformation) [191L]. Yliniemela et al. [8] suggested that there are several reasons for choosing conformers not based on Boltzmann distribution and conformational energies. First, molecular mechanical or semiempirical conformational energies are not very accurate. Second, solvent and physiological environment effects cannot be properly accounted for. Third, even a non-optimal conformer will be somewhat populated if the energy is not too high above the global minimum energy.

One selected bioactive conformation per compound is normally used in CoMFA. However, several studies were pursued with multiple sets of conformations, and CoMFA was used to select the probable bioactive conformations [45L,154L,254L, 275L]. For example, van Steen et al. [254L] investigated their hydrophobic and hydrophilic interaction site concept with two hypotheses about the way that the N4-substituents of phenylpiperazine derivatives interact with the $5\text{-}HT_{1A}$ receptor. The first hypothesis was that by all compounds adapting one conformation, both interaction sites can be reached by all compounds. The second hypothesis was that the N4-substituents with different hydrophobic character adopt a different conformation for each of the interaction sites. Thus, different N4-substituents were oriented according to one of the two possible directions corresponding to the hydrophobic or hydrophilic interaction site, depending on the chemical properties of the N-substituent. For hydrophilic oxygen-containing substituents, a third orientation was used. Unfortunately, none of the models gave very high statistics, and the authors could not select one as the preferred set.

Similar results were obtained in the study of two classes of acetylcholinesterase inhibitors, N-benzylpiperidine benzisoxazoles (NBPBs) and 1-benzyl-4-[2-(N-benzoylamino)ethyl]-piperidines (NBEPs) [249L]. Two conformations for the NBEPs were

Table 8 *CoMFA results of two classes of acetylcholinesterase inhibitors*

Alignment	N	L	Q^2	R^2	s	R^2_{pred}
I	57	6	0.672	0.936	0.316	0.817
II	57	6	0.722	0.962	0.245	0.868

examined. Alignment I brought the amide carbonyl of NBEPs close to the isoxazole oxygen of NBPBs, thus maximizing the similarity of the electrostatic fields. Alignment II made the same carbonyl group point in the opposite direction so as to maximize the steric similarity between the two classes. They used 57 compounds for the training set, and 20 compounds for the test set.

Although alignment II gave slightly better statistics (Table 8), the authors concluded that in the absence of experimental data both alignments were plausible, especially considering that the active site of the enzyme is relatively large and, thus, several binding sites may be available for substrates and inhibitors.

Carrieri et al. [42L] selected the bioactive conformation from a previous QSAR. A QSAR analysis developed from 25 hippurates as inhibitors of papain was as follows:

$$\log 1/K_m = 0.57\sigma + 1.03\,\pi'_3 + 0.61MR_4 + 3.80$$
$$n = 25,\ R^2 = 0.823,\ S = 0.208,\ Q^2 = 0.740$$

where K_m is the Michaelis-Menten binding constant, MR_4 is the molar refractivity of the para substituent, σ is the Hammett electronic substituent constant of the meta and para substituents and π'_3 is the hydrophobic constant referring only to the more hydrophobic of the two meta substituents. An arbitrary π'_3 value of 0 was assigned to the hydrophilic meta substituent based on a hypothesis that only the hydrophobic meta substituents fit into the hydrophobic pocket of the enzyme, so that hydrophilic meta substituents were assumed to project toward the surrounding aqueous solvent. To test the hypothesis, two separate CoMFA models were derived, the first one being consistent with the above QSAR equation ('split' alignment in which meta hydrophobic and hydrophilic substituents pointed toward different directions) and the second one overlapping all the meta substituents. The 'split' alignment (Q^2 = 0.73) yielded a better CoMFA model than the other alignment (Q^2 = 0.24), with accurate prediction of six test compounds (rms residuals = 0.26). A similar approach was taken in other studies, even if there was no known QSAR [9L,148L,209L,254L].

2.4.2. Alignment

An increasing number of experimentally determined ligand-bound macromolecular structures is becoming available. The availability of structures of ligand–macromolecule complexes of all the compounds of a dataset can avoid ambiguity in alignments. This was the case for glucose analog inhibitors of glycogen phosporylase b [64L,270L]. To align these inhibitors, it was sufficient to match the protein backbone atoms in the corresponding complexes. However, such experimental structures are typically available for only a few complexes, and the bound conformations of the remaining ligands must be deduced theoretically. Congeneric series are usually modelled with the conformation and orientation of the known compound. Such a procedure was applied to numerous cases: triazine [104L] and benzylpyrimidine [67L] inhibitors of dihydrofolate reductase, amino acid ester substrates of α-chymotrypsin [39L], N-benzoyl- and N-methansulfonyl phenylglycinate substrates of papain [42L], 2-heterosubstituted statine inhibitors of HIV-1 protease [152L], disoxaril analogs binding to the capsid protein 1 of human rhinovirus-14 [13L], structurally diverse acetylcholinesterase inhibitors [48L], and non-congeneric inhibitors of HIV-1 protease [266L].

An alignment was also produced by using a theoretically derived 3D model of the target as demonstrated by Gamper et al. [96L,194L]. In this study, a set of 27 chemically diverse haptens were docked with a computer program into a model of the monoclonal antibody IgE(Lb4). Since most of the ligands exhibited more than one plausible binding geometry, they examined several alignments of a subset of nine representative compounds. Each alignment, consisting of a different combination of conformations and orientations, was independently submitted to PLS. The models with highest Q^2 values were further considered and served as references to align the remaining ligands.

Many times, an appropriate macromolecular structure is not available. For such cases, different alignment approaches have been used [9]. Pharmacophores are most often used as the basis of alignment [10,30L,95L,183L,302L]. There are a number of approaches for pharmacophore identification [5]. Sometimes, however, common pharmacophore elements were absent as in polycyclic aromatic hydrocarbons which were aligned on their principal moment of inertia [58L,272L]. In other studies, alignments were based on electrostatic and steric complementarity [37L,49L,79L,117L, 260L, 265L].

Quite often, several CoMFA models were derived for the same training set using different alignment rules. Alternate alignments were obtained using different active conformations and/or different types of superposition procedures (usually rms fitting about atoms or field fitting). However, it is difficult or even impossible to predict whether any particular superposition method will be more suited for a given set of molecules. Therefore, based on the CoMFA results, choice of such an alignment or conformation used was considered justified [302L]. However, it is not always possible to choose a particular alignment based on the CoMFA results [154L].

The selection of either the bioactive conformation or the superposition may be influenced by the choice of the other, and the two aspects are sometimes considered simultaneously. Alternative conformations and/or alignments of even only a few molecules often influence CoMFA results. Additional examples and discussions on this subject are presented below in sections 2.10 and 2.10.1.

2.5. Interaction energy fields

Besides the standard steric and electrostatic fields, a number of other fields have been used alone or in combination with the standard fields in different studies.

2.5.1. *Hydrophobic fields*

Since the nature of hydrophobic interactions and their importance in drug–receptor interactions have long been recognized, a question was posed with respect to CoMFA: do the steric interactions account for the majority of energy derived from interacting two hydrophobic groups? Abraham and Kellogg addressed this question in the previous volume of this book [2L]. They developed the HINT program to evaluate ligand docking and protein folding and used it to calculate hydrophatic fields for CoMFA [1L].

Kim et al. [142L] employed the GRID H_2O probe to model the hydrophobic proerties of 48 benzodiazepine analogs binding to the benzodiazepine receptor. The results were fully consistent with a previous study based on a mixed Hansch-CoMFA approach in which the lipophilicity of the substituents was described through the π constant [11]. The GRID-CoMFA method improved the statistics and afforded coefficient contour maps for the hydrophobic effects.

2.5.2. *Molecular lipophilicity potentials*

For nearly a decade, several people have been interested in the application of molecular lipophilicity potentials (MLP) in QSAR. Different definitions have been proposed [2L,90L,94L]. Gaillard et al. applied MLP for calculating log P and used it as an additional field in CoMFA [93L,94L,246L]. Gaillard et al. claimed that MLP encodes hydrogen bonds and hydrophobic interactions not adequately described by the steric and electrostatic fields and that it also includes an entropy component [95L].

2.5.3. *E-state fields*

Kellogg et al. [128L] suggested that electrotopological state (E-state) and hydrogen electrotopological state (HE-state) fields can be used alone or in combination with the steric, electrostatic and/or hydropathic (HINT) fields in CoMFA [1L]. These fields were constructed from a nonempirical index that incorporates electronegativity, the inductive influence of neighboring atoms and the topological state into a single atomistic descriptor. The E-state fields were calculated for non-hydrogen atoms and derived from the counts of valence and σ bonding electrons in a hydrogen-suppressed chemical graph representing a molecule. The index was formulated to encode information about the electronegativity, π and lone-pair electron content, topological status and the environment of an atom within a molecule. On the other hand, the HE-state fields were calculated for all heavy atoms in a molecule that are bonded to a hydrogen. Kellogg et al. indicated that the E-state and HE-state fields are complementary; the most significant difference between the E-state and HE-state fields is that the E-state is localized on and around heavy (non-hydrogen) atoms, while the HE-state is localized on and around the hydrogens.

Table 9 *CoMFA results of a corticosteroid-binding globulin (CBG) dataset*

Fields	N	L	R^2	Q^2
Steric + electrostatic	21	3	0.950	0.736
E-state + HE-state	21	3	0.979	0.803

As an illustration and application of the E-state and HE-state fields, Kellogg et al. used a corticosteroid-binding globulin (CBG) dataset. The results of their CoMFA study are shown in Table 9. They reported that the best CoMFA model obtained from this dataset was from both the E-state and HE-state fields compared to any other combination of steric, electrostatic, hydrophatic, E-state and HE-state fields.

2.5.4. *Molecular orbital fields*

In a CoMFA study of cytochrome P450-mediated metabolism of chlorinated volatile organic compounds, Waller et al. [262L] supplemented the standard CoMFA steric and electrostatic fields with three molecular orbital fields (the electron density of HOMO, LUMO and frontier orbital field). The most consistent model was obtained from the combination of steric, electrostatic, LUMO and HINT hydropathicity fields. However, the complex nature of the molecular orbital fields precluded the generation of contour plots from these models. Waller and Marshall [260L] also reported the use of the fields arising from the charge distribution on the molecular orbitals (HOMO) in a CoMFA study with angiotensin-converting enzyme inhibitors and thermolysin inhibitors. Navajas et al. [194L] used the LUMO field in correlation with mutagenic activity of furanone analogs.

2.5.5. *Atom-based indicator variable*

Can the steric interaction energies commonly used in CoMFA be replaced by variables indicating the presence of an atom of a particular molecule in a predefined volume within the region enclosing the ensemble of superimposed molecules [151L]? Such atom-based indicator vectors were used as steric fields in subsequent PLS analyses with and without electrostatic fields. Kroemer and Hecht [151L] applied this method to five training sets (80 compounds each) and five test sets (60 compounds) randomly selected from 256 dihydrofolate reductase inhibitors and obtained models with varying degrees of R^2 and Q^2 values. However, the atom-based indicator variable method gave better results than the standard CoMFA.

2.5.6. *Van der Waals intersection volume*

The steric potentials used in CoMFA increase sharply for interatomic distances smaller than van der Waals contact interdistances. This produces large variations in the steric energy with slight displacement of atoms along the 1 or 2 Å CoMFA lattice. Taking into account the appreciable flexibility in torsion angle and local conformational changes of both the receptor and the ligand, interatomic distances never become appreciably less than van der Waals contact distance [234L]. Assuming that the steric potential energy increase beyond van der Waals contact interdistances is roughly proportional to the volume of intersection of the van der Waals envelope between the ligand and the receptor molecule, Muresan et al. [190L] proposed that the intersection volume of van der Waals envelopes of ligand molecules and probe atoms could be used as a measure of steric interactions. They suggested that these interaction volumes vary smoothly with interatomic distances, and that the large variations in steric potential associated with receptor grid interdistances will thus be greatly reduced.

2.5.7. Comparative molecular similarity indices analysis (CoMSIA)

Another approach to avoid the sharp increase in steric potentials was introduced by Klebe et al. [146L]. In the CoMSIA approach, molecular similarity indices between a probe atom and the molecule at each lattice position were used. For a steroid dataset, similar statistical results were obtained from the CoMSIA or the standard CoMFA approach.

2.5.8. Comparative molecular moment analysis (CoMMA)

All the above fields are calculated from superimposed 3D structures. On the other hand, Comparative Molecular Moment Analysis (CoMMA) [233L] utilizes descriptors calculated from individual 3D structure independent of the orientation and location of the molecules in 3D space. These descriptors are related to molecular shape and charge, such as the three principal moments of inertia, magnitude of dipole moment and the magnitude of principal quadrupole moment.

Detailed discussions on CoMSIA and CoMMA can be found in the chapters by G. Klebe and B.D. Silverman in this volume, respectively.

2.6. Grid spacing and lattice positions

Two aspects are of special concern in placing the lattice points around the molecules: the size of the spacing and the location of the grid box. The effects of grid offset and lattice positions have been investigated by various people [37L,47L,64L,91L,117L, 129L,141L,150L,206L,289L].

As noted in the chapter on R^2-guided region selection, Cho and Tropsha [47L] observed that Q^2 values were sensitive to the overall orientation of rigidly aligned molecules. When they systematically rotated several molecular aggregates in the three-dimensional coordinate system, the resulting CoMFA Q^2 values differed by as much as 0.5. They reasoned that in CoMFA the steric and electrostatic fields are sampled on such a coarse grid that these fields are inadequately represented. Kim *et al.* [322L] observed similar results.

In a study on the inhibition of glycogen phosphorylase b by glucose analogs, Cruciani and Watson [64L] observed that important information could be lost when the grid spacing was too large or the probes were inadequately described. Examination of the values of R^2 and Q^2 of different models showed that if the grid spacing was increased from 1 Å to 2 Å, both the fitting and the predicting capability dropped dramatically. They claimed that the 2 Å spacing was too large for sensitive and highly directional interactions, such as those found in multiple hydrogen bonds, to be adequately defined. On the other hand, the 1 Å spacing using the GRID phenolic OH probe was sufficient for eliminating noisy variables while retaining only relevant information by means of the GOLPE approach.

In a study of human immunodeficiency virus (I) protease inhibitors, Oprea et al. [206L] compared the CoMFA electrostatic contour maps with the molecular electrostatic potential (MEP) contours. They found that the CoMFA individual field was not able to distinguish the subtle changes in the overall fields. For example, the deep negative potential created by a carbonyl moiety surrounded by weak positive charges of two NH moieties was located by the MEP field. However, the averaging effect of the 2 Å grid caused the CoMFA field to show only positive contours in that region. They successfully reproduced the MEP values using a 1 Å CoMFA grid.

In a correlation study of hydrogen-bond basicity (log K_β) with computed molecular electrostatic potential for 23 aromatic heterocycles, Kenny [129L] investigated how effectively the electrostatic potential predicts hydrogen-bond basicity when it is computed at a distance *r* from the site of the nitrogen lone pair. The value of *r* corresponding to electrostatic potential local minima ranged from 1.21 Å to 1.28 Å, and the optimal fit for the CoMFA correlation of log K_β was 1.4 Å. He reported that the electrostatic potential fits log K_β most effectively when it was calculated within the van der Waals radius of the nitrogen. He indicated that in a standard CoMFA with 2 Å spacing and commonly used sp^3 carbon probe the lattice points do not correspond to the electrostatic potential minima. These findings may explain the often observed better performance of CoMFA models derived without dropping electrostatic energies sampled at sterically 'bad' points or within the common van der Waals volume of the superimposed molecules.

In a study of six different structural classes of insecticides that act at the GABA receptor, Calder et al. [37L] initially used a 2 Å grid spacing. However, although the 4-substituents were symmetric, the CoMFA electrostatic coefficient contour maps in this region of the 4-substituent were markedly asymmetric. The Q^2 value from a 2 Å grid spacing was only slightly smaller than that from 0.75 Å. However, attempting to interpret this asymmetric field could mislead the chemist in designing new compounds. When the grid spacing was reduced to 0.75 Å, this field asymmetry in the region of the 4-substituent disappeared.

Folkers [91L] reported that the GRID methyl probe was very efficient at a 2 Å grid spacing for describing steric bulk effects, whereas the water probe was more adequate for analyzing H bonding at higher resolutions (1 Å). Horwitz et al. [117L] reported the Q^2 value being more stable when the grid resolution was set to 1 Å (values comprised between 0.629 and 0.647) compared with the grid spacing of 2 Å (values from 0.570 to 0.654).

Although these results clearly suggested that for a detailed CoMFA study a 1 Å grid spacing is preferred over a 2 Å grid spacing, about two-thirds of the studies listed in Table 10 used 2 Å grid spacing. Many of the other studies in Table 10 with missing grid spacing information may have also been done with a default 2 Å grid setting. Only one-fifth of the studies were done using a 1 Å grid spacing. A probable reason for this is because many other studies also showed that lattice spacings of 1 Å or 2 Å yielded similar results in terms of Q^2 values. For example, Tomkinson et al. [248L], Tong et al. [249L], Kroemer and Hecht [150L] and Debnath et al. [76L] reported a small improvement in the correlation switching from a 2 Å to 1 Å spacing. However, the gain in Q^2 value was not large enough to justify the substantial increase in computing time and model complexity. Akamatsu et al. [289L] reported that use of 1 Å, 1.5 Å or 2 Å grid spacing yielded almost equivalent model quality in their CoMFA study.

Some authors [95L,148L,246L] have proposed a 1.5 Å spacing, probably as a compromise between an accurate description of the molecules and the need to keep the

Table 10 *Summary of the CoMFA models in applications*[a,b]

Equation no.	Reference no.[c]	Target property	Target property form used in CoMFA	N^g	Type of compounds	Grid (Å)	Probe group	Additional probes or parameters used	Fit statistics[d] L^i	R^2	s	CV statistics[e] Q^2	s_{cv}	Forecast statistics[f] M^j	R^2	s^k_{pred}	Comments[h]
1	[150]	Inhibition of dihydrofolate reductase (DHFR)	$-\log IC_{50}$	80	Triazines	2	$C^{+}_{sp^3}$		4			0.86			0.48–0.60		
2	[150]	Inhibition of DHFR	$-\log IC_{50}$	80	Triazines	2	$C^{+}_{sp^3}$		5			0.62		70	0.60–0.64		
3	[151]	Inhibition of DHFR from *M. lufu*	$-\log IC_{50}$	80	Triazines	0.75–2	$C^{+}_{sp^3}$	Indicator variable instead of steric fields						60			Many models
4	[91]	Inhibition of DHFR	?	10	Benzylpyrimidines	2	H^+		3			0.98	0.17				
5	[102]	Inhibition of DHFR	$\log (1/K_{i\text{-}app})$	35	Triazines	2	$C^{+}_{sp^3}$		4	0.90	0.32	0.76	0.49				
6	[102]	Inhibition of DHFR	$\log (1/K_{i\text{-}app})$	35	Triazines	2	$C^{+}_{sp^3}$		4	0.87	0.35	0.75	0.50				QSAR-guided variable selection
7	[67]	Inhibition of DHFR from *C. albicans*	$\log (1/IC_{50})$	10	Trimethoprims	2	$C^{+}_{sp^3}$		1			0.78					
8	[102]	Inhibition of DHFR from chicken liver	$\log (1/K_i)$	35	Triazines	2	$C^{+}_{sp^3}$	π	4	0.90	0.32	0.76	0.49				
9	[102]	Inhibition of DHFR from chicken liver	$\log (1/K_i)$	35	Triazines	2	$C^{+}_{sp^3}$	π	4	0.87	0.36	0.75	0.50				Variable selection
10	[67]	Inhibition of DHFR from *E. coli*	$\log (1/IC_{50})$	9	Trimethoprims	2	$C^{+}_{sp^3}$		5			0.74					2 alignments, 3 excluded
11	[231]	Inhibition of DHFR from *E. coli*	$\log (1/IC_{50})$	7	Benzylpyrimidines				3			0.85					3 excluded
12	[231]	Inhibition of DHFR from *E. coli*	$\log (1/IC_{50})$	10	Benzylpyrimidines				5			0.86					3 excluded
13	[67]	Inhibition of DHFR from *E. coli*	$\log (1/IC_{50})$	10	Trimethoprims	2	$C^{+}_{sp^3}$		5			0.85					
14	[67]	Inhibition of DHFR from *M. lufu*	$\log (1/IC_{50})$	10	Trimethoprims	2	$C^{+}_{sp^3}$		2			0.47					
15	[79]	Inhibition of angiotensin converting enzyme (ACE)	pIC_{50}	23	Carboxylates	2	$C^{+}_{sp^3}$		7	0.99	0.14	0.81					
16	[293]	Inhibition of ACE	$\log (1/I_{50})$	29	Dipeptides	?	?		2	0.89		0.79		29		0.41	GRID, INLR
17	[293]	Inhibition of ACE	$\log (1/I_{50})$	29	Dipeptides	?	?		3	0.87		0.76		29		0.45	GRID
18	[293]	Inhibition of ACE	$\log (1/I_{50})$	29	Dipeptides	?	?		2	0.92		0.77		29		0.34	D-optimal design, INLR
19	[293]	Inhibition of ACE	$\log (1/I_{50})$	29	Dipeptides	?	?		3	0.88		0.72		29		0.45	D-optimal design
20	[56]	Inhibition of ACE	$\log (1/I_{50})$	58	Dipeptides	2	H_2O,	COO^-, CO, NH_2, NH_3^+, $C^{+}_{sp^3}$	3			0.89					

Table 10 (continued)

Equation no.	Reference no.[c]	Target property	Target property form used in CoMFA	N^g	Type of compounds	Grid (Å)	Probe group	Additional probes or parameters used	Fit statistics[d] L^i	R^2	s	CV statistics[e] Q^2	s_{cv}	Forecast statistics[f] M^j	R^2	s^k_{pred}	Comments[h]
21	[163]	Inhibition of ACE	?	30	Dipeptides	?	?					0.73					$q^2 = 0.87$ for IVS-PLS
22	[260]	Inhibition of ACE	pIC_{50}	68	Diverse carboxylates, phosphates and thiols	2	$C^+_{sp^3}$	HOMO field, $DG_{solvation}$	7	0.87	0.82	0.73		20 19 17	0.66 0.71 0.33		Diverse Phosphate Thiols
23	[79]	Inhibition of ACE	pIC^{50}	68	Diverse compounds	2	$C^+_{sp^3}$	indicator variables	4	0.83	0.90	0.68	1.26	20 19 10 17	0.53 0.57 0.60 0.20	0.77 0.52 0.61 1.12	
24	[79]	Inhibition of ACE	pIC_{50}	12	Phosphates	2	$C^+_{sp^3}$		2	0.83	0.63	0.46					
25	[79]	Inhibition of ACE	pIC_{50}	30	Thiols	2	$C^+_{sp^3}$		4	0.79	0.90	0.26					
26	[293]	Inhibition of ACE	log $(1/IC_{50})$	58	Dipeptides	?	?		3	0.88	0.45	0.72					D-optimal design
27	[293]	Inhibition of ACE	log $(1/IC_{50})$	58	Dipeptides	?	?		2	0.92	0.39	0.76					D-optimal design, INLR
28	[47]	Inhibition of HIV protease	?	59	HIV protease inhibitors	2	$C^+_{sp^3}$		5			0.38					
29	[47]	Inhibition of HIV protease	?	59	HIV protease inhibitors	1	$C^+_{sp^3}$		7			0.80					$q^2_{cutoff} = 0.4$, q^2-*GRS*
30	[206,266]	Inhibition of HIV protease	pIC_{50}	59	Hydroxyethylamines, statines, norstatines, keto amides, and dihydroxyethylenes	2	$C^+_{sp^3}$		6	0.98	0.15	0.77	0.55	18	0.66	0.63	
31	[152]	Inhibition of HIV-1 protease	log $(1/K_i)$	100	Statines	2	$C^+_{sp^3}$		5	0.84		0.59	0.49	67	0.57		8 excluded, 0.26 for 75 compounds, 2 designed, slightly underpredicted
32	[221]	Inhibition of HIV-1 protease	log $(1/IC_{50})$	14	Flavones	2	$C^+_{sp^3}$		9	1.00	0.04	0.81					1 excluded
33	[221]	Inhibition of HIV-1 protease	log $(1/IC_{50})$	12	Flavones	2	$C^+_{sp^3}$		8	1.00		0.77					
34	[76]	Inhibition of HIV-1 (gp120)	log $(1/EC_{50})$	20	Porphyrins	2	$C^+_{sp^3}$		4	0.98	0.11	0.59	0.49				1 excluded
35	[144]	Inhibition of HIV-1 protease	log $(1/EC_{50})$	28	Arylquinolineamines	2	$C^+_{sp^3}$		5	0.92	0.18	0.61	0.39				
36	[207]	Inhibition of HIV-1 protease	pIC_{50}	59	Transition state isosters	2	$C^+_{sp^3}$		6	0.98	0.15	0.78	0.55	34	0.68		2 excluded

Table 10 (continued)

Equation no.	Reference no.[c]	Target property	Target property form used in CoMFA	N^g	Type of compounds	Grid (Å)	Probe group	Additional probes or parameters used	Fit statistics[d] L^i	R^2	s	CV statistics[e] Q^2	s_{cv}	Forecast statistics[f] M^j	R^2	s^k_{pred}	Comments[h]
37	[109]	Inhibition of HIV-1 reverse transcriptase	log (1/*C*)	40	Phenylthiothymines	2	$C^+_{sp^3}$,	$O^+_{sp^3}$, , H^+	7	0.98	0.22	0.70	0.88				
38	[183]	Inhibition of monoamine oxidase-A	log IC_{50}	25	Indoles and isatins	?	$C^+_{sp^3}$		10	0.92	0.25	0.74					
39	[246]	Inhibition of monoamine oxidase-A	pIC_{50}	18	Isoquinolines	1.5	$C^+_{sp^3}$	MLP	2	0.77		0.51		3 (0.17)			Steric + MLP
40	[246]	Inhibition of monoamine oxidase-A	pK_i	31	Isoquinolines	1.5	$C^+_{sp^3}$		2	0.77		0.51					Steric only
41	[148]	Inhibition of monoamine oxidase-A and B	pIC_{50}	39	Indenopyridazin-5-ones	1.5	$C^+_{sp^3}$	MLP	3	0.93		0.75		2			o:f = 4.85:4.69; 6.16:5.80
42	[148]	Inhibition of monoamine oxidase-B	pIC_{50}	24	Indenopyridazin-5-ones	1.5	$C^+_{sp^3}$	MLP	3	0.98		0.84		2			o:f = 4.85:4.87; 6.16:5.88
43	[7]	Inhibition of monoamine oxidase-B	log (1/IC_{50})	39	Indolepyridazines	?	?	MLP	3	0.93		0.75					1 designed
44	[183]	Inhibition of monoamine oxidase-B	log IC_{50}	21	Indoles and isatins	?	$C^+_{sp^3}$		10	0.99	0.11	0.60					
45	[42]	Hydrolysis by papain	log (1/K_m)	21	Phenyl N-benzoyl glycinates	2	$C^+_{sp^3}$		7	0.98	0.08	0.73		6		0.26	
46	[42]	Hydrolysis by papain	log (1/K_m)	12	Phenyl N-methanesulfonyl glycinates	2	$C^+_{sp^3}$		1	0.82	0.15		0.52				
47	[139]	Hydrolysis by papain	log (1/K_m)	17	m-substituted N-benzoylglycine esters	2	H_2O,	H^+	4	0.86	0.17		0.18				
48	[139]	Hydrolysis by papain	log (1/K_m)	37	N-benzoylglycine esters	2	CH_3,	H_2O, H^+	4	0.96	0.22		0.25				
49	[139]	Hydrolysis by papain	log (1/K_m)	21	p-substituted N-benzoylglycine esters	2	CH_3,	H^+	4	0.98	0.18		0.21				
50	[141]	Hydrolysis by papain	log (1/K_m)	22	Phenyl hippurates	2	H_2O		4	0.95	0.10		0.30				
51	[130]	Hydrolysis by papain	log (1/K_m)	22	3-pyridyl-m & p-Phenyl hippurates	2	H_2O		4	0.95	0.13		0.35				
52	[130]	Hydrolysis by papain	log (1/K_m)	8	Phenyl hippurates	2	H_2O		2	0.84	0.15		0.20				
53	[344]	Inhibition of glycogen phosphorylase b	ln K_i	47	Glucose analogs	1	OH	(phenolic)	4	0.93		0.81					MECE (ligand + H_2O)

Table 10 (continued)

Equation no.	Reference no.[c]	Target property	Target property form used in CoMFA	N^g	Type of compounds	Grid (Å)	Probe group	Additional probes or parameters used	Fit statistics[d] L^i	Fit statistics[d] R^2	Fit statistics[d] s	CV statistics[e] Q^2	CV statistics[e] s_{cv}	Forecast statistics[f] M^j	Forecast statistics[f] R^2	Forecast statistics[f] s^k_{pred}	Comments[h]
54	[344]	Inhibition of glycogen phosphorylase b	$\ln K_i$	47	Glucose analogs	1	OH (phenolic)		4	0.95		0.79					
55	[64]	Inhibition of glycogen phosphorylase b	$\ln K_i$	36	Glucose analogs	1	OH (phenolic)		3	0.92	0.48	0.80	0.78				GOLPE
56	[270]	Inhibition of glycogen phosphorylase b	$\ln K_i$		Glucose analogs	1	OH (phenolic)		4	0.92	0.54	0.76	0.98				GOLPE
57	[260]	Inhibition of thermolysin	pK_i	61	Diverse compounds	2	$C^+_{sp^3}$		9	0.85	0.83	0.65		11	0.17		
58	[79]	Inhibition of thermolysin	pK_i	61	Diverse compounds	2	$C^+_{sp^3}$		11	0.98	0.31	0.70	1.30	11	0.29	0.35	
59	[145]	Inhibition of thermolysin	$-\ln K$	20	Diverse compounds	2	$C^+_{sp^3}$		6	0.99	0.41	0.77	2.34				Many correlations with different alignments
60	[249]	Inhibition of acetylcholinesterase (AChE)	$\log IC_{50}$	57	Piperidine, N-benzylpiperidines	2	$C^+_{sp^3}$		6	0.94	0.32	0.67		20	0.82		
61	[48]	Inhibition of acetylcholinesterase (AChE)	$-\log IC_{50}$	60	Physostigmine, tetrahydroacridine, edrophonium and others	1	$C^+_{sp^3}$		3	0.80	0.74	0.62	1.03				
62	[48]	Inhibition of acetylcholinesterase (AChE)	$-\log IC_{50}$	60	Physostigmine, tetrahydroacridine, edrophonium and others	1	$C^+_{sp^3}$		7	0.93	0.45	0.73	0.89				q^2-GRS
63	[224]	Inhibition of human placental aromatase	pIC_{50}	29	Fadrozoles	2	$C^+_{sp^3}$		3	0.95	0.18	0.76	0.40				
64	[203]	Inhibition of human placental aromatase	pK_i	33	Steroids	2	Fe^{+2}		5	0.89		0.56					
65	[203]	Inhibition of human placental aromatase	pK_i	50	Steroids	2	Fe^{+2}		1	0.67		0.62					
66	[24]	Inhibition of serine protease thermitase	pK_i	19	Peptide chloromethyl ketones	2	$C^+_{sp^3}$		6	0.99	0.20	0.93	0.50	1			o:f = 7.78:8.16; 2 excluded
67	[24]	Inhibition of serine protease thermitase	pK_i	18	Peptide methyl ketones	2	$C^+_{sp^3}$		5	0.99	0.16	0.86	0.67				1 excluded
68	[13,222]	Inhibition of human rhinovirus-14	pEC_{50}	17	Alkylisoxazoles	2	$C^+_{sp^3}$	$\log k'$	5	0.97	0.11	0.72	0.38	6, 7		0.53	(4cmpd) ae (average absolute error)
69	[145]	Binding to human rhinovirus 14 (HRV-14)	$-\ln MIC$	8	HRV-14 inhibitors	2	$C^+_{sp^3}$		3	0.98	0.28	0.53	1.46				Many correlations with different alignments

Table 10 (continued)

Equation no.	Reference no.[c]	Target property	Target property form used in CoMFA	N^g	Type of compounds	Grid (Å)	Probe group	Additional probes or parameters used	Fit statistics[d] L^i	Fit statistics[d] R^2	Fit statistics[d] s	CV statistics[e] Q^2	CV statistics[e] s_{cv}	Forecast statistics[f] M^j	Forecast statistics[f] R^2	Forecast statistics[f] s^k_{pred}	Comments[h]
70	[106]	Substrate of N-methyltransferase	?	30	Benzylamines	2	$C^+_{sp^3}$		?	0.87	0.61	0.69	0.96				
71	[106]	Substrate of N-methyltransferase	?	15	Phenylethanolamines	2	$C^+_{sp^3}$		?	0.92	0.99	0.44	0.88				
72	[106]	Substrate of N-methyltransferase	?	21	Phenylethylamines	2	$C^+_{sp^3}$		?	0.81	0.77	0.57	1.14				
73	[143]	Substrates of elastase	$\log (1/K_m)$	89	Dipeptides	?	?		2	0.54	0.20	0.41					
74	[143]	Substrates of elastase	$\log K_{cat}$	89	Dipeptides	?	?		4	0.84	0.21	0.75					
75	[143]	Substrates for elastase	$\log (K_{cat}/K_m)$	89	Dipeptides	?	?		3	0.75	0.29	0.63					
76	[39,40,203]	Hydrolysis by α-chymotrypsin	$\log (1/K_m)$	71	N-acyl-L-aminoacid esters	2	$C^+_{sp^3}$		3	0.91	0.49	0.87	0.58				
77	[201,40]	Hydrolysis by α-chymotrypsin	$\log (1/K_m)$	12	N-acyl-L-aminoacid esters	2	$C^+_{sp^3}$		2	0.96	0.33	0.83	0.70	59	0.79	0.67	Cluster analysis
78	[39],	Hydrolysis by α-chymotrypsin	$\log (1/K_m)$	12	N-acyl-L-aminoacid esters	2	$C^+_{sp^3}$		1			0.65	1.03	59	0.77	0.73	Factorial design
79	[91]	Inhibition of hydrolysis by trypsin hydrolase	?	12	Benzylamines	2	CH_3		6			0.82	0.27				
80	[145]	Binding to renin	ΔH	12	Diverse compounds	2	$C^+_{sp^3}$		5	0.99	0.87	0.83	3.73				1 excluded, many correlations with different alignments
81	[145]	Binding to renin	ΔS	12	Diverse compounds	2	$C^+_{sp^3}$		5	0.99	3.66	0.84	13.04				1 excluded, many correlations with different alignments
82	[226]	Inhibition of bilirubin UDP-glucuronosyltransferase	$\log (1/IC_{50})$	18	Phenolphthalein and triphenylalkyl carboxylic acids	2	$C^+_{sp^3}$		4	0.97		0.72	0.27				
83	[226]	Inhibition of bilirubin UDP-glucuronosyltransferase	$\log (1/IC_{50})$	10	Phenolphthaleins	2	$C^+_{sp^3}$		5	0.98		0.85	0.34				
84	[216]	Binding to cytochrome P4502 A5	pK_i	15	Coumarins	2	$C^+_{sp^3}$	LUMO field	3	0.97		0.72	0.48				
85	[114]	Binding to squalene epoxide	?	22	Benzylamines	2	$C^+_{sp^3}$		2	0.95	0.12	0.50	0.38				

Table 10 (continued)

Equation no.	Reference no.[c]	Target property	Target property form used in CoMFA	N^g	Type of compounds	Grid (Å)	Probe group	Additional probes or parameters used	Fit statistics[d] L^i	R^2	s	CV statistics[e] Q^2	s_{cv}	Forecast statistics[f] M^j	R^2	s^k_{pred}	Comments[h]
86	[86]	Enantioselectivity of ester hydrolysis by lipase *Candida rugosa*	ln *E*	22	Norbornanes	1	H^+		5	0.99	0.18	0.52					
87	[139]	Equilibrium of enzyme–substrate complex formation	$\log (k_1/k_2)$	8	Phenyl-b-D-glucosides	2	H_2O,	H^+	2	0.94	0.11		0.24				
88	[139]	Inhibition of carbonic anhydrase	$\log (1/K_i)$	18	Benzenesulfonamides	2	H_2O,	H^+	4	0.96	0.18		0.22				
89	[25]	Inhibition of dipeptidyl peptidase IV	pK_i	24	Dipeptides and aminoacid pyrrolides	2	$C^+_{sp^3}$		5	0.99	0.10	0.90	0.40				
90	[208]	Inhibition of human synovial fluid phospholipase A2	% activity	26	?	?	?		2	0.89	7.10	0.78	9.81				GOLPE
91	[83]	Inhibition of kynureninase	pK_i	13	S-aryl-L-cysteine S, S-dioxides	2	$C^+_{sp^3}$		2	0.90	0.41	0.5					
92	[158]	Inhibition of prolyl endopeptidase	$\log (1/IC_{50})$	10	Prolinal or thioprolinal diamides	2	$C^+_{sp^3}$		2	0.82	0.41	0.43	0.63	3		0.48	
93	[91,184]	Inhibition of herpes simplex virus 1 thymidine kinase	?	24	Phenylguanines	2	$C^+_{sp^3}$		5			0.74					
94	[86]	Enantioselectivity of ester hydrolysis by lipase *Candida rugosa*	ln *E*	22	Norbornanes	1	H^+		5	0.99	0.18	0.52					
95	[23]	Hydrolysis by *Candida rugosa* lipase		10	2-Arylpropionic esters	?	$C^+_{sp^3}$		5	1.00	0.01	0.68		2			$L = 2$ is better with $n = 12$
96	[324]	Glyoxalase I catalysed rearrangement	$\log (K_{cat}/K_m)$	7	Glutathione ketoaldehyde hemiacetals	2	H^+		2	0.94	0.06	0.31	0.20				Distance matrix
97	[324]	Oxidation rate by phosphoglyceraldehyde dehydrogenase	log *K*	3	Aromatic aldehydes	2	H^+		3	0.94	0.05	0.23	0.20				Distance matrix
98	[324]	Reaction rate by thiaminase	log *K*	12	Anilines	2	H^+		4	0.93	0.29	0.30	0.91				Distance matrix
99	[17]	Induction of aryl hydrocarbon hydroxylase (AHH)	pEC_{50}	20	Polychlorinated dibenzofurans	1	$C^+_{sp^3}$		5	0.95	0.42	0.80	0.90				
100	[17]	Induction of aryl hydrocarbon hydroxylase (AHH)	*pEC*50	20	Polychlorinated dibenzofurans	1	$C^+_{sp^3}$		2			0.90	0.50				

Table 10 (continued)

Equation no.	Reference no.[c]	Target property	Target property form used in CoMFA	N^g	Type of compounds	Grid (Å)	Probe group	Additional probes or parameters used	Fit statistics[d] L^i	R^2	s	CV statistics[e] Q^2	s_{cv}	Forecast statistics[f] M^j	R^2	s^k_{pred}	Comments[h]
101	[199]	?	log *RBA*	19	Steroids	1.5	$C^+_{sp^3}$		4			0.45		30	0.43		
102	[199]	?	log *RBA*	20	Steroids	1.5	$C^+_{sp^3}$		5			0.45		29	0.29		
103	[263]	Binding to androgen receptor	pK_i	21	Steroids, diverse compounds	2	$C^+_{sp^3}$		4	0.99	0.24	0.79	1.01	7		0.58	Electrostatic
104	[199]	Binding to corticosteroid binding globulin	log $(1/K)$	21	Steroids	1.5	$C^+_{sp^3}$		2			0.79		10	0.56		
105	[199]	Binding to corticosteroid binding globulin	log $(1/K)$	21	Steroids	1.5	$C^+_{sp^3}$		2			0.90		10	0.48		
106	[35]	Binding to corticosteroid binding globulin	?	21	Steroids	2	?		2	0.90	0.40	0.69	0.69				SAMPLS
107	[128]	Binding to corticosteroid-binding globulin	?	21	Steroids	1	$C^+_{sp^3}$		3	0.95		0.74					
108	[128]	Binding to corticosteroid-binding globulin	?	21	Steroids	1	$C^+_{sp^3}$	E-state, HE-state	3	0.98		0.80					
109	[47]	Binding to corticosteroid-binding globulin	?	21	Steroids	1	$C^+_{sp^3}$					0.79					q^2cutoff = 0.5, q^2-GRS
110	[47]	Binding to corticosteroid-binding globulin	?	21	Steroids	2	$C^+_{sp^3}$					0.66					q^2cutoff = 0.4
111	[205]	Binding to progesterone receptor	$-\log IC_{50}$	22	Steroids	?	?		7	0.98	0.25	0.32	1.43				
112	[253]	Binding to progesterone receptor	*RBA*	43	Steroids	?	?		4	0.88		0.27		13			
113	[22]	H_1 antagonistic activity	pA_2	15	Thiazolidinones	1	$C^+_{sp^3}$,	π	3	0.91	0.37	0.55	0.82				1 excluded
114	[81]	H_1 agonistic activity	pD_2	34	Histamines	1.5	?	LIP_m	5	0.88	0.29	0.54	0.57				
115	[3]	Intrinsic activity of 5-HT_{1A}	*IA*	20	Diverse compounds	2	$C^+_{sp^3}$		5	0.94	0.10	0.48					
116	[11]	Binding to 5-HT receptor	pK_i	24	Quinolines	?	?		1	0.83	0.49	0.70	0.65				
117	[95]	Binding to 5-HT_{1A} receptor	pK_i	101	Arylpiperazines	1.5	$C^+_{sp^3}$,	MLP	4	0.88	0.43	0.64		12			
118	[95]	Binding to 5-HT_{1A} receptor	pK_i	30	(Aryloxy) propanolamines	1.5	$C^+_{sp^3}$	MLP	2	0.87	0.47	0.59		4			
119	[254]	Binding to 5-HT_{1A} receptor	$-\log K_i$	43	Benzodioxinyl-piperazines	1	$C^+_{sp^3}$		4	0.94	0.21	0.82					

Table 10 (continued)

Equation no.	Reference no.[c]	Target property	Target property form used in CoMFA	N^g	Type of compounds	Grid (Å)	Probe group	Additional probes or parameters used	Fit statistics[d] L^i	R^2	s	CV statistics[e] Q^2	s_{cv}	Forecast statistics[f] M^j	R^2	s^k_{pred}	Comments[h]
120	[4]	Binding to 5-HT_{1A} receptor	pK_i	45	Tetrahydropyridinnl) indoles	2	$C^+_{sp^3}$		4	0.85	0.29	0.45		6		0.46	
121	[4]	Binding to 5-HT_{1A} receptor	pK_i	15	Tetrahydropyridinnl) indoles	2	$C^+_{sp^3}$		2	0.72	0.26	0.52		12		0.60	
122	[47]	Binding to 5-HT_{1A} receptor	?	20	5-HT_{1A} receptor ligands	2	$C^+_{sp^3}$		7			0.74					
123	[47]	Binding to 5-HT_{1A} receptor	?	20	5-HT_{1A} receptor ligands	1	$C^+_{sp^3}$		4			0.81					q^2cutoff = 0.4, q^2-GRS
124	[95]	Binding to 5-HT_{1A} receptor	pK_i	60	Aminotetralins	1.5	$C^+_{sp^3}$,	MLP	1	0.44	0.77	0.36					Steric
125	[95]	Binding to 5-HT_{1A} receptor	pK_i	131	Arylpiperazines and (Aryloxy) propanolamines	1.5	$C^+_{sp^3}$,	MLP	3	0.83	0.58	0.66					
126	[95]	Binding to 5-HT_{1A} receptor	pK_i	185	Arylpiperazines, (Aryloxy) propanolamines, and Tetrahydropyridyl-indoles	1.5	$C^+_{sp^3}$,	MLP	3	0.81	0.54	0.61					
127	[160]	Binding to 5-HT_{1A} receptor	pK_i	17	β-blockers		$C^+_{sp^3}$		5	0.99	0.13			4			
128	[95]	Binding to 5-HT_{1A} receptor	pK_i	54	Tetrahydropyridyl-indoles	1.5	$C^+_{sp^3}$,	MLP	4	0.84	0.28	0.43					
129	[160]	Binding to 5-HT_{1B} receptor	pK_i	17	β-blockers		$C^+_{sp^3}$		5	1.00	0.25				1		
130	[4]	Binding to 5-HT_2 receptor	pK_i	45	Tetrahydropyridinnl) indoles	2	$C^+_{sp^3}$		5	0.84	0.37	0.41			6	0.36	
131	[4]	Binding to 5-HT_2 receptor	pK_i	15	Tetrahydropyridinnl) indoles	2	$C^+_{sp^3}$		5	0.93	0.29	0.62			12	0.65	
132	[30]	Binding to 5-HT_3 receptor	log IC_{50}	39	Piperazino pyrrolo thieno pyrazines	2	$C^+_{sp^3}$		5	0.86	0.76	0.46	1.48				H^+ gave identical results, Gasteiger-Hückel charges gave poor results
133	[94]	Binding to α_1-adrenergic receptor	?	38	Aryltetrahydropyrazol-opyridines	1	$C^+_{sp^3}$		5	0.92		0.76					

Table 10 (continued)

Equation no.	Reference no.[c]	Target property	Target property form used in CoMFA	N^g	Type of compounds	Grid (Å)	Probe group	Additional probes or parameters used	Fit statistics[d] L^i	Fit statistics[d] R^2	Fit statistics[d] s	CV statistics[e] Q^2	CV statistics[e] s_{cv}	Forecast statistics[f] M^j	Forecast statistics[f] R^2	Forecast statistics[f] s^k_{pred}	Comments[h]
134	[94]	Binding to α_1-adrenergic receptor	?	33	Diverse compounds	2	MLP		2	0.84		0.22					
135	[57]	Binding to α_1-adrenergic receptor	pA	14	Prazosin analogues	?	CH	(aromatic)	2	0.96	0.11	0.73	0.32				GOLPE
136	[174]	Binding to α_2-adrenergic receptor	pK_i	46	Phenethylamine phenols	2	$C^+_{sp^3}$		3			0.80	0.66	22	0.80	0.77	
137	[232]	Binding to A_3 adenosine receptor (rat brain)	pK_i	37	Benzyladenosine uronamides	2	$C^+_{sp^3}$		5	0.80		0.63	0.56	5		0.01[k]	
138	[19]	Binding to angiotensin II (AT_1) receptor	pIC_{50}	28	Diverse compounds	1	$C^+_{sp^3}$		2	0.92	0.29	0.80	0.47				GOLPE
139	[168]	Binding to angiotensin II (AT_1) receptor	pIC_{50}	28	Diverse compounds	1	$C^+_{sp^3}$, Cl		2	0.92	0.29	0.80	0.47	5		0.84	GOLPE
140	[168]	Binding to angiotensin II (AT_1) receptor	pIC_{50}	18	Diverse compounds	1	$C^+_{sp^3}$, Cl		2	0.95	0.24	0.79	0.50	5		0.58	GOLPE
141	[72]	Binding to angiotensin II AT_1/AT_2 receptor	$\log(1/IC_{50})$	131	Quinazolinones	?	?		?	0.94	0.36	0.81					
142	[72,220]	Angiotensin II antagonism	$\log(1/IC_{50})$	50	Diverse compounds	2	$C^+_{sp^3}$		2	0.75	0.56	0.64	0.66	39	12		For prediction, see ref. [72]
143	[209]	Binding to benzodiazepine receptor	pIC_{50}	26	Aryl-dihydropyridazinoindol-ones	2	$C^+_{sp^3}$		4	0.95	0.20	0.56					
144	[284]	Binding to benzodiazepine receptor	IC_{50}?	34?	5-Thienyl- and 5-furyl benzodiazepines	?	?		?	0.84							
145	[101]	Binding to benzodiazepine receptor (mitochondrial)	pIC_{50}	42	Arylpyrrolobenzothiazepines	2	$C^+_{sp^3}$		4	0.94	0.33	0.76	0.65	7		0.54	
146	[153]	Binding to benzodiazepine receptor	pIC_{50}	37	Benzodiazepines	2	$C^+_{sp^3}$		7	0.98		0.77					Different charges
147	[142]	Binding to benzodiazepine receptor	$\log(1/IC_{50})$	48	Benzodiazepines	2	H_2O, H^+		7	0.96	0.18		0.40				
148	[12]	Binding to benzodiazepine receptor	?	26	Phenylquinolines	2	$C^+_{sp^3}$		3	0.89	0.18	0.41	0.99				
149	[9]	Binding to benzodiazepine receptor	pIC_{50}	24	Indenopyridazines	2	$C^+_{sp^3}$		3	0.89	0.31	0.57		8		0.48	
150	[275]	Binding to diazepam-insensitive benzodiazepine receptors	$pK_i(DI)$	38	Benzodiazempams	2	$C^+_{sp^3}$		7	0.96	0.18	0.73	0.45				
151	[275]	Binding to diazepam-insensitive benzodiazepine receptors	$pK_i(DS)$	38	Benzodiazempams	2	$C^+_{sp^3}$		11	0.99	0.09	0.70	0.59				

Table 10 (continued)

Equation no.	Reference no.[c]	Target property	Target property form used in CoMFA	N^g	Type of compounds	Grid (Å)	Probe group	Additional probes or parameters used	Fit statistics[d] L^i	R^2	s	CV statistics[e] Q^2	s_{cv}	Forecast statistics[f] M^j	R^2	s^k_{pred}	Comments[h]
152	[275]	Difference in binding to diazepam-sensitive and insensitive benzodiazepine receptors	$pK_i(DI)$ – $pK_i(DS)$	38	Benzodiazempams	2	$C^+_{sp^3}$		6	0.80	0.19	0.79	0.56				
153	[32]	Binding to cholecystokinin-A	?	14	Benzodiazepine and glutamic acids	?			5	0.99	0.39	0.45					
154	[31]	Binding to cholecystokinin-B	log ($1/IC_{50}$)	20	Benzodiazepine and glutamic acids	2	$C^+_{sp^3}$		5	1.00	0.09	0.57		12		1.06 [k]	
155	[185]	Binding to dopamine D_1 receptor	pK_i	10	Phenyltetrahydroiso-quinolines and tetrahydrodibenzoqu-inolizines	2	$C^+_{sp^3}$		2	1.00	0.08	0.58	0.73				
156	[174]	Binding to dopamine D_1 receptor	pK_i	61	Phenethylamine phenols	2	$C^+_{sp^3}$		4		1.13	0.51	0.79				Many equations with detailed description
157	[174]	Binding to dopamine D_1 receptor	pK_i	225	Phenethylamine phenols	2	$C^+_{sp^3}$		4			0.50	0.89				
158	[241]	Binding to dopamine D_1 receptor	pK_i	16	Tetrahydroprotober-berines	2	$C^+_{sp^3}$		5	0.98	0.16	0.71					
159	[241]	Binding to dopamine D_2 receptor	pK_i	13	Tetrahydroprotober-berines	2	$C^+_{sp^3}$		3	0.98	0.08	0.70					
160	[174]	Binding to dopamine D_2 receptor	pK_i	26	Phenethylamine phenols	2	$C^+_{sp^3}$		4	0.89	0.45	0.54	0.85				
161	[337]	Binding to dopamine D_3 receptor	log K_i	30	Arylamides	1	$C^+_{sp^3}$,	Ca^{+2}, H_2O	2	0.80		0.65					GOLPE
162	[29]	Binding to dopamine D_{1A} receptor	log ($1/K_d$)	11	DA agonists	2	$C^+_{sp^3}$		2	0.96		0.83	0.46	6			
163	[269]	Binding to epididymal retinoic acid binding protein	–log K	18	Retinoids	2	$N^+_{sp^3}$		5	0.99	0.03	0.75					
164	[200]	Binding to GABA receptor	?	27	Diverse compounds	?	?		4	0.93	0.25	0.36					
165	[200]	Binding to GABA receptor	?	27	Pyrimidines				4	0.95	0.75	0.44					
166	[193]	Binding to melatonin receptor	pK_i	28	Melatonin	2	$O^+_{sp^3}$,	H^+	5	0.97		0.76		3			
167	[193]	Binding to melatonin receptor	pK_i	31	Melatonin	2	$O^+_{sp^3}$,	H^+	5	0.96		0.66					

Table 10 (continued)

Equation no.	Reference no.[c]	Target property	Target property form used in CoMFA	N^g	Type of compounds	Grid (Å)	Probe group	Additional probes or parameters used	Fit statistics[d] L^i	Fit R^2	Fit s	CV statistics[e] Q^2	CV s_{cv}	Forecast statistics[f] M^j	Forecast R^2	Forecast s^k_{pred}	Comments[h]
168	[352]	Binding to melatonin receptor	pK_i	44	Diverse compounds	2	$C^+_{sp^3}$		5	0.97	0.25	0.80	0.61	9	0.76		4 excluded
169	[197]	Binding to muscarinic receptor	pK_i	22	Benzofurans	2	$C^+_{sp^3}$,	π	3	0.84	0.31	0.60					
170	[191]	Binding to neuromodulatory σ_3 receptor	pIC_{50}	34	Phenylaminotetralins	2	$C^+_{sp^3}$		5	0.94	0.28	0.59	0.73				2 excluded
171	[98]	Binding to nicotine receptor	pK_i	28	Aminomethylpyridines, naphthyridines, and misc. compounds	2	$C^+_{sp^3}$		2	0.85		0.61					
172	[177]	Binding to σ_1 receptor	$-\log K_i$	19	N-benzyl-N-normetazocines	2	$C^+_{sp^3}$					0.39					
173	[348]	Binding to μ opioid receptor	$\log (1/ED_{50})$	13	Methylfentanyl derivatives	?	?		5	1.00	0.05	0.72		1		0.70 [k]	
174	[248]	Inhibition of ADP-induced platelet aggregation (P_2t purinoceptor antagonism)	pIC_{50}	9	ADP analogs	1	$C^+_{sp^3}$		2	0.92		0.71	1.2	1			o:f = 6.1:6.7
175	[283]	Human olfactory detection	log 1/ (threshold value)	9	Pyrazines	2	$C^+_{sp^3}$		2	0.97	0.29	0.76		5			
176	[283]	Human olfactory detection	log 1/ (threshold value)	12	Furanones and others	2	$C^+_{sp^3}$		3	0.98	0.86	0.77		5			13 conformers were used
177	[261]	Binding to Ah (dioxin) receptor	pIC_{50}	64	Dioxins and furans	2	$C^+_{sp^3}$		7	0.88	0.62	0.71	0.97				
178	[261]	Binding to Ah (dioxin) receptor	pIC_{50}	78	Dioxins, furans, and biphenyls	2	$C^+_{sp^3}$		10	0.90	0.56	0.71	0.97				
179	[261]	Binding to Ah (dioxin) receptor	pIC_{50}	76	Dioxins, furans and indolocarbazoles	2	$C^+_{sp^3}$		9	0.88	0.57	0.59	1.06				
180	[261]	Binding to Ah (dioxin) receptor	pIC_{50}	69	Dioxins, furans and naphthalenes	2	$C^+_{sp^3}$		8	0.89	0.59	0.70	0.95				
181	[261]	Binding to Ah (dioxin) receptor	pIC_{50}	94	Dioxins, furans, biphenyls, naphthalenes and indolocarbazoles	2	$C^+_{sp^3}$		8	0.86	0.61	0.62	1.02				5 excluded
182	[149], [219]	Binding to aryl hydrocarbon receptor	log *au*	50	Chlorinated dibenzo-p-dioxin and dibenzofurans	2	$C^+_{sp^3}$		7	0.95	0.80	0.72	1.90	3			No predictivity

Table 10 (continued)

Equation no.	Reference no.[c]	Target property	Target property form used in CoMFA	N^g	Type of compounds	Grid (Å)	Probe group	Additional probes or parameters used	Fit statistics[d] L^i	R^2	s	CV statistics[e] Q^2	s_{cv}	Forecast statistics[f] M^j	R^2	s^k_{pred}	Comments[h]
183	[277]	Binding to androgen receptor	log $RA(A)$	18	Androstan-17b-ols	2	$C^{+}_{sp^3}$,	CLOGP	4	0.93	0.39		?				5 excluded
184	[277]	Binding to androgen receptor	log $RA(A)$	12	Estratrien-17b-ols	2	$C^{+}_{sp^3}$		3	0.88	0.30		?				1 excluded
185	[97]	Binding to estradiol receptor	ln RBA	44	Halogenated estradiols	1	$C^{+}_{sp^3}$		11	0.99	0.16	0.89	0.47				
186	[277]	Binding to estrogen receptor	log $RA(E)$	18	Androstan-17b-ols	2	$C^{+}_{sp^3}$,	CLOGP	4	0.84	0.33		?				5 excluded
187	[265]	Binding to estrogen receptor	pK_i	58	Diverse compounds	2	$C^{+}_{sp^3}$	HINT	3	0.88	0.69	0.59					
188	[277]	Binding to estrogen receptor	log $RA(E)$	11	Estratrien17b-ols	2	$C^{+}_{sp^3}$,	CLOGP	2	0.79	0.60		?				3 excluded
189	[264]	Binding to estrogen receptor	pC_{50}	11	o-chlorinated hydroxybiphenyls	1	$C^{+}_{sp^3}$		2	0.97	0.27	0.54	1.12				3 ionizable compounds not used, 3 unknown compounds predicted
190	[85]	Binding to Jurkat human lymphoblastic T-cell line receptor	log $(1/IC_{50})$	9	Lactoferrins and transferrins	2	$C^{+}_{sp^3}$,	Dipole moments	2	0.94	0.07	0.59	0.17				
191	[210]	Binding to LTD_4 receptor	pK_i	24	Diverse compounds	2	$C^{+}_{sp^3}$		5	0.99	0.09	0.79	0.57				10 outliers, multiple conformers per molecule
192	[271]	Binding to vertebrate skeletal muscle ryanodine receptor	ln K_D	19	Ryanoids	?	?		?	0.92				2			
193	[250]	Binding to RNA	pK_D	10	Xanthines	1.5	$C^{+}_{sp^3}$		3	0.96	0.28	0.34	1.09				
194	[250]	Binding to RNA	pK_D	10	Xanthines	1.5	$C^{+}_{sp^3}$		3	0.93	0.35	0.73	0.70				GOLPE
195	[96]	Binding to antibody IgE(Lb4)	log K_A	27	Haptens	1	$C^{+}_{sp^3}$		6	0.99		0.81					4 alignments
196	[68]	Calcium-channel agonistic potency	–log potency	36	Benzoylpyrrole-carboxylates	1	OH^-		4	0.91							
197	[68]	Calcium-channel agonistic potency	–log potency	12	Benzoylpyrrole-carboxylates	1	OH^-		2	0.97							
198	[69]	Calcium-channel agonistic potency	log EC_{50}	36	Pyrrole-3-carboxylates	1	OH^-		4	0.86		(0.72)					
199	[302]	Binding to calcium channel	pIC_{50}	26	pyrrolobenzothiazines	1	$C^{+}_{sp^3}$		5	0.97	0.21	0.70	0.65	10	0.68		Average error = 0.26 prediction of different classes

Table 10 (continued)

Equation no.	Reference no.[c]	Target property	Target property form used in CoMFA	N^g	Type of compounds	Grid (Å)	Probe group	Additional probes or parameters used	Fit statistics[d] L^i	R^2	s	CV statistics[e] Q^2	s_{cv}	Forecast statistics[f] M^j	R^2	s^k_{pred}	Comments[h]
200	[374]	Modulation of calcium channel function (ryanodine receptor)	% fraction. conduction	10	Ryanodines	1	$C^+_{sp^3}$,		2	0.98		0.76		2			
201	[374]	Modulation of calcium channel function (ryanodine receptor)	% fraction. conduction	12	Ryanodines	1	$C^+_{sp^3}$,		2	0.96		0.73					
202	[46]	Potassium channel openers	pEC_{50}	18	Diverse compounds	?	?		5	0.99	0.26	0.76					
203	[202]	Vasorelaxant activity (potassium channel opening)	pEC_{50}	13	Benzopyran-4-carbothioamides	1	$C^+_{sp^3}$	π, π^2	6	1.00	0.11	0.57					
204	[267]	Inhibition of astrocytic chloride transport	pIC_{50}	18	Oxoisoindolines	2	$C^+_{sp^3}$		2	0.91	0.20	0.82					
205	[302]	Binding to rat cortex homogenate (calcium channel blocker)	pIC_{50}	26	Pyrrolobenzothiazines	1	$C^+_{sp^3}$		5	0.97	0.21	0.70	0.65	^{3}10	0.86;	0.26;	close analogs
206	[45]	Inhibition of binding to dopamine transporter	$\log(1/IC_{50})$	12	Tropane-2b-carboxylic acid esters	2	$C^+_{sp^3}$		3	0.94	0.23	0.76		13			
207	[45]	Inhibition of binding to dopamine transporter	$\log(1/IC_{50})$	12	Tropane-2b-carboxylic acid esters	2	$C^+_{sp^3}$		4	0.96	0.17	0.73					
208	[44]	Binding to dopamine transporter	?	25	Cocaines	?	?		3	0.93	0.20						
209	[154]	Inhibition of endothelin-1 binding to A10 rat thoracic aorta smooth muscle cells	$\log(1/IC_{50})$	36	Aryl sulfonamides	1	$C^+_{sp^3}$		6	0.94	0.30	0.70	0.69				
210	[154]	Inhibition of endothelin-1 binding to A10 rat thoracic aorta smooth muscle cells	$\log(1/IC_{50})$	32	Aryl sulfonamides	1	$C^+_{sp^3}$		3	0.57	0.79			4			
211	[113]	Inhibition of growth hormone release	$pGHI$	34	Somatostatin analogs	2	$C^+_{sp^3}$		4	0.86	0.54	0.54	1.22	6	0.88		3 alignments
212	[113]	Inhibition of growth hormone release	$pGHI$	64	Somatostatin analogs	2	$C^+_{sp^3}$		5	0.92	0.50	0.51	1.02	6	0.93		3 alignments
213	[130]	Inhibition of valinomycin-induced uptake of K^+	$\log(1/K_i)$	11	Procaine analogues	2	H_2O		2	0.94	0.16		0.27				
214	[138]	Antimicrobial potency	$\log MIC$	9–12	Alkylbenzyldimethyl-ammonium chlorides	2	H_2O		2	0.97–0.99	0.07–0.10		0.16–0.27				
215	[138]	Antimicrobial potency	$\log MKC$	10–12	Alkylbenzyldimethyl-ammonium chlorides	2	H_2O		2	0.98–1.00	0.07–0.17		0.18–0.28				

Table 10 (continued)

Equation no.	Reference no.[c]	Target property	Target property form used in CoMFA	N^g	Type of compounds	Grid (Å)	Probe group	Additional probes or parameters used	Fit statistics[d] L^i	R^2	s	CV statistics[e] Q^2	s_{cv}	Forecast statistics[f] M^j	R^2	s^k_{pred}	Comments[h]
216	[138]	Antimicrobial potency	log *C*	6	Alkylbenzyldimethyl-ammonium chlorides	2	H_2O		2	0.99–1.00	0.08–0.16		0.16–0.54				
217	[285]	Inhibition of gram-negative organism	?	17	Quinolines	2	$C^+_{sp^3}$		4	0.91	0.30	0.49					
218	[285]	Inhibition of gram-negative organism	?	17	Quinolines	2	$C^+_{sp^3}$		4	0.91	0.28	0.53					
219	[285]	Inhibition of gram-negative organism	?	17	Quinolines	2	$C^+_{sp^3}$		4	0.89	0.30	0.58					
220	[286]	Inhibition of gram-negative organism	?	19	Quinolines	2	$C^+_{sp^3}$		5	0.97	0.16	0.46					
221	[285]	Inhibition of gram-positive organism	?	20	Quinolines	2	$C^+_{sp^3}$		5	0.99	0.05	0.56					
222	[287]	Inhibition of gram-positive organism (*E. coli*)	?	60	Quinolines	1	$C^+_{sp^3}$		6	0.89	0.32	0.69		6			
223	[287]	Inhibition of gram-positive organism (*S. aureus*)	?	60	Quinolines	1	$C^+_{sp^3}$		8	0.91	0.28	0.62					
224	[324]	Bacteriostatic activity	log (1/*C*)	38	Sulfanilamides	2	$C^+_{sp^3}$		9	0.94	0.26	0.67	0.61				Distance matrix
225	[75]	Genotoxicity (antibacterial)	log *SOSIP*	40	Nitrofurans		$C^+_{sp^3}$,	H^+	4	0.91	0.31		0.63	6		0.50[l]	
226	[75]	Genotoxicity (antibacterial)	log SOSIP	44	Nitrofurans	2	$C^+_{sp^3}$,	H+	5	0.96	0.20		0.41				
227	[239]	Antifungal activity against *Candida albicans*; Inhibition of p-450$_{DM}$	–log (*MIC*$_{90}$-X/ *MIC*$_{90}$-ref)	40	Bifonazoles	1	$C^+_{sp^3}$		6	0.95	0.20	0.57	0.60	16	0.69		2 alignments
228	[49]	Ability to form intracellular covalent topoisomerase II-DNA complexes	ln(*PCPDCF* + 1)	60	Demethylepipo-dophyllotoxins	1	$C^+_{sp^3}$		4	0.87	0.45	0.58	0.82	41	0.24	0.42	q^2*cutoff* = 0.1, q^2-GRS, one excluded
229	[49]	Ability to form intracellular covalent topoisomerase II-DNA complexes	ln(*PCPDCF* + 1)	101	Demethylepipo-dophyllotoxins	1	$C^+_{sp^3}$		5			0.40	0.79				
230	[49]	Ability to form intracellular covalent topoisomerase II-DNA complexes	ln(*PCPDCF* + 1)	101	Demethylepipo-dophyllotoxins	1	$C^+_{sp^3}$,	C_{sp^3}, H^+, $O^-_{sp^3}$	5	0.84	0.40	0.58	0.66				q^2cutoff = 0.2, q^2-GRS
231	[43,299]	Inhibition of topoisomerase I (anticancer/antitumor agents)	p*IC*$_{50}$	32	Camptothecins	?	$C^+_{sp^3}$		4	0.87	0.37	0.73	0.53	3			
232	[66]	Binding to microtubules	log (1/ratio)	50	Taxols	2	$C^+_{sp^3}$		4	0.96	0.16	0.84	0.30	47			
233	[117]	Antitumor activity	Arbitrary rating	32	Thioxanthen-9-ones	2	$C^+_{sp^3}$	log *P*	?	0.94		0.72					SEAL superposition

Table 10 (continued)

Equation no.	Reference no.[c]	Target property	Target property form used in CoMFA	N^g	Type of compounds	Grid (Å)	Probe group	Additional probes or parameters used	Fit statistics[d] L^i	R^2	s	CV statistics[e] Q^2	s_{cv}	Forecast statistics[f] M^j	R^2	s^k_{pred}	Comments[h]
234	[118]	Cytotoxicity against HCT-8 cells	log (1/IC_{50})	38	Pyrazoloacridines	1	$C^+_{sp^3}$			0.95		0.58					SEAL superposition, various lattices
235	[118]	Cytotoxicity against l1 210 cells	log (1/IC_{50})	44	Pyrazoloacridines	2	$C^+_{sp^3}$			0.96		0.66					SEAL superposition, various lattices
236	[111]	Induction of human leukemia cell differentiation	log C	18	Amides and ureas	2	$C^+_{sp^3}$		4	0.98	0.10	0.78	0.37	5		0.56	
237	[176]	Inhibition of 6C3HED lymphosarcoma	L	96	Sulfonylureas	2	$C^+_{sp^3}$,	log P	5	0.94		0.66					
238	[276]	Inhibition of L1 210 cell line	log (1/C)	19	Pyridopyrimidines	?	?		?	?		?		5			$o = 7.20/c = 7.98$; $o = 7.70/c = 7.84$; $o = 5.19/c = 5.13$; $o = 7.54/c = 8.00$; $o = 6.30/c = 7.00$
239	[329]	Inhibition of spermidine transport into L1 210 cell	pK_i	57	Polyamines	?	$C^+_{sp^3}$		5	0.91		0.66		11			
240	[194]	Mutagenecity	ln ($TA100$)	16	Furanones	0.25	?										
241	[227]	Mutagenecity	log $TA98$	31	Nitroaromatics	?	?	log P, (log P)^2	?	1.00		0.70					
242	[217,218]	Mutagenicity (TA100)	ln $TA100$	16	Hydroxyfuranones	2	$C^+_{sp^3}$,	E_{LUMO}	3	0.97		0.92		5			
243	[38]	Mutagenicity	log $TA98$	163	Nitroaromatics	2	$C^+_{sp^3}$,	log P, E_{LUMO}	4	0.83	0.84	0.78	0.94	16		0.90	3 excluded
244	[347]	Mutagenicity (TA100)	log $TA100$	12	Hydrazines	2	$C^+_{sp^3}$,	E_{LUMO}, LUMO $C^-_{sp^3}$	2	0.97		0.92					
245	[140]	Acute toxicity	log (1/LD_{100})	11	Alkanes	2	H_2O		2	0.94	0.09	0.89	0.17				
246	[140]	Acute toxicity	log (1/LD_{50})	6	Alkanes	2	H_2O		2	0.98	0.06	0.71	0.27				
247	[140]	Acute toxicity	log (1/LD)	17	Alkanes	2	H_2O		3	0.96	0.08	0.94	0.09				
248	[140]	Acute toxicity	log (1/LD)	17	Alkanes	2	H2O		5	0.99	0.05	0.96	0.05				
249	[323]	Hepatotoxicity	log ALT	11	Thiobenzamides	2	$C^+_{sp^3}$		3	0.96	0.13	0.51	0.44				Distance matrix
250	[293]	Toxicity on *B. calyciflorus*	?	21	Sulfactants	?	?		3	0.95		0.86		9		0.63	
251	[293]	Toxicity on *B. calyciflorus*	?	21	Sulfactants	?	?		2	0.91		0.85		9		0.68	

Table 10 (continued)

Equation no.	Reference no.[c]	Target property	Target property form used in CoMFA	N^g	Type of compounds	Grid (Å)	Probe group	Additional probes or parameters used	Fit statistics[d] L^i	R^2	s	CV statistics[e] Q^2	s_{cv}	Forecast statistics[f] M^j	R^2	s^k_{pred}	Comments[h]
252	[293]	Toxicity on *T. platurus* and *B. calyciflorus*	?	21	Sulfactants	?	?		3	0.95		0.86		9		0.78	
253	[293]	Toxicity on *T. platurus* and *B. calyciflorus*	?	21	Sulfactants	?	?		2	0.91		0.85		9		0.85	
254	[289]	Insecticidal activity	pLD_{50}	18	Dioxatricyclododec-10-ones	2	$C^+_{sp^3}$,	log P	2	0.76	0.28	0.52	0.40				2 excluded
255	[289]	Insecticidal activity	pLD_{50}	11	Dioxatricyclododec-10-ones	2	$C^+_{sp^3}$,	log P	1	0.89	0.31	0.79	0.42				2 excluded
256	[289]	Insecticidal activity	pLD_{50}	31	Dioxatricyclododec-10-enes	2	$C^+_{sp^3}$,	log P	4	0.81	0.33	0.47	0.56				2 excluded
257	[192]	Insecticidal activity	pEC_{50}	37	Dibenzoylhydrazines	2	$C^+_{sp^3}$,	log P	3	0.77	0.35	0.45	0.54				
258	[192]	Insecticidal activity	pEC_{50}	43	Dibenzoylhydrazines	2	$C^+_{sp^3}$,	log P	3	0.89	0.25	0.47	0.55				
259	[5]	Insecticidal activity (knockdown activity against houseflies)	?	27	Dioxatricycloalkene and others	?	$C^+_{sp^3}$,	log P	3	0.83	0.32		0.55				
260	[6]	Insecticidal activity (knockdown activity against houseflies)	log ($1/KD_{50}$)	17	m-substituted benzyl esters	2	$C^+_{sp^3}$	π	3	0.98	0.17		0.43				
261	[6]	Insecticidal activity (knockdown activity against houseflies)	log ($1/KD_{50}$)	19	N-methylol esters of imides and lactams	2	$C^+_{sp^3}$	π	4	0.95	0.21	0.42	0.70				
262	[6]	Insecticidal activity (knockdown activity against houseflies)	log ($1/KD_{50}$)	51	N-methylol esters of imides and lactams and benzyl esters	2	$C^+_{sp^3}$		6	0.97	0.15		0.50				
263	[6]	Insecticidal activity (knockdown activity against houseflies)	log ($1/KD_{50}$)	14	Benzyl esters	2	$C^+_{sp^3}$		3	0.88	0.14		0.28				
264	[6]	Insecticidal activity (knockdown activity against houseflies)	log ($1/KD_{50}$)	51	Pyrethroids	2	$C^+_{sp^3}$		6	0.97	0.15		0.50				
265	[37]	Insecticidal activity (non-competitive block GABA, *in vivo*)	log ($1/LD_{50}$)	36	Bicyclooctanes, dithianes, picrotoxins, lindanes, cyclodienes, silatranes	0.75	?		5	0.89		0.43					

Table 10 (continued)

Equation no.	Reference no.[c]	Target property	Target property form used in CoMFA	N^g	Type of compounds	Grid (Å)	Probe group	Additional probes or parameters used	Fit statistics[d] L^i	R^2	s	CV statistics[e] Q^2	s_{cv}	Forecast statistics[f] M^j	R^2	s^k_{pred}	Comments[h]
266	[37]	Insecticidal activity (non-competitive block GABA, *in vivo*)	log ($1/IC_{50}$)	36	Bicyclooctanes, dithianes, picrotoxins, lindanes, cyclodienes, silatranes	0.75	?		5	0.92		0.44					
267	[51]	Herbicidal potency	SRS_{NL}	48	3-pyrazolyltrifluorotolyl ethers	?	?		5	0.86	6.14	0.36	13.2				
268	[51]	Herbicidal potency	SRS_{BL}	48	3-pyrazolyltrifluorotolyl ethers	?	?		3	0.77	8.75	0.41	14.1				
269	[52]	Herbicidal potency	pGR_{50}	20	Pyrazole olefinic nitriles	2	$C^+_{sp^3}$		2	0.89	0.17	0.89					
270	[139]	Local anesthetic action	log ($1/ED_{50}$)	8	Diethylaminoethyl benzoates	2	H_2O,	H^+	3	0.94	0.19		0.28				
271	[15]	Antimalarial potency for D-6 clone	(IC_{50})	42	Artemisinin analogs	2	?		5	0.97	0.26	0.86		14		1.08	
272	[289]	Inhibition of TBPS binding to rat brain membranes	pIC_{50}	22	Dioxatricyclododec-10-ones	2	$C^+_{sp^3}$		4	0.91	0.21	0.50	0.50				2 excluded
273	[289]	Inhibition of TBPS binding t rat brain membranes	pIC_{50}	19	Dioxatricyclododec-10-ones	2	$C^+_{sp^3}$		4	0.84	0.35	0.37	0.71				2 excluded
274	[195]	Binding to α-cyclodextrin	log K	16	Barbiturates	1	$C^+_{sp^3}$	HINT	6	0.99		0.75					
275	[195]	Binding to β-cyclodextrin	log K	16	Barbiturates	1	$C^+_{sp^3}$	HINT	4	0.96		0.65					
276	[88]	Binding to cellulose fibres	A	49	Anthraquinone vat dyes	?	$C^+_{sp^3}$		5	0.99		0.84					
277	[340]	Binding to cellulose fibres	$-Dm^o$	27	Azo dyes	2	$C^+_{sp^3}$		2	0.92		0.78					
278	[87]	Binding to cellulose fibres	A	21	Bisazo dyes	2	$C^+_{sp^3}$		5	0.99	0.40	0.70		2			Multiple conformations were used for same molecule. Excellent predictivity
279	[179]	Glycine conjugation	log A	19	Aromatic, aliphatic carboxylic acids	2	$C^+_{sp^3}$,	MLP	3	0.94		0.52					
280	[200]	Inhibition of ergosterol biosynthesis	?	20	Pyrimidines	?	?		4	0.99	0.33	0.91					GOLPE-like
281	[200]	Inhibition of ergosterol biosynthesis	?	20	Diverse compounds	?	?		4	0.99	0.33	0.91					GOLPE-like

Table 10 (continued)

Equation no.	Reference no.[c]	Target property	Target property form used in CoMFA	N^g	Type of compounds	Grid (Å)	Probe group	Additional probes or parameters used	Fit statistics[d] L^i	R^2	s	CV statistics[e] Q^2	s_{cv}	Forecast statistics[f] M^j	R^2	s^k_{pred}	Comments[h]
282	[47]	Inhibition of initiation of protein biosynthesis	(pIC_{50})	7	Cephalotaxine esters	2	$C^+_{sp^3}$		1			0.46					
283	[47]	Inhibition of initiation of protein biosynthesis	(pIC_{50})	7	Cephalotaxine esters	1	$C^+_{sp^3}$		1			0.65					q^2cutoff = 0.0, q^2-GRS
284	[125]	Inhibition of liver microsomal metabolism	ln $1/K_i$	27	Diverse compounds	1	$C^+_{sp^3}$		5	0.99	0.17	0.70					q^2-GRS
285	[182]	Inhibition of mitochondrial respiration	?	19	Isooquinolines	1.5	$C^+_{sp^3}$		2	0.86		0.66					Steric
286	[262]	Metabolic rate constant	CL_{int}	12	Chlorinated volatile organic compounds	2	$C^+_{sp^3}$,	LUMO, HINT	4	0.95	5.35	0.53					
287	[166]	Ploydimethylsiloxane membrane flux	log J_{ssi}	101	Aromatic and heteroaromatic compounds	2	$C^+_{sp^3}$,	Ring indicators, log MFS_i, f_{chex}, I_{HB}	6	0.97	0.21	0.95	0.29				6 excluded
288	[104]	Inhibition of Hill reaction	pI_{50}	77	Anilides	2	$C^+_{sp^3}$,	π	7	0.89	0.41	0.75	0.61	297 90		0.63 0.77	20 excluded 18 excluded
289	[278]	Rate constants of nucleophilic reaction	log k	6	Alkoxides	2	$C^+_{sp^3}$		4	1.00		0.90					Reactant complexes
290	[278]	Rate constants of nucleophilic reaction	log k	18	Aryloxides	2	$C^+_{sp^3}$		5	0.99		0.82					Transition state structures
291	[278]	Rate constants of nucleophilic reaction	log k	18	Aryloxides	2	$C^+_{sp^3}$		4	0.96		0.85					Reactant complexes
292	[278]	Rate constants of nucleophilic reaction	log k	6	Arylsulfides	2	$C^+_{sp^3}$		3	0.98		0.83					Reactant complexes
293	[281]	Rate of nucleophilic substitution reaction in MeCN	log k	15	Benzyl benzenesulfonate	2	$C^+_{sp^3}$		5	0.99	0.07	0.94					
294	[281]	Rate of nucleophilic substitution reaction in MeOH	log k	15	Benzyl benzenesulfonate	2	$C^+_{sp^3}$		5	0.98	0.08	0.92					
295	[279]	Rate of [3,3]-sigmatropic rearrangement	log k'	11	Allylic esters	2	$C^+_{sp^3}$		3	0.99	0.04	0.90					
296	[136]	Rate constant of elimination reaction of arenesulphonates	log k	4	Phenylethyl arenesulphonates	2	H^+		1	0.97	0.11						
297	[136]	Rate constant of elimination reaction of arenesulphonates	log k	4	Phenylethyl arenesulphonates	2	H^+		1	0.95	0.14						

Table 10 (continued)

Equation no.	Reference no.[c]	Target property	Target property form used in CoMFA	N^g	Type of compounds	Grid (Å)	Probe group	Additional probes or parameters used	Fit statistics[d] L^i	R^2	s	CV statistics[e] Q^2	s_{cv}	Forecast statistics[f] M^j	R^2	s^k_{pred}	Comments[h]
298	[136]	Rate constant of elimination reaction of arenesulphonates	log k	4	Phenylethyl arenesulphonates	2	H^+		1	0.92	0.14						
299	[136]	Rate constant of elimination reaction of arenesulphonates	log k	4	Phenylethyl arenesulphonates	2	H^+		1	0.94	0.13						
300	[136]	Rate constant of elimination reaction of arenesulphonates	log k	4	Phenylethyl arenesulphonates	2	H^+		1	0.94	0.14						
301	[136]	Rate constant of elimination reaction of arenesulphonates	log k	4	Phenylethyl arenesulphonates	2	H^+		1	0.74	0.24						
302	[136]	Rate constant of elimination reaction of arenesulphonates	log k	4	Phenylethyl arenesulphonates	2	H^+		1	0.84	0.19						
303	[238]	Brown's steric parameter	E_R	36	Phosphine ligands	1	$C^+_{sp^3}$		3	0.95	27.0	0.77	61.0				
304	[238]	[^{13}C] Carbonyl chemical shift	Δ_{CO}	36	Phosphine ligands	2	$C^+_{sp^3}$		2	0.92	0.34	0.70	0.72				
305	[281]	LUMO energy	E_{LUMO}	15	Benzyl benzenesulfonate	2	$C^+_{sp^3}$		5	0.99	0.05	0.97					
306	[272]	Formation enthalpy	$\Delta_f H^\circ_m$	16	Polycyclic aromatic hydrocarbons	2	$C^+_{sp^3}$		6	1.00		0.62					
307	[272]	Sublimation enthalpy	$\Delta_{sub} H^\circ_m$	30	Polycyclic aromatic hydrocarbons	2	$C^+_{sp^3}$		3	0.95		0.82		5			Average deviation of 5%
308	[136]	Dissociation constant	pK_a	6	Benzoic acids	2	H+		1	0.93	0.11	0.68	0.23				
309	[136]	Dissociation constant	pK_a	6	Phenyl acetic acids	2	H+		1	0.98	0.03	0.95	0.05				
310	[323]	Hammett σ	σ	11	Thiobenzamides	2	$C^+_{sp^3}$		3	0.96	0.08	0.76	0.19				
311	[324]	Hammett σ	σ	7	Glutathione ketoaldehyde hemiacetals	2	H^+		2	0.97	0.08	0.89	0.16				
312	[324]	Hammett σ	σ	3	Aromatic aldehydes	2	H^+		3	0.97	0.10	0.83	0.26				
313	[324]	Hammett σ	σ	12	Anilines	2	H^+		4	0.84	0.23	0.67	0.43				
314	[324]	Hammett σ	σ	7	Glutathione ketoaldehyde hemiacetals	2	H^+		3	0.98	0.07	0.84	0.21				Distance matrix

Table 10 (continued)

Equation no.	Reference no.[c]	Target property	Target property form used in CoMFA	N^g	Type of compounds	Grid (Å)	Probe group	Additional probes or parameters used	Fit statistics[d] L^i	R^2	s	CV statistics[e] Q^2	s_{cv}	Forecast statistics[f] M^j	R^2	s^k_{pred}	Comments[h]
315	[324]	Hammett σ	σ	11	Aromatic aldehydes	2	H^+		4	0.97	0.11	0.69	0.37				Distance matrix
316	[324]	Hammett σ	σ	12	Anilines	2	H^+		4	0.95	0.12	0.36	0.45				Distance matrix
317	[196]	Hydrophobicity	log P	10	Cytosine nucleosides	?	$O^{-4}_{sp^3}$		3	0.97	0.08	0.78					
318	[196]	Hydrophobicity	log k'	10	Cytosine nucleosides	?	$O^{-4}_{sp^3}$		3	0.95	0.08	0.54					
319	[237]	Distribution coefficient	log D	38	Amino acids	1	H_2O		3	0.90	0.42	0.65	0.79				GRID
320	[258]	Partition coefficient	log P	15	Dioxins, biphenyls and benzofurans	1	$C^+_{sp^3}$		4	1.00	0.09	0.90	0.52				
321	[132]	Partition coefficient	log P	19	Carbamates	2	H_2O		5	0.96	0.12	0.41	0.58				GRID
322	[132]	Partition coefficient	log P	19	Carbamates	2	H_2O		4	0.94	0.20	0.38	0.59				GRID
323	[137]	Partition coefficient	log P	59	Diverse compounds	2	H_2O		8	0.95	0.21	0.84	0.35				GRID
324	[133]	Partition coefficient	log P	17	Furrans	2	H_2O		5	0.98	0.12	0.55	0.50				GRID
325	[137]	Partition coefficient	log P	19	Pyrazines	2	H_2O		4	0.93	0.19	0.67	0.38				GRID
326	[137]	Partition coefficient	log P	19	Pyridines	2	H_2O		4	0.97	0.14	0.61	0.42				GRID
327	[112]	Partition coefficients	log D (pH 7.40)	28	Diazine analogs of ridogrel	1	H_2O		5	0.92	0.18	0.71					GRID
328	[280]	Partition coefficients	log P	36	Orthopramides	2	$O.spc^-$		5	0.96	0.13	0.85		6			
329	[280]	Partition coefficients	log P	10	Orthopramides	2	$O.spc^-$		2	1.00	0.09	0.60		6			
330	[237]	Solubility	–log S	19	Amino acids	2, 1	H_2O, $C^+_{sp^3}$		2	0.91	0.20	0.80	0.30				GRID
331	[132]	Capacity factor	log k'	19	Carbamates	2	H_2O		5	0.96	0.11	0.38	0.36				
332	[132]	Capacity factor	log k'	19	Carbamates	2	H_2O		4	0.94	0.12	0.35	037				
333	[112]	Capacity factor	log $k'w$	12	Diazine analogs of ridogrel	1	H_2O		5	0.97	0.09	0.74					
334	[137]	Capacity factor	log k'	6	Diverse compounds	2	H_2O		2	0.97	0.08	0.80	0.19				
335	[137]	Capacity factor	log k'	6	Diverse compounds	2	H_2O		2	1.00	0.05	0.87	0.28				
336	[137]	Capacity factor	log k'	6	Diverse compounds	2	H_2O		2	0.99	0.08	0.92	0.12				
337	[137]	Capacity factor	log k'	59	Diverse compounds	2	H_2O		8	0.93	0.16	0.77	0.26				

Table 10 (continued)

Equation no.	Reference no.[c]	Target property	Target property form used in CoMFA	N^g	Type of compounds	Grid (Å)	Probe group	Additional probes or parameters used	Fit statistics[d] L^i	Fit R^2	Fit s	CV statistics[e] Q^2	CV s_{cv}	Forecast statistics[f] M^j	Forecast R^2	Forecast s^k_{pred}	Comments[h]
338	[133]	Capacity factor	$\log k'$	17	Furans	2	H_2O		5	0.98	0.09	0.54	0.43				
339	[137]	Capacity factor	$\log k'$	19	Pyrazines	2	H_2O		4	0.91	0.15	0.52	0.32				
340	[137]	Capacity factor	$\log k'$	19	Pyridines	2	H_2O		4	0.97	0.08	0.61	0.26				
341	[133]	Capacity factor	$\log k'$	54	Triazines	2	H_2O		5	0.96	0.10	0.86	0.17	14			
342	[58]	Chromatographic retention index	$\log I$	31	Polycyclic aromatic hydrocarbons	2	$C^+_{sp^3}$		3	0.95	0.16	0.86					
343	[258]	HPLC retention time	$\log T_r$	24	Dioxins, biphenyls and benzofurans	1	$C^+_{sp^3}$		7	0.99	0.04	0.85	0.18				
344	[8]	Enantioseparation	$\log \alpha$	21	Alkyl aryl sulfoxides	1.5	$C^+_{sp^3}$		5	0.99	0.01	0.56					
345	[41]	Enantioseparation	$\log \alpha$	18	Aryloxy acetic acid methyl esters	2	$C^+_{sp^3}$		4	0.97	0.01	0.63		1			$o = 1.89$; $c = 1.31$
346	[10]	Enantioseparation	$\log \alpha$	25	Phenyloxadiazolines	2	$C^+_{sp^3}$		1	0.74	0.60	0.02					
347	[238]	Inorganic reaction	ΔG^*	16	Phosphine ligands	1	Ca^{2+}		2	0.88	0.83	0.27	2.20				
348	[238]	Inorganic reaction	ΔG°	11	Phosphine ligands	1	$C^+_{sp^3}$		2	0.92	0.78	0.45	2.10				
349	[238]	Inorganic reaction	ΔG^*	10	Phosphine ligands	2	$C^+_{sp^3}$		1	0.75	3.40	0.39	5.30				
350	[238]	Inorganic reaction	ΔH°	8	Phosphine ligands	1	Br^-		1	0.94	5.70	0.33	19.0				
351	[238]	Inorganic reaction	ΔH°	19	Phosphine ligands	1	$C^+_{sp^3}$		2	0.96	5.40	0.64	16.0				
352	[238]	Tolman cone angle	Q_T	36	Phosphine ligands	1	$C^+_{sp^3}$		3	0.95	5.00	0.76	12.0				
353	[181]	?	$\log (1/IC_{50})$	25	Diverse compounds	?	$C^+_{sp^3}$		?	0.96	0.17	0.53		5			
354	[89]	?	?	19	Pyridazinones	2	$C^+_{sp^3}$		2	0.78	0.15	0.58	0.20				
355	[89]	?	?	37	Sulmazoles, and isomazoles	2	$C^+_{sp^3}$		2	0.90	0.25	0.67	0.45	3			
356	[179]	Binding to CoA	$\log A$	19	Aromatic and aliphatic carboxylic acids	2	$C^+_{sp^3}$,	MLP	5	0.98		0.66		2			
357	[92]	Binding to major his to compatibility complex	?	8	Nonapeptides	1	$C^+_{sp^3}$		3	0.98	0.64	0.70	2.55				1 excluded
358	[91]	Binding to major his to compatibility complex	?	10	Nonapeptides	2	$C^+_{sp^3}$		2			0.53	2.67				Steric only

Table 10 (continued)

Equation no.	Reference no.[c]	Target property	Target property form used in CoMFA	N^g	Type of compounds	Grid (Å)	Probe group	Additional probes or parameters used	Fit statistics[d] L^i	R^2	s	CV statistics[e] Q^2	s_{cv}	Forecast statistics[f] M^j	R^2	s^k_{pred}	Comments[h]
359	[293]	Characterization of pulp using near infrared reflectance spectroscopy (NIR)	?	28	?	?	?		5	0.93		0.86		11		6.23	D-optimal design
360	[293]	Characterization of pulp using NIR	?	28	?	?	?		3	0.94		0.90		11		3.80	D-optimal design, INLR
361	[293]	Fat content of beef or pork meat	?	70	?	?	?		5	0.97		0.97		33		0.99	D-optimal design
362	[293]	Fat content of beef or pork meat	?	70	?	?	?		12	0.99		0.95		33		0.90	D-optimal design, INLR
363	[293]	Water content of beef or pork meat	?	70	?	?	?		4	0.95		0.95		33		0.81	D-optimal design
364	[293]	Water content of beef or pork meat	?	70	?	?	?		11	0.98		0.94		33		0.65	D-optimal design, INLR

[a] Many times several CoMFA models were presented. In such a case, the one (usually the 'best') model was selected based not on the original authors' suggestion, but on the present authors' judgment.
[b] Some columns are missing because they were not reported by the authors.
[c] Reference numbers refer to the references in the chapter 'List of CoMFA References' by K.H. Kim.
[d] Statistics of the final CoMFA model.
[e] Statistics of the cross-validation test.
[f] Statistics of the forecasts. Different authors used different definition for forecasted R^2.
[g] N = number of compounds used in the CoMFA model.
[h] Abbreviations: GOLPE = Generating Optimal Linear PLS Estimations method; Q^2-GRS = Q^2-guided Region Selection method; INLR = Implicit Nonlinear Latent Variable Regression method; SAMPLS = SAMple-distance Partial Least Squares method; GRID = GRID force field and probe; IVS = Interactive Variable Selection; MECE = Minimum Equivalent Chemical Entities; MLP = Molecular Lipophilicity Potential; HOMO = Highest Occupied Molecular Orbital; LUMO = Lowest Unoccupied Molecular Orbital; HINT = Hydropathic INTeractions; E-state = Electrotopological state; HE-state = Hydrogen Electrotopological state; and SEAL = Steric and Electrostatic Alignment method; o : f = observed : forcasted, c = calculated.
[i] L = optimum number of components for the final CoMFA model.
[j] M = number of compounds predicted.
[k] RMSE or SEP.
[l] Calculated by the authors.

number of variables low. Brusniak et al. [29L] tried lattice spacings of 1 Å, 2 Å and 3 Å, and obtained Q^2 values of 0.72 (2), 0.83 (2) and 0.74 (3). The performance of the very coarse 3 Å spacing, which is certainly unusual in the literature, was surprisingly good.

Studies have shown that the magnitude of the effects on Q^2 values varied from relatively little [141L,142L] to as much as 0.5 [47L], due to the difference in the orientation of aligned molecules with respect to the grid box. It was observed that the large variation in Q^2 values sometimes decreased as the grid spacing changed from 2 Å to 1 Å. On the other hand, the decrease in the grid spacing may increase the noise level in PLS analysis, and may yield a lower Q^2 value. It was observed that such variation is more pronounced with a dataset of diverse structures than with a dataset of less diverse structures [47L]. The decrease in grid spacing increased the probability of placing the probe atom in a region where the steric and electrostatic field changes best correlated with biological activity.

2.7. *Scaling and intercorrelation*

2.7.1. *Scaling of energy fields*

The results of PLS analysis depend on the variance of the variables. If the original properties have been measured on the same relative scale, such as interaction energies in kcal/mol, there is less or no problem with high variance properties.

One of the often used variable weighting methods is the block scaling realized in SYBYL CoMFA (through the 'CoMFA_STD' keyword). This method ensures the same statistical importance to the steric and electrostatic fields, as well as additional parameters such as log *P*, each viewed as a 'block' of independent information. Lack of block scaling has, in some cases, dramatically worsened the results [102L].

Cruciani and Watson [64L] applied different scaling methods to the energy values calculated from a single probe. Their results showed that the R^2 of the fitted model was generally not affected by different data pretreatment, whereas greater effects were seen on the Q^2 of the cross-validated model. On the basis of their results, they concluded that the most appropriate pretreatment procedure was autoscaling on a subset of variables selected using a D-optimal algorithm to eliminate a reasonable amount of noise.

In a study of 43 N4-substituents of phenylpiperazine derivatives interacting with the 5-HT_{1A} receptor, van Steen et al. [254L] examined the contribution of the steric and electrostatic field descriptors toward the CoMFA models they had developed. For three alignment sets, the cross-validated and conventional R^2 values were lower when both fields were used compared to when only the steric field was used. The electrostatic fields had a negative effect on the overall cross-validated and conventional R^2 values and, thus, the contribution of the electrostatic field was of minor importance in comparison with the steric field. However, the CoMFA model derived from both fields indicated that it contained 53% steric and 47% electrostatic contributions. These calculations were performed using the CoMFA standard column scaling. When no scaling was applied, however, the ratio for the steric and electrostatic contributions was found to be 98% and 2%, respectively. These results indicated that scaling of energy fields influences the CoMFA results significantly, and the results from no scaling were in better agreement with the results obtained from the separate steric and electrostatic fields.

Kroemer et al. [153L] also examined how much CoMFA results were affected by different scaling procedures in a study with 37 ligands of the benzodiazepine receptor. They used two different scaling options: CoMFA standard scaling and no scaling. When they used HF/STO-3G/MPA fields, the contribution of the electrostatic components was 49% with scaling, whereas it was 7% without scaling: the former was a two-component model with $Q^2 = 0.62$ and $R^2 = 0.79$, whereas the latter was a four-component model with $Q^2 = 0.65$ and $R^2 = 0.87$.

We conclude that autoscaling may assign too much significance to those variables with only small variation and may not reflect real structural variations.

2.7.2. Scaling of other than energy fields

Sometimes one has to resort to external parameters because the molecular mechanics force fields used to calculate standard CoMFA descriptors are not parameterized for certain interactions and do not model important enthalpic and enthropic phenomena. DePriest et al. [79L] investigated a series of angiotensin-converting enzyme (ACE) inhibitors by using, in addition to the standard steric and electrostatic fields, indicator variables multiplied by 10, 100, 1000 or 10 000 to account for the chemical function (carboxylate, phosphate, hydroxamate and sulfur) directly bound to the zinc in the active site. The Zn indicator variable multiplied by 10 improved significantly the external predictivity of the model.

Davis et al. [68L,69L] performed a detailed study on the effects of scaling with macroscopic descriptors such as CLOGP and CMR. Depending on the relative scaling of the energy fields versus the macroscopic descriptors, the overall PRESS changed from 0.29 to 0.65.

2.7.3. Intercorrelations

Besides the problem of weighting effects, there can be the problem of intercorrelation when one includes variables other than the energy fields in CoMFA. For example, in the study of intrinsic knockdown activity of benzyl chrysanthemates, tetramethrins and related imido- and lactam-N-carbonyl esters against house flies, Akamatsu et al. [6L] tried to include a π_{alc} term to monitor the hydrophobic influence of substituents. They found that this term was playing a minor role, and inclusion of a π term in the CoMFA model was not statistically supported. They found a high correlation between the π_{alc} term and the CoMFA steric (SFT) and electrostatic (EFT) energy fields terms, as shown below:

$$\pi = 0.557 + [\mathrm{SFT}] + [\mathrm{EFT}],\ L = 6,\ n = 51,\ s = 0.247,\ R = 0.980$$

Because of such a collinearity, they argued that the separation of the π_{alc} term from the [SFT] and [EFT] terms was incomplete and that fractions of the π_{alc} term were included within the [SFT] and [EFT] terms. It is well known in classical QSAR that any variables that show collinearity should not be used together in the same correlation. Inclusion of such terms can yield a misleading QSAR model and make the interpretation of a QSAR difficult. Inclusion of such terms in 3D QSAR would result in similar consequences.

A series of thiazolidinones acting as H_1-antagonists was analyzed by Bolognese et al. [22L], using a combined Hansch and CoMFA approach. The following QSAR equation explained the effects of 3- and 4-phenyl substituents on the potency:

$$pA_2 = -3.49\ F_3 + 1.66\ \pi_4 - 0.92\ L_4 + 9.67,\ n = 15,\ R^2 = 0.804$$

In the above equation, F_3 is the field constant of the 3-substituents, and π_4 and L_4 are the hydrophobic constants of 4-substituents and the Verloop's length parameter of the 4-substituents, respectively. In the CoMFA study, steric and electrostatic fields as well as π_4 were used. Besides the negative steric contours of the resulting CoMFA model, which were consistent with the negative coefficient of L_4 in the classical QSAR shown above, positive steric contours were also observed. The positive contours resulted from a collinearity between the π_4 and the steric field of the 4-substituent.

Greco et al. [104L] circumvented the problem of collinearity between the steric field and scalar hydrophobic parameters with the knowledge of preliminary QSAR studies. Since the classical QSAR suggested that the steric properties of the varying substituents were irrelevant, they included the hydrophobic constants for the m- and p-phenyl substituents, but completely eliminated the steric and electrostatic fields at these positions. The variables used in CoMFA were the steric field of the m- and p-unsubstituted moiety and the π_3 and π_4 constants multiplied by proper weighting factors.

Intercorrelation between energy fields is to be suspected when models from different fields for a given set have comparable statistics and graphical results [95L]. In such cases, a tentative interpretation of the results is still possible, but the predictive ability of the model is questionable. The only solution to this problem is changing the composition of the training set, if possible, to break the undesired collinearity. Further aspects on the subject of intercorrelation is discussed in section 2.9, below.

2.8. *Variable selection*

Although there is a small risk of chance correlation in PLS, it is well known that including irrelevant variables into the independent parameter columns causes detrimental effects on the selection of a CoMFA model by PLS [50L]. Therefore, it would be beneficial to select only those variables that have significant effects on the biological activity to be correlated. Different approaches used in recent CoMFA are described below.

2.8.1. *Generating optimal linear PLS estimations (GOLPE)*

The Generating Optimal Linear PLS Estimations (GOLPE) procedure [17L,55L] evaluates the effects of individual variables on the model predictivity and extracts only those variables that improve the model predictivity. The procedure may be divided into three steps. First, a normal linear PLS model is applied using all the variables. This is followed by a variable preselection using a D-optimal design procedure. At this step, redundancy in the energy data matrix is reduced, and a sufficient collinearity among the remaining variables is maintained. In the second step, a matrix that contains variable combinations according to a fractional factorial design is built. At this step, dummy

avoid making predictions for a new compound that lies outside the boundaries of the training set [124L]. Then, what constitutes an ideal test set? Oprea et al. [207L] suggested that an ideal test set should include molecules (i) tested in the same conditions employed for the training set, (ii) falling within the lattice region occupied by the training set molecules and (iii) exhibiting well-distributed values of the target property, yet not exceeding those of the training set by more than 10% in order to avoid risky extrapolations.

2.10.1. Efforts to improve predictivity of CoMFA models

Aside from attempting to improve the predictiveness of CoMFA models by variable reduction, others have proposed different approaches. Kroemer and Hecht [150L] developed an automated procedure which systematically reorients those molecules that are underpredicted by the model. In this procedure, each compound was excluded once, and its activity was predicted by the CoMFA model derived from the remaining compounds. If the activity of the excluded compound is calculated to be lower than the observed activity, the compound is translated along the three principal axes of a Cartesian coordinate system by a user-specified increment to create a number of new orientations located at the points of a cube with the initial position of the compound in its center. The new alignments with the smallest residual activity are kept. From this position, the molecule is then rotated around the three axes of the coordinate system. Subsequently, the increments for rotation and translation are set to half of the original value, and the translation followed by the rotation procedure is continued until the final orientation of the molecule is chosen. If necessary, the whole process is repeated several times for the entire set until the final model is chosen.

In their study with two independent sets of 80 dihydrofolate reductase inhibitors and a test set of 70 compounds, they used 0.1 Å for the translation increment and 1° for the rotation increment. Two cycles were performed yielding a maximum translation of 0.3 Å along one direction and a maximum rotation of 3° around one axis. The results obtained using an sp^3 carbon probe with +1 charge with 2 Å grid spacing are shown in Table 14.

A three-step procedure was used for the alignment and subsequent prediction of the test molecules. First, the similarities between the template molecule and the reoriented molecule were determined with respect to the molecular fields. Second, the six most similar alignments were selected. The activities of the six orientations were predicted and the mean activity was calculated.

Table 14 *CoMFA results of dihydrofolate reductase inhibitors*

	Dataset A			Dataset B		
Analysis	L	Q^2	R^2_{pred}	L	Q^2	R^2_{pred}
Initial	4	0.58	0.44	4	0.33	0.55
After reorientation	4	0.86	0.48–0.60[a]	5	0.80	0.60–0.64[a]

[a] Field-fitted, corrected.

Table 15 *CoMFA results of dihydrofolate reductase inhibitors with randomized activities*

	Randomized activity A			Randomized activity B		
Analysis	L	Q^2	R^2_{pred}		Q^2	R^2_{pred}
Initial	1	–0.10		2	–0.17	
After reorientation	5	0.62	–0.45, –0.46[a]	5	0.60	–0.40, –0.42[a]

[a] Field-fitted, corrected.

The Q^2 values for both datasets were largely improved by the realignment process: from 0.58 and 0.33 to 0.86 and 0.80 for the dataset A and B, respectively (Table 14). However, the predictivity of the model (R^2_{pred}) improved only moderately: from 0.44 to 0.48–0.60 for the dataset A and from 0.55 to 0.60–0.64 for the dataset B.

However, this procedure gave a model from two sets of randomized activities, with an improved Q^2 but negative R^2_{pred} values. These results, tabulated in Table 15, showed that the Q^2 value alone was not a good measure for the predictivity of the model, and that the realignment procedure created false models. (See the discussion above in section 2.9.)

2.10.2. Measure of predictivity

The issue of how the predictive R^2 (R^2_{forecast}) should be defined is still in debate, although this subject was discussed in the previous volume [61L]. There is disagreement about what to use for the Y_{mean} in deriving R^2_{forecast} from the equation:

$$R^2_{\text{forecast}} = 1 - \text{PRESS/SD}$$

where PRESS is the predictive sum of squared residuals for the test set molecules, and SD is the sum of the squared deviation of the test set target property Y about Y_{mean}. Some authors compute Y_{mean} from the training set Y values, whereas others derive Y_{mean} from the Y values of the test set.

Sometimes, a large difference is observed between the two predictive R^2-based indices [203L]: $R^2_{\text{forecast(test)}}$ obtained using the test set mean activity value Y_{mean} and $R^2_{\text{forecast(train)}}$ obtained using the training set Y_{mean}. Such a discrepancy between the two predictive R^2 indices is due to a different distribution in the activity of the test set compared to the activity of the training set.

There are important implications whenever the Y variance of the test set is not similar to that of the training set. If the activities of the test set molecules fall within a small interval, $R^2_{\text{forecast(test)}}$ will always underestimate the predictive performance of the model. In this case, provided that predictions are accurate, $R^2_{\text{forecast(train)}}$ will be large only if the observed activities cluster far from the Y_{mean} of the training set. If the test molecules exhibit activities all close to the Y_{mean} of the training set, both $R^2_{\text{forecast(test)}}$ and $R^2_{\text{forecast(train)}}$ will be exceedingly small, even if the predictions are accurate as judged by their average or standard error.

Regarding the use of $R^2_{\text{forecast(test)}}$ for prediction, how does one calculate the $R^2_{\text{forecast(test)}}$ when a single compound is to be predicted? In this case, the $R^2_{\text{forecast(test)}}$ value becomes minus infinity!

In the light of these complications, and awaiting theoretically more solid definition of predictive R^2, the use of standard error of prediction or other similar dimension-dependent indices is suggested as they are independent of the variance of both the training set and the test set. In contrast to the standard error of predictions, R^2 indices ($R^2_{\text{forecast(train)}}$ or $R^2_{\text{forecast(test)}}$) offer the advantage of not being dimension-dependent. Unfortunately, they are too heavily influenced by the distribution of the actual Y values within the test set.

3. Examples of CoMFA Applications

There are over 350 CoMFA models described in almost 200 publications since 1993. Table 10 summarizes these CoMFA models. Several datasets have been studied by many different authors to investigate different procedures and methods. The dataset that has been used most often is the steroid datasets of Cramer III et al. [12] (see the chapter by E. Coats in this volume).

Started as a method to derive 3D QSAR for ligand–macromolecule interactions that can be used when there is no three-dimensional macromolecular structure available, the use of CoMFA progressed into diverse applications. The most numerous applications of CoMFA have been with the ligands acting on various enzymes and receptors. The methods have also been used in the fields of agrochemistry — pesticides, insecticides or herbicides. In addition, the methods have been applied for the correlation of physico-chemical parameters such as pK_a or Hammett σ values and for the development of new descriptors that can be used in classical QSAR studies; such applications include partition coefficients, capacity factors, enantioseparation factors and C13 chemical shifts. Both thermodynamic and kinetic data have also been correlated using the CoMFA approach. These applications are loosely divided into nine groups below, and each group is briefly summarized.

3.1. Enzyme inhibitors and substrates

Almost 100 CoMFA models have been reported of compounds that act on an enzyme. The enzymes involved are too numerous to list, and the ligands associated with these studies are as numerous and diverse as the enzymes. Some of the most frequently studied enzymes are dihydrofolate reductase, angiotensin converting enzyme, HIV protease, monoamine oxidase and papain.

3.2. Binding affinities to various receptors

There are almost 100 CoMFA models involved with binding affinities of various receptors, including steroid, adrenergic, 5-hydroxytryptamine, angiotensin, benzodiazepine, cholecystokinin, dopamine, GABA, melatonin, nicotine, σ, hormone and other receptors.

3.3. Antibacterial and antifungal activities

Quinolines [285L–287L], sulfanilamides [324L], nitrofurans [75L], and alkylbenzyl-dimethylammonium chlorides [138L] were studied for their antibacterial activities, whereas oxocyclododecylsulfonamides [377L] and bifonazoles [239L] were investigated for their antifungal activities.

3.4. Anticancer activities

Numerous studies were aimed at improving anticancer activities of various compounds: the antitumor activity or cytotoxicity of thioxanthen-9-ones, pyrazoloacridines, amides and ureas, sulfoynlureas, pyridopyrimidines and polyamines against various cell lines [117L,118L,276L,329L,377L]. The ability to form intracellular covalent topoisomerase II–DNA complexes of demethylepipodophyllotoxins was also investigated [49L].

3.5. Toxic activities

The acute toxicities of alkanes [140L], the genotoxicities of nitrofurans [75L], the hepatotoxicities of thiobenzamides and the toxicities on *Thamnocephalus platurus* and *Brachionus calyciflorus* of non-ionic sulfactants were analyzed in different CoMFA studies. The genotoxicity study of Debnath et al. [75L] was aimed at antibacterial potency.

The mutagenicity activities of furanones, nitroaromatics, hydroxyfuranones and hydrazines were also correlated [38L,194L,217L,218L,227L,347L].

3.6. Agrochemical activities

CoMFA models were derived for the herbicidal potency of pyrazolyltrifluorotolyl ethers and pyrazole olefinic nitriles [51L], and the insecticidal activity of various compounds [5L,6L,37L,192L,289L]. Several series required log P or π as an additional parameter in the CoMFA models [5L,6L,289L].

3.7. Physico-chemical parameters

The CoMFA methodology has been applied not only to correlate various physico-chemical parameters (dissociation constants, Hammett's σ electronic constants [136L,323L,324L], steric and hydrophobic parameters), but also to correlate chemical reactivities and reaction rate constants [278L,281L]. The earlier works were summarized in the previous volume of this book by K.H. Kim [135L].

Among others, the use of CoMFA for the calculation of partition coefficients and capacity factors are of special interest. Since the CoMFA method was originally devised to correlate the drug–receptor interactions, it was questioned whether the method could be used to correlate global molecular properties such as partition coefficients, molar volume or *in vivo* data. However, there are now ample examples showing that the method can be used to correlate such global molecular properties. The hydrophobic parameters studied encompass not only the octanol–water partition coefficients (*log* P) of pyrazines [137L], pyridines [137L], triazine [133L], furan [133L] and benzyl N,N-dimethylcarbamates [132L], as well as a set of furan, benzene, pyrrole, 1-methylpyrrole, benzofuran, indole, 1-methylindole [131L] and orthopramides [280L], but also the capacity factors (log k') obtained from reversed-phase high-performance liquid chromatography (RP-HPLC) of mostly the same sets of compounds. This approach applies not only to congeneric series, but also to a mixed set of noncongeneric series [131L], distribution coefficients (log D) of diazine analogs of ridogrel and amino acids [112L,237L], respectively, hydrophobicity of cytosine nucleosides [196L], the water solubility of amino acids [237L], partition coefficients and solubilities of amino acid derivatives [237L].

Waller [258L] also used the CoMFA methodology to calculate partition coefficients of structural isomers, which many conventional methods do not distinguish.

Altomare et al. [8L,10L,41L] successfully correlated the HPLC enantioseparation factor of alkyl aryl sulfoxides, aryloxy acetic acid methyl esters and aryloxadiazolines on chiral stationary phases. With a similar aim but on a quite different system, Faber et al. [86L] used CoMFA to correlate the enantioselectivity in the hydrolysis of substrates by *Candida rugosa* lipase.

Brown's steric parameter [238L], carbon-13 chemical shifts of phosphine compounds [238L] and LUMO energy [281L] have also been correlated using CoMFA.

3.8. *Thermodynamic or kinetic data of reactions*

Yoo et al. [278L,279L,281L], Kim [136L] and Folkers et al. [91L] correlated the rate constants of various reactions. Steinmetz [238L] applied CoMFA to correlate various parameters of inorganic reactions with phosphorous ligands.

Welsh et al. [272L] used CoMFA to calculate the sublimation enthalpy ($\Delta_{sub}H_m^\circ$) and formation enthalpy ($\Delta_f H_m^\circ$) of polycyclic aromatic hydrocarbons (PAHs).

3.9. *Development of substituent descriptors*

One unique application of the CoMFA approach is on the characterization and derivation of transferable substituent descriptors that can be used in QSAR. For example, van de Waterbeemd et al. [252L] derived substituent parameters called 3D principal properties (3D PPs) from the steric and electrostatic CoMFA fields for 59 common organic substituents. In a similar approach, Cocchi and Johansson [56L] derived principal properties of amino acids.

4. Miscellaneous Aspects of CoMFA Applications

4.1. *Multiple binding modes*

The binding mode of the compounds that interact with a macromolecule is frequently assumed to be similar. Although in many instances this seems to be a plausible working hypothesis, results from X-ray crystallography often reveal that some compounds, even very close analogs, bind with alternative orientations in the binding site or bind to different site points within the same binding region [13,14].

4.2. *Agonists and antagonists in the same model*

The issue of whether receptor agonists and antagonists can be included into one model or should be kept separate has been addressed by several authors. Minor et al. [185L] discarded agonists from a CoMFA model derived from dopamine D_1 antagonists based on the assumption that the binding modes of agonists versus antagonists were different. Myers et al. [191L] also removed two mispredicted compounds from a CoMFA model built up on σ_3 ligands; they justified the omission based on their antagonistic profiles which could, in turn, imply an orientation at the receptor different from those of the remaining analogs.

On the other hand, agonists (like triazolam) and antagonists (like flumazenil) of the diazepam-sensitive benzodiazepine receptor were merged into the same training set by Wong et al. [275L]. Martin et al. [15] combined previously established CoMFA models for receptor affinity D_1 agonist and antagonists because the cross-validation statistics improved in the combined model. Gaillard et al. [95L] analyzed several chemically diverse classes of serotonin 5-HT_{1A} ligands without making distinctions between agonists and antagonists. In the same paper, the authors mentioned a theoretically derived model of ligand– receptor interaction [16] where the binding sites of agonists and antagonists overlapped partially.

Agarwal and Taylor [3L] used CoMFA to correlate the intrinsic activity (*IA*) of ligands which was defined as the ratio of the maximal effect produced by a ligand to that produced by a full agonist. A structurally diverse set of 5-HT_{1A} receptor ligands with *IA* data determined by the inhibition of 5-HT sensitive forskolin-stimulated adenylate cyclase was used. $IA = 1$ was assigned for a full agonist, $IA = 0$ for a full antagonist and $0 < IA < 1$ for a partial agonist. The CoMFA results suggest that agonist and antagonist ligands can share parts of a common binding site on the receptor, with a primary agonist binding region that is also occupied by antagonists and a secondary binding site accommodating the excess bulk present in many antagonists and partial agonists. They suggested that the secondary binding site may inhibit conformational changes in the receptor that are associated with agonist activity when both binding sites are fully occupied.

It seems reasonable to merge agonists and antagonists together into one CoMFA if preliminary CoMFA models developed separately for the two classes yield similar results in terms of statistics and coefficient contour maps.

4.3. *Receptor selectivity*

CoMFA has been successfully applied to highlight 3D properties responsible for ligand selectivity between different receptors. A series of tetrahydropyridinylindole agonists of the serotonin 5-HT_{1A} and 5-HT_2 receptors have been investigated by Agarwal et al.

[4L]. Separate CoMFA models for the two receptor subtypes were developed, and the resulting coefficient contour maps were compared visually.

A more effective procedure to capture the determinants of receptor selectivity was proposed by Wong et al. [275L] in a study with imidazo-1,4-diazepine derivatives tested on diazepam-insensitive (*DI*) and diazepam-sensitive (*DS*) benzodiazepine receptors. The negative logarithm of the ratio between *DI* and *DS* K_i values (*pDI–pDS*) was used as dependent variable. In this case, interpretation of the resulting CoMFA contour maps was straightforward.

For most compounds that Wong et al. [275L] investigated, the conformations and orientations of the ligands were assumed to be identical at both receptors. However, the azido group at the 8-position was thought to be arranged in different conformations at the DI and DS receptors ('anti' and 'syn', respectively). Based on the contour plots, the CoMFA model for receptor selectivity appears to be derived from the 'anti' conformation for the azido substituent.

4.4. Nonlinear relationships

In classical QSAR studies, nonlinear relationships are often observed with both *in vivo* and *in vitro* biological activity data. Such relationships provided some of the most useful information in classical QSAR: the optimum value of the physico-chemical property such as log P_o in the structure–activity relationships.

Several approaches are proposed for describing a nonlinear relationship in CoMFA. A nonlinear method called Implicit Nonlinear Latent Variable Regression (INLR) is very similar to ordinary PLS models, except that it has a curved inner relation such as a quadratic or cubic polynomial or spline [292L,293L]. Kimura et al. [143L] used a quadratic PLS (QPLS) model to derive nonlinear models for biological activities (log ($1/K_m$), log k_{cat} and log (k_{cat}/K_m)) of synthetic substrates for elastase. They showed that significantly improved models were obtained from the QPLS method judged by their Q^2 values.

A large list of nonlinear PLS approaches has been cited in a recent paper by Berglund and Wold [290L]. Recently, PLS analysis of distance matrices was described to describe nonlinear relationships [17,116L,175L,323L].

4.5 Lateral validations

Lateral validation refers to the method of validating a new QSAR by comparing it with other QSAR equations. This method was originally used by Hansch in classical QSAR. The possibility of supporting a new CoMFA by lateral validation was recently investigated [136L]; this included comparative studies of the dissociation constants of benzoic acids and phenylacetic acids and the rate constants for the elimination reaction of substituted arenesulfonates. The results indicated that the coefficients of the PLS regression equation in CoMFA contain useful information and they can be used in the lateral validation or lateral comparison of single-component models. However, a comparison of the coefficients in CoMFA studies is deterred by the fact that the optimum number of components for a CoMFA model varies depending on the constitution of compounds included in the analysis, as well as various adjustable parameters in the CoMFA procedures.

4.6. Predictivity of CoMFA

One goal of a CoMFA study is to predict the potency of new compounds before their synthesis. Table 10 lists about 90 examples where a CoMFA model was used for the prediction of test set compounds. Table 10 shows that the activities of more than 1700 compounds in different test sets have been predicted by various CoMFA models. A similar table compiled up to early 1994 contained 25 CoMFA models, and they were used to predict more than 290 compounds in various test sets. The average predicted error for these compounds was 0.70 which corresponds to 0.98 kcal/mol. It is not easy to estimate the average error of all predicted compounds in Table 10, and the magnitude of errors depends on the target property used. A rough estimate of the average predicted error for receptor and enzyme studies appears to be 0.6 to 0.7. Most of the compounds predicted, however, were close analogs, congeners or even homologs of molecules employed to derive the corresponding CoMFA model. Thus, the average estimate of predictivity of CoMFA model overestimates the real predictivity of CoMFA models when exploited in a 'real lead' optimization process.

4.7. Reporting CoMFA results

Many CoMFA publications do not include sufficient information, such as the optimum number of components for the model chosen, the probe, the grid size, the statistical indices such as Q^2 or s_{cv} for the cross-validation test, the type of compounds studied, the number of compounds used or the compounds left out from the model derivation. Sometimes the only information presented of a CoMFA study was the CoMFA coefficient contour maps or R^2 of the fit. Some models were derived without describing the precise form of the biological property (e.g. IC_{50}, ln $1/IC_{50}$ or log $1/IC_{50}$). Table 10 shows that many CoMFA studies are missing some of the crucial information.

Sometimes, the information presented in the paper is confusing. For example, the optimum number of components described for the cross-validation and the final model are not the same and sometimes the statistical indices reported in the table or the figure are not the same as those in the text.

Most of the studies that did not provide the information might have been performed using the default settings. Sometimes the CoMFA study was a re-evaluation of a previous study, or the objective of the study was not developing a CoMFA model itself, but investigating various aspects of the CoMFA procedures. However, inclusion of critical data would be beneficial to the readers. Some of these publications were proceedings of a conference and could not include detailed information.

In classical QSAR, it has been standard to present the calculated (fitted) activity values along with the observed values and their deviations. However, in most CoMFA studies, this has not been practiced. Calculated activity values from the model and their deviations from the observed values may provide important additional information about the model. There may be a small number of compounds showing larger deviations, or every compound may show a similar deviation without a particular outlier. Without the calculated activity values using the chosen model, such information is completely lost.

Recommendations [134L,173L,244L] for CoMFA studies and publications have been published in several places including the Appendix in the previous volume of this book [245L]. If these procedures were followed, many of the common mistakes could have been avoided. We urge the authors of CoMFA papers to consider these recommendations as a checklist for the publication.

While most studies report a single or a few CoMFA models, Cho and Tropsha [47L] claimed that reporting the single value of Q^2 and associated CoMFA fields is not adequate, because the results of CoMFA are sensitive to the overall orientation of molecular aggregates with respect to the location of the grid box. Thus, they suggested that a range of possible Q^2 values should be presented instead of one number.

5. Concluding Remarks

In the first volume of this book, limitations in CoMFA and practical problems in PLS analyses were discussed in detail [91L,155L]. Three years have passed since that time, and the number of CoMFA applications increased from about 50 [243L] to over 350 since the last volume of this book. How much have those limitations and problems been solved since then? What are the limitations and shortcomings of the method at the present time? What are the advances achieved during the last three years?

Significant advances have been made in the areas of series design and selection of training set, variable selection and describing nonlinear relationships. However, many limitations and problems in CoMFA still remain unsolved. The optimum number of components and Q^2 still vary significantly depending on adjustable parameters, and inconsistent results are often obtained. It is difficult to compare the results of different CoMFA studies. Sometimes it is also difficult even to reproduce the literature results because of so many adjustable variables involved in the study and lack of all relevant information described in the paper. Application of lateral validation for a new CoMFA model seems to be pessimistic at the present time. No significant breakthrough has been achieved regarding the choice of probe groups, location of grid box, scaling of different fields or external parameters added, and the intercorrelations among different descriptors. The situations regarding the choice of lattice spacing, standard cutoff values, atomic charges and number of compounds per component in a CoMFA model have hardly changed. The results of CoMFA are, in most cases, still sensitive to the overall orientation of molecular aggregates with respect to the location of the grid box.

Several aspects in CoMFA have achieved some advances but still need further improvement. They include the description of hydrophobic interactions, selection of the best CoMFA model based on its predictivity and use of various PLS plots. CoMFA has been applied to much broader areas including the separation of enantiomers and description of global properties such as capacity factors and partition coefficients. Improvement in the predictability of a CoMFA model is also greatly desired.

Perhaps one of the most significant advances in recent CoMFA applications is the use of ligand–macromolecule complex structures as more three-dimensional macromolecular structures are becoming available. This approach is extending to include the three-dimensional structures obtained by homology modelling. (See the chapter by K.H. Kim in this volume.) Inclusion of such information has been useful not only for the selection of bioactive conformations, alignments and docking of new ligands, but also in the interpretation of CoMFA results. Inclusion of the active site water molecules in CoMFA is also noteworthy. Another point to note among the recent applications is that a greater number of studies utilized multiple conformations and alignments, and often the choice of particular conformation or alignment was considered to be justified based on the CoMFA results.

As any other QSAR approach, exploiting a CoMFA model to design novel, more potent compounds is the primary goal. This important issue has received less emphasis in the literature of the last six years than it deserves. This might be partially due to the fact that designing new compounds based on the coefficient contour maps is not a trivial practice. The Leapfrog module of SYBYL was devised for such a purpose, but the efficiency of this algorithm has not yet been documented in the literature.

There is no doubt that the methodology of CoMFA for 3D QSAR will be advanced further in the coming years. The applications of CoMFA are expected to encompass even broader areas. And, eventually, the method will lead to or contribute significantly to the design and development of new therapeutic, agricultural and pesticidal agents.

References

(See the chapter by Ki Hwan Kim for references ending with letter ‘L’.)

1. Lin, C.T., Pavlik, P.A. and Martin, Y.C., *Use of molecular fields to compare series of potentially bioactive molecules designed by scientists or by computer*, Tetrahed. Comput. Methodol., 3 (1990) 723–738.
2. Wermuth, C.-G. and Langer, T., *Pharmacophore identification*, In Kubinyi, H. (Ed.) 3D QSAR in drug design: Theory, methods and applications, ESCOM, Leiden, The Netherlands, 1993, pp. 117–136.
3. Horwitz, J.P., Massova, I., Wiese, T.E., Besler, B.H. and Corbett, T.H., *Comparative molecular field analysis of the antitumor activity of 9H-thioxanthen-9-one derivatives against pancreatic ductal carcinoma 03*, J. Med. Chem., 37 (1994) 781–786.
4. Kim, K.H. and Martin, Y.C., *Direct prediction of linear free energy substituent effects from 3D structures using comparative molecular field analysis: 1. Electronic effects of substituted benzoic acids*, J. Org. Chem., 56 (1991) 2723–2729.
5. Marshall, G.R., *Binding-site modeling of unknown receptors*, In Kubinyi, H. (Ed.) 3D QSAR in drug design: Theory, methods and applications, ESCOM, Leiden, The Netherlands, 1993, pp. 80–116.
6. Klebe, G., *Structural alignment of molecules*, In Kubinyi, H. (Ed.) 3D QSAR in drug design: Theory, methods and applications, ESCOM, Leiden, The Netherlands, 1993, pp. 173–199.
7. Golender, V.E. and Vorpagel, E.R., *Computer-assisted pharmacophore identification*, In Kubinyi, H. (Ed.) 3D QSAR in drug design: Theory, methods and applications, ESCOM, Leiden, The Netherlands, 1993, pp. 137–149.
8. Yliniemela, A., Konschin, H., Neagu, C., Pajunen, A., Hase, T., Brunow, G. and Teleman, O., *Design and synthesis of a transition state analog for the ene reaction between maleimide and 1-alkenes*, J. Am. Chem. Soc., 117 (1995) 5120–5126.

9. Itai, A., Tomioka, N., Yamada, M., Inoue, A. and Kato, Y., *Molecular superposition for rational drug design*, In Kubinyi, H. (Ed.) 3D QSAR in drug design: Theory, methods and applications, ESCOM, Leiden, The Netherlands, 1993, pp. 200–225.
10. Martin, Y.C., Bures, M.G., Danaher, E.A., DeLazzer, J., Lico, I. and Pavlik, P.A., *A fast new approach to pharmacophore mapping and its application to dopaminergic and benzodiazepine agonists*, J. Comput.-Aid. Mol. Design, 7 (1993) 83–102.
11. Greco, G., Novellino, E., Silipo, C. and Vittoria, A., *Study of benzodiazepines receptor sites using a combined QSAR-CoMFA approach*, Quant. Struct.-Act. Relat., 11 (1992) 461–477.
12. Cramer III, R.D., Patterson, D.E. and Bunce, J.D., *Comparative molecular field analysis (CoMFA): 1. Effect of shape on binding of steroids to carrier proteins*, J. Am. Chem. Soc., 110 (1988) 5959–5967.
13. Mattos, C., Rasmussen, B., Ding, X., Petsko, G.A. and Ringe, D., *Analogous inhibitors of elastase do not always bind analogously*, Nature Struct. Biol., 1 (1994) 55–58.
14. Mattos, C., Ringe, D., *Multiple binding modes*, In Kubinyi, H. (Ed.) 3D QSAR in drug design: Theory, methods and applications, ESCOM, Leiden, The Netherlands, 1993, pp. 226–254.
15. Martin, Y.C., Lin, C.T. and Wu, J., *Application of CoMFA to D1 dopaminergic agonists: A case study*, In Kubinyi, H. (Ed.) 3D QSAR in drug design: Theory, methods and applications, ESCOM, Leiden, The Netherlands, 1993, pp. 643–660.
16. Kuipers, W., van Wijngaaden, I. and Ijzerman, A.P., *A model of the serotonin 5-HT1A receptor: Agonist and antagonist binding sites*, Drug Des. Discuss., 11 (1994) 231–249.
17. Kubinyi, H., *QSAR: Hansch analysis and related approaches*, VCH, Weinheim, Germany, 1993.

List of CoMFA References, 1993–1997

Ki Hwan Kim
Department of Structural Biology, D46Y AP10-2, Pharmaceutical Products Division, Abbott Laboratories, 100 Abbott Park Road, Abbott Park, IL 60064-3500, U.S.A.

From its first publication in 1988 to 1992, the sum of published CoMFA papers was approximately 80. Between 1993 and 1996, that amount nearly tripled. In addition, there are numerous CoMFA-related papers, such as those dealing with the interaction energy fields, nonlinearity, superposition, conformational analysis, molecular similarity, PLS algorithms, neural networks, molecular diversity and various 3D QSAR approaches. If all of these were to be included, the list of references would be very long. Only some of these publications are included in this list.

The CoMFA references included in the list resulted from an exhaustive search of the papers published in 1993 through September 1997. A majority of the references was found by the keyword searches of 'CoMFA' and '3D QSAR', as well as a citation search to the original 1988 CoMFA publication of Cramer III et al. All volumes of the journal of *Quantitative Structure–Activity Relationships* published since 1993 were also manually searched to find additional references. Several individuals were also contacted by personal communications for the papers that have been published in rare places or are currently in print.

The reference list includes regular publications, as well as review papers, the proceedings of conferences, theses and worldwide web publications. The language used in the publication was not restricted to English; however, only a few were written in other languages. The list does include some papers closely related to CoMFA procedures which do not contain CoMFA results; it includes those papers that employed non-traditional fields, principal component analysis or similarity matrices. However, no effort was made to include an exhaustive listing of papers on such related topics. Conference abstracts were usually excluded unless they were part of a regular journal page. A list of the 1997 CoMFA-related papers is appended at the end of this list and included in the conference abstracts.

References that contain CoMFA results are specifically marked with a star symbol (*) after the corresponding reference number, except some of the 1997 references. The relevant CoMFA information for these studies can be found in Table 10 of the chapter by Ki Hwan Kim et al. in this volume.

The help of Mrs. Ruth Swanson, of the Abbott Library Information Services, for the initial computer searching of the Chemical Abstracts is greatly appreciated. Special thanks also go to Dr. Hugo Kubinyi who helped me update the 1997 list at the last moment and to many fellow scientists who sent me reprints or preprints.

Despite my efforts to include all the relevant CoMFA references published between 1993 and 1997, it is possible that some have been omitted. The author sincerely apologizes to the authors of such papers.

H. Kubinyi et al. (eds.), 3D QSAR in Drug Design, Volume 3. 317–38.

(a) List of CoMFA References, 1993–1996

1. Abraham, D.J. and Kellogg, G.E., *The effect of physical organic properties on hydrophobic fields*, J. Comput.-Aided Mol. Design, 8 (1994) 41–49.
2. Abraham D.J. and Kellogg, G.E., *Hydrophobic fields*, In Kubinyi, H. (Ed.) 3D QSAR in drug design: Theory, methods and applications, ESCOM, Leiden, The Netherlands, 1993, pp. 506–522.
3. *Agarwal, A. and Taylor, E.W., *3-D QSAR for intrinsic activity of 5-HT1A receptor ligands by the method of comparative molecular field analysis*, J. Comput. Chem., 14 (1993) 237–245.
4. *Agarwal, A., Pearson, P.P., Taylor, E.W., Li, H.B., Dahlgren, T., Herslof, M., Yang, Y.H., Lambert, G., Nelson, D.L., Regan, J.W. and Martin, A.R., *3-dimensional quantitative structure–activity relationships of 5-HT receptor binding data for tetrahydropyridinylindole derivatives — a comparison of the Hansch and CoMFA methods*, J. Med. Chem. 36 (1993) 4006–4014.
5. *Akamatsu, M., Fujita, T., Ozoe, Y., Mochida, K., Nakamura, T. and Matsumura, F., *3D QSAR of insecticidal dioxatricycloalkene and its related compounds*, In Wermuth, C.-G. (Ed.) Trends in QSAR and Molecular Modeling, Proceedings of the 9th European Symposium on Structure–Activity Relationships: QSAR and Molecular Modeling, ESCOM, Leiden, The Netherlands, 1993, pp. 525–526.
6. *Akamatsu, M., Nishimura, K., Osabe, H., Ueno, T. and Fujita, T., *Quantitative structure–activity studies of pyrethroids: 29. Comparative molecular-field analysis (3-dimensional) of the knockdown activity of substituted benzyl chrysanthemates and tetramethrin and related imido- and lactam-N-carbonyl esters*, Pesticide Biochem. Physiol., 48 (1994) 15–30.
7. *Altomare, C., Carotti, A., Carta, V., Kneubuhler, S., Carrupt, P.A. and Testa, B., *Modeling of new pyridazine inhibitors of MAO-B using QSAR and CoMFA approaches*, In Sanz, F., Giraldo, J. and Manaut, F. (Eds.) QSAR and molecular modeling: Concepts, computational tools and biological applications, Proceedings of the 10th European Symposium on Structure–Activity Relationships: QSAR and Molecular Modeling, Barcelona, Spain, September 4–9, 1994, J.R. Prous Science Publishers, Barcelona, 1995, pp. 463–465.
8. *Altomare, C., Carotti, A., Cellamare, S., Fanelli, F., Gasparrini, F., Villani, C., Carrupt, P.A. and Testa, B., *Eantiomeric resolution of sulfoxides on a DACLH_DNB chiral stationary phase — a quantitative structure–enantioselective retention relationship (QSERR) study*, Chirality, 5 (1993) 527–537.
9. *Altomare, C., Campagna, F., Carta, V., Cellamare, S., Carotti, A., Genchi, G. and De Sarro, G., *Synthesis, benzodiazepine receptor affinity and anticonvulsant activity of 5-H-indeno[1,2-c]pyridazine derivatives*, 49 (1994) 313–323.
10. *Altomare, C., Cellamare, S., Carotti, A., Barreca, M.L., Chimirri, A., Monforte, A.M., Gasparrini, F., Villani, C., Cirrilli, M. and Mazza, F., *Substituent effects on the enantioselective retention of anti-HIV 5-aryl-delta(2)-1,2,4-oxadiazolines on R,R-dach-DNB chiral stationary-phase*, Chirality, 8 (1996) 556–566.
11. *Anzini, M., Cappelli, A., Vomero, S., Giorgi, G., Langer, T., Hamon, M., Merahi, N., Emerit, B.M., Cagnotto, A., Skorupska, M., Mennini, T. and Pinto, J.C., *Novel, potent, and selective 5-HT3 receptor antagonists based on the arylpiperazine skeleton: Synthesis, structure, biological activity, and comparative molecular field analysis studies*, J. Med. Chem., 38 (1995) 2692–2704.
12. *Anzini, M., Cappelli, A., Vomero, S., Langer, T. and Bourguignon, J.-J., *CoMFA analysis of ligands of the mitochondrial benzodiazepine receptor: A versatile tool for the design of new lead compounds*, In Sanz, F., Giraldo, J. and Manaut, F. (Eds.) QSAR and Molecular Modeling: Concepts, Computational Tools and Biological Applications, Proceedings of the 10th European Symposium on Structure–Activity Relationships: QSAR and Molecular Modeling, Barcelona, Spain, September 4–9, 1994, J.R. Prous Science Publishers, Barcelona, 1995, pp. 470–472.
13. *Artico, M., Botta, M., Corelli, F., Mai, A., Massa, S. and Ragno, R., *Investigation on QSAR and binding mode of a new class of human rhinovirus-14 inhibitors by CoMFA and docking experiments*, Bioorg. Med. Chem., 4 (1996) 1715–1724.
14. Avery, M.A., Gao, F., Mehrotra, S., Chong, W.K. and Jennings-White, C., *The organic and medicinal chemistry of artemisinin and analogs*, Res. Trends Trivandrum: India, (1993) 413–468.

15. *Avery, M.A., Gao, F.G., Chong, W.K.M., Mehrotra, S. and Milhous, W.K., *Structure–activity relationships of the antimalarial agent artemisinin: 1. Synthesis and comparative molecularfield analysis of C-9 analogs of artemisinin and 10-deoxoartemisinin*, J. Med. Chem., 36 (1993) 4264–4275.
16. Baroni, M., Clementi, S., Crucianai, G., Kettanehwold, N. and Wold, S., *D-optimal designs in QSAR*, Quant. Struct.-Act. Relat., 12 (1993) 225–231.
17. *Baroni, M., Costantino, G., Cruciani, G., Riganelli, D., Valigi, R. and Clementi, S., *Generating optimal linear PLS estimations (GOLPE): An advanced chemometric tool for handling 3D-QSAR problems*, Quant. Struct.-Act. Relat., 12 (1993) 9–20.
18. Baroni, M., Costantino, G., Cruciani, G., Riganelli, D., Valigi, R. and Clementi, S., *Multivariate data modeling of new steric, topological and CoMFA-derived substituent parameters*, In Wermuth, C.-G. (Ed.) Trends in QSAR and Molecular Modeling 92, Proceedings of the 9th European Symposium on Structure–Activity Relationship. QSAR and Molecular Modeling, ESCOM, Leiden, The Netherlands, 1993, pp. 256–259.
19. *Belvisi, L., Bravi, G., Catalano, G., Mabilia, M., Salimbeni, A. and Scolastico, C., *A 3D QSAR CoMFA study of non-peptide angiotensin II receptor antagonists*, J. Comput.-Aided Mol. Design, 10 (1996) 567–582.
20. Benigni, R. and Guiliani, A., *Analysis of distance matrices for studying data structures and separating classes*, Quant. Struct.-Act. Relat., 12 (1993) 397–401.
21. Benigni, R., *EVE, a distance based approach for discriminating nonlinearly separable groups*, Quant. Struct.-Act. Relat., 13 (1994) 406–411.
22. *Bolognese, A., Diurno, M.V., Greco, G., Greco, G., Grieco, P., Mazzoni, O., Novellino, E., Perissutti, E. and Silipo, C., *Quantitative structure–activity relationships in a set of Thiazolidin-4-ones acting as H1-histamine antagonists*, J. Receptor Signal Transduct. Res., 15 (1995) 631–641.
23. *Botta, M., Cernia, E., Corelli, F., Manetti, F. and Soro, S., *Probing the substrate specificity for lipases: A CoMFA approach for predicting the hydrolysis rates of 2-arylpropionic esters catalyzed by Candida rugosa lipase*, Biochim. Biophys. Acta, 1296 (1996) 121–126.
24. *Brandt, W., Lehmann, T., Willkomm, C., Fittkau, S. and Barth, A., *CoMFA investigations on two series of artificial peptide inhibitors of the serine protease thermitase*, Int. J. Pep. Prot. Res., 46 (1995) 73–78.
25. *Brandt, W.L.T., Thondorf, I., Born, I., Schutkowski, M., Rahfield, J.-U.N.K. and Barth, A., *A model of the active site of dipeptidyl peptidase IV predicted by comparative molecular field analysis and molecular modeling simulations*, Int. J. Pept. Protein Res., 46 (1995) 494–507.
26. Briens, F.B.R., Rault, S. and Robba, M., *Applicability of CoMFA in ecotoxicology: A critical study on chlorophenols*, Ecotoxicol. Environ. Saf., 31 (1995) 37–48.
27. Briens, F.B.R., Rault, S. and Robba, M., *Comparative molecular field analysis of chlorophenols: Application in ecotoxicology*, SAR QSAR Environ. Res., 2 (1994) 147–157.
28. Bro, R., *Multiway calibration: Multilinear PLS*, J. Chemom., 10 (1996) 47–61.
29. *Brusniak, M.-Y.K., Pearlman, R.S., Neve, K.A. and Wilcox, R.E., *Comparative molecular field analysis-based prediction of drug affinities at recombinant D1A dopamine receptors*, J. Med. Chem., 39 (1996) 850–859.
30. *Bureau, R., Lancelot, J.C., Prunier, J. and Rault, S., *Conformational analysis and 3D QSAR study on novel partial agonists of 5-HT3 receptors*, Quant. Struct.-Act. Relat., 15 (1996) 373–381.
31. *Bureau, R., Rault, S. and Robba, M., *Comparative molecular field analysis of CCK-B antagonists*, Eur. J. Med. Chem., 29 (1994) 487–494.
32. *Bureau, R., Rault, S., Pilo, J.-C. and Robba, M., *Comparative molecular field analysis of CCK-A antagonists using field fit as alignment technique*, In Wermuth, C.-G., (Ed.) Trends in QSAR and Molecular Modeling 92, Proceedings of the 9th European Symposium on Structure–Activity Relationships: QSAR and Molecular Modeling, ESCOM, Leiden, The Netherlands, 1993, pp. 522–524.
33. Burke, B.J. and Hopfinger, A.J., *Molecular similarity*, In Kubinyi, H. (Ed.) 3D QSAR in drug design: Theory, methods and applications, ESCOM, Leiden, The Netherlands, 1993, pp. 276–306.
34. Burke, B.J., Dunn III, W.J. and Hopfinger, A., *Construction of a molecular shape analysis — three-dimensional quantitative structure–analysis relationship for an analog series of pyridobenzodiazepinone inhibitors of muscarinic 2 and 3 receptors*, J. Med. Chem., 37 (1994) 3775–3788.

35. *Bush, B.L. and Nachbar, Jr, R.B., *Sample-distance partial least squares: PLS optimized for many variables, with application to CoMFA*, J. Comput.-Aided Mol. Design, 7 (1993) 587–619.
36. Bush, B.L., Nachbar, Jr., R.B. and Sheridan, R.P., *SAMPLS: Sample-distance partial lease squares (PLS) for many variables, with application to CoMFA*, In Sanz, F., Giraldo, J. and Manaut, F. (Eds.) QSAR and molecular modeling: Concepts, Computational Tools and Biological Applications, Proceedings of the 10th European Symposium on Structure–Activity Relationships: QSAR and Molecular Modeling, Barcelona, Spain, September 4–9, 1994, J.R. Prous Science Publishers, Barcelona, 1995, pp. 415–419.
37. *Calder, J.A., Wyatt, J.A., Frenkel, D.A. and Casida, J.E., *CoMFA validation of the superposition of 6 classes of compounds which block GABA receptors noncompetitively*, J. Comput.-Aid. Mol. Design, 7 (1993) 45–60.
38. *Caliendo, G., Fattorusso, C., Greco, G., Novellino, E., Perissutti, E. and Santagada, V. *Shape-dependent effects in a series of aromatic nitro compounds acting as mutagenic agents on T. typhimurium TA98*, SAR QSAR Environ. Res., 4 (1995) 21–27.
39. *Caliendo, G., Greco, G., Novellino, E., Perissutti, E. and Santagada, V., *Combined use of factorial design and comparative molecular field analysis (CoMFA): A case study*, Quant. Struct.-Act. Relat., 13 (1994) 249–261.
40. *Caliendo, G., Greco, G., Novellino, E., Persissutti, E. and Santagada, V., *An integrated approach to CoMFA and cluster analysis for series design*, In Sanz, F., Giraldo, J. and Manaut, F. (Eds.) QSAR and Molecular Modeling: Concepts, Computational Tools and Biological Applications, Proceedings of the 10th European Symposium on Structure–Activity Relationships: QSAR and Molecular Modeling, Barcelona, Spain, September 4–9, 1994, J.R. Prous Science Publishers, Barcelona, 1995, 473–477.
41. *Carotti, A., Altomare, C., Cellamare, S., Monforte, A., Bettoni, G., Loiodice, F., Tangari, N. and Tortorella, V., *LFER and CoMFA studies on optical resolution of alpha-alkyl a-aryloxy acetic acid methyl esters on DACH-DNB chiral stationary phase*, J. Comput.-Aid. Mol. Design, 9 (1995) 131–138.
42. *Carrieri, A., Altomare, C., Barreca, M.L., Contento, A., Carotti, A. and Hansch, C., *Papain catalyzed hydrolysis of aryl esters: A comparison of the Hansch, docking and CoMFA methods*, Farmaco, 49 (1994) 573–585.
43. Carrigan, S.W., *Molecular modeling studies and comparative molecular field analysis of 20-(S)-camptothecin analogs*, University of Georgia, Athens, GA, U.S.A. 1996.
44. *Carroll, F.I.M.., Lewin, A.H., Boja, J.W., and Kuhar, M.J., *Pharmacophore development of (-)-cocaine analogs for the dopamine, serotonin, and norepinephrine uptake sites using a QSAR and CoMFA approach*, In Wermuth, C.-G. (Ed.) Trends in QSAR and Molecular Modeling 92, Proceedings of the 9th European Symposium on Structure–Activity Relationships: QSAR and Molecular Modeling, ESCOM, Leiden, The Netherlands, 1993, pp. 530–531.
45. *Carroll, F.I., Mascarella, S.W., Kuzemko, M.A., Gao, Y., Abraham, P., Lewin, A.H., Boja, J.W. and Kuhar, M.J., *Synthesis, Ligand Binding, and QSAR (ComFA and Classical) Study of 3.beta.-(3′-Substituted phenyl)-,3.beta.-(4′-Substituted phenyl)-, and 3.beta.-(3′,4′-Disubstituted phenyl)tropane-2.beta.-carboxylic Acid Methyl Esters*, J. Med. Chem., 37 (1994) 2865–2873.
46. *Chen, H., Zhou, J., Xie, G. and Pang, S. *The studies on pharmacophore model of K+ channel opener*, ACTA Physico-Chimica Sinica (Wuli Huaxue Huebao), (1997), in press.
47. *Cho, S.J. and Tropsha, A., *Cross-validated R2-guided region selection for comparative molecular field analysis: A simple method to achieve consistent results*, J. Med. Chem., 38 (1995) 1060–1066.
48. *Cho, S.J., Garsia, M.L.S., Bier, J. and Tropsha, A. *Structure-based alignment and comparative molecular field analysis of acetylcholinesterase inhibitors*, J. Med. Chem., 39 (1996) 5064–5071.
49. *Cho, S.J., Tropsha, A., Suffness, M., Cheng, Y.-C. and Lee, K.-H., *Antitumor agents: 163. Three-dimensional quantitative structure–activity relationship study of 4'-O-demethylepipodophyllotoxin analogs using the CoMFA /q2-GRS approach*, J. Med. Chem., 39 (1996) 1383–1395.
50. Clark, M. and Cramer III, R.D., *The probability of chance correlation using partial least squares (PLS)*, Quant. Struct.-Act. Relat., 12 (1993) 137–145.
51. *Clark, R.D., *Synthesis and QSAR of herbicidal 3-pyrazolyl α,α,α-trifluorotolyl ethers*, J. Agr. Food Chem., 44 (1996) 3643–3652.
52. *Clark, R.D., Parlow, J.P., Brannigan, L.H., Schnur, D.M. and Duewer, D.L., *Applications of scaled rank-sum statistics in herbicide QSAR*, In Hansch, C. and Fujita, T. (Eds.) Classical and three-

dimensional QSAR in agrochemistry, ACS Symposium series Vol. 606, American Chemical Society, Washingotn, DC., 1995, pp. 264–281.
53. Clementi, S., Cruciani, G., Baroni, M. and Costantino, G., *Series design*, In Kubinyi, H. (Ed.) 3D QSAR in drug design: Theory, methods and applications, ESCOM, Leiden, The Netherlands, 1993, pp. 567–582.
54. Clementi, S., Cruciani, G., Fifi, P., Riganelli, D., Valigi, R. and Musumarra, G., *A new set of principal properties for heteroaromatics obtained by GRID*, Quant. Struct.-Act. Relat., 15 (1996) 108–120.
55. Clementi, S., Cruciani, G., Riganelli, D. and Valigi, R., *GOLPE: Merits and drawbacks in 3D-QSAR*, In Sanz, F., Giraldo, J. and Manaut, F. (Eds.) QSAR and molecular modeling: Concepts, computational tools and biological applications, Proceedings of the 10th European Symposium on Structure–Activity Relationships: QSAR and Molecular Modeling, Barcelona, Spain, September 4–9, 1994, J.R. Prous Science Publishers, Barcelona, 1995, pp. 408–414.
56. Cocchi, M. and Johansson, E., *Amino acids characterization by GRID and multivariate data analysis*, Quant. Struct.-Act. Relat., 12 (1993) 1–8.
57. *Cocchi, M., Cruciani, G., Menziani, M.C. and De Benedetti, P.G., *Use of advanced chemometric tools and comparison of different 3D descriptors in QSAR analysis of prazosin analogs a_1-adrenergic antagonists*, In Wermuth, C.-G. (Ed.) Trends in QSAR and Molecular Modeling 92, Proceedings of the 9th European Symposium on Structure–Activity Relationships: QSAR and Molecular Modeling, ESCOM, Leiden, The Netherlands, 1993, pp. 527–529.
58. *Collantes, E.R., Tong, W., Welsh, W.J. and Zielinski, W.L., *Use of moment of inertia in comparative molecular field analysis to model chromatographic retention of nonpolar solutes*, Anal. Chem., 68 (1996) 2038–2043.
59. Cramer III, R.D., *Partial least squares (PLS): Its strengths and limitations*, Perspect. Drug Discovery Design, 1 (1993) 269–278.
60. Cramer III, R.D., Clark, R.D., Patterson, D.E. and Ferguson, A.M., *Bioisosterism as a molecular diversity descriptor: Steric fields of single 'topomeric' conformers*, J. Med. Chem., 39 (1996) 3060–3069.
61. Cramer III, R.D., DePriest, S.A., Patterson, D.E. and Hecht, P., *The developing practice of comparative molecular field analysis*, In Kubinyi, H. (Ed.) 3D QSAR in drug design: Theory, methods and applications, ESCOM, Leiden, The Netherlands, 1993, pp. 443–485.
62. Crippen, G.M., *Intervals and the deduction of drug binding site models*, J. Comput. Chem., 16 (1995) 486–500.
63. Crucian, B., Clementi, S. and Baroni, M., *Variable selection in PLS analysis*, In Kubinyi, H. (Ed.) 3D QSAR in drug design: Theory, methods and applications, ESCOM, Leiden, The Netherlands, 1993, pp. 551–564.
64. *Cruciani, G. and Watson, K.A., *Comparative molecular field analysis using GRID force-field and GOLPE variable selection methods in a study of inhibitors of glycogen phosphorylase b*, J. Med. Chem., 37 (1994) 2589–2601.
65. Cruciani, G., Riganelli, D., Valigi, R., Clementi, S. and Musumara, G., *Grid characterization of heteroaromatics*, In Sanz, F., Giraldo, J. and Manaut, F. (Eds.) QSAR and Molecular Modeling: Concepts, Computational Tools and Biological Applications, Proceedings of the 10th European Symposium on Structure–Activity Relationships: QSAR and Molecular Modeling, Barcelona, September 4–9, 1994, J.R. Prous Science Publishers, Barcelona, 1995, pp. 493–495.
66. *Czaplinski, K.-H. and Grunewald, G.L., *A comparative molecular field analysis derived model of the binding of taxol analogues to microtubules*, Bioorg. Med. Chem., 4 (1994) 2211–2216.
67. *Czaplinski, K.-H., Haensel, W., Wiese, M. and Seydel, J.K., *New benzylpyrimidines: Inhibition of DHFR from various species — QSAR, CoMFA and PC analysis*, Eur. J. Med. Chem., 30 (1995) 779–787.
68. *Davis, A.M., Gensmantel, N.P. and Marriott, D.P., *Use of the GRID program in the 3-D QSAR analysis of a series of calcium channel agonists*, In Wermuth, C.-G. (Ed.) Trends in QSAR and molecular modeling 92, Proceedings of the 9th European Symposium on Structure–Activity Relationships: QSAR and Molecular Modeling, ESCOM, Leiden, The Netherlands, 1993, pp. 517–518.
69. *Davis, A.M., Gensmantel, N.P., Jahansson, E. and Marriott, D.P., *The use of the GRID program in the 3-D QSAR analysis of a series of calcium-channel agonists*, J. Med. Chem., 37 (1994) 963–972.

70. De Jong, S. *PLS fits closer than PCR*, J. Chemom., 7 (1993) 551–557.
71. De Jong, S. *SIMPLS: An alternative approach to partial least squares regression*, Chemometr. Intell. Lab. Sys., 18 (1993) 251–263.
72. *de Laszlo, S.E., Glinka, T.W., Greenlee, W.J., ball, R., Nachbar, R.B. and Prendergast, K. *The design, binding affinity prediction and synthesis of macrocyclic angiotensin II AT1 and AT2 receptor antagonists*, Bioorg. Med. Chem. Lett., 6 (1996) 923–928.
73. Dean, P.M., *Molecular similarity*, In Kubinyi, H. (Ed.) 3D QSAR in drug design: Theory, methods and applications, ESCOM, Leiden, The Netherlands, 1993, pp. 150–172.
74. Debnath, A.K., Jiang, S. and Neurath, A.R., Molecular modeling of the V_3 loop of the HIV-1 envelope glycoprotein gp120 reveals possible binding pocket for porphyrins. In Sanz, F., Giraldo, J. and Manaut, F. (Eds.) QSAR and Molecular Modeling: Concepts, Computational Tools and Biological Applications, Proceedings of the 10th European Symposium on Structure–Activity Relationships: QSAR and Molecular Modeling, Prous Science Pub., Barcelona, Spain, 1995, pp. 585–587.
75. *Debnath, A.K., Hansch, C., Kim, K.H. and Martin, Y.C., *Mechanistic interpretation of the genotoxicity of nitrofurans as antibacterial agents using quantitative structure–activity relationships (QSAR) and comparative molecular field analysis (CoMFA)*, J. Med. Chem., 36 (1993) 1007–1016.
76. *Debnath, A.K., Jiang, S., Strick, N., Lin, K., Haberfield, P. and Neurath, A.R., *Three-dimensional structure–activity analysis of a series of porphyrin derivatives with anti-HIV-1 activity targeted to the V3 loop of the gp120 envelope glycoprotein of the human immunodeficiency virus type 1*, J. Med. Chem., 37 (1994) 1099–1108.
77. Deng, Q.L., Cao, B. and Lai, L.H., *Receptor mapping by comparative molecular-field analysis of phospholipase A(2) inhibitors*, J. Chinese Chem. Soc., 42 (1995) 739–744.
78. Deng, Q.L., Cao, B., Lai, L.H. and Tang, Y.Q., *Comparative molecular field analysis (CoMFA) study on known inhibitors of phospholipase A2*, Yaoxue Xuebao, 30 (1995) 428–34.
79. *DePriest, S.A., Mayer, D., Naylor, C.B. and Marshall, G.R., *3D-QSAR of angiotensin-converting enzyme and thermolysin inhibitors — a comparison of CoMFA models based on deduced and experimentally determined active-site geometries*, J. Am. Chem. Soc., 115 (1993) 5372–5384.
80. Diana, G.D., Nitz, T.J., Mallamo, J.P. and Treasurywala, A.M., *Antipicornavirus compounds: Use of rational drug design and molecular modeling*, Antivir. Chem. Chemother., 4 (1993) 1–10.
81. *Dove, S., Kuhne, R. and Schunack, W., *H1 agonistic 2-heteroaryl and 2-phenylhistamines: CoMFA and possible receptor binding sites*, In Sanz, F., Giraldo, J. and Manaut, F. (Eds.) QSAR and Molecular Modeling: Concepts, Computational Tools and Biological Applications, Proceedings of the 10th European Symposium on Structure–Activity Relationships: QSAR and Molecular Modeling, Barcelona, Spain, September 4–9, 1994, J.R. Prous Science Publishers, Barcelona, 1995, pp. 427–432.
82. Doweyko, A.M., *Three-dimensional pharmacophores from binding data*, J. Med. Chem., 37 (1994) 1769–1778.
83. *Dua, R.K., Taylor, E.W. and Phillips, R.S., *A-aryl-L-cysteine S, S,-dioxides — design, synthesis, and evaluation of a new class of inhibitors of kynureninase*, J. Am. Chem. Soc. 115 (1993) 1264–1270.
84. Dunn III, W.J., Hopfinger, A.J., Catana, C. and Duraiswami, C., *Solution of the conformation and alignment tensors for the binding of trimethoprim and its analogs to dihydrofolate reductase: 3D-quantitative structure–activity relationship studying using molecular shape analysis, 3-way partial least squares regression, and 3-way factor analysis*, J. Med. Chem., 39 (1996) 4825–4832.
85. *Elass, A., Vergoten, G., Legrand, D., Mazurier, J., Elass-Rochard, E. and Spik, G., *Processes underlying interactions of human lactoferrin with the jurkat human lymphoblastic T-cell line receptor*, Quant. Struct.-Act. Relat., 15 (1996) 102–107.
86. *Faber, N.M., Griengl, H., Honig, H. and Zuegg, J., *On the prediction of the enantioselectivity of Candida rugosa lipase by comparative molecular field analysis*, Biocatalyst, 9 (1994) 227–239.
87. *Fabian, W.M.F. and Timofei, S., *Comparative molecular field analysis (CoMFA) of dye-fiber affinities: Part 2. Symmetrical bisazo dyes*, Theochem, 362 (1996) 155–162.
88. *Fabian, W.M.F., Timofei, S. and Kurunczi, L., *Comparative molecular field analysis (CoMFA), semi-empirical (AM1) molecular orbital and multiconformational minimal steric difference (MTD) calculations of anthraquinone dye-fiber affinities*, Theochem, 340 (1995) 73–81.

89. *Feng, J. and Zhou, J., *Comparative molecular field analysis of inotropic compounds and pyridazinone*, ACTA Physico-Chimica Sinica (Wuli Huaxue Xuebao), 11 (1995) 206–210.
90. Floersheim, P., Nozulak, J. and Weber, H.P., *Experience with comparative molecular field analysis*, In Wermuth, C.-G. (Ed.) Trends in QSAR and Molecular Modeling 92, Proceedings of the 9th European Symposium on Structure–Activity Relationships: QSAR and Molecular Modeling, ESCOM, Leiden, The Netherlands, 1993, pp. 227–232.
91. *Folkers, G., Merz, A. and Rognan, D., *CoMFA: Scope and limitations*, In Kubinyi, H. (Ed.) 3D QSAR in drug design: Theory, methods and applications, ESCOM, Leiden, The Netherlands, 1993, pp. 583–618.
92. *Folkers, G., Merz, A. and Rognan, D., *CoMFA as a tool for active site modeling*, In Wermuth, C.-G. (Ed.) Trends in QSAR and Molecular Modeling 92, Proceedings of the 9th European Symposium on Structure–Activity Relationships: QSAR and Molecular Modeling, ESCOM, Leiden, The Netherlands, 1993, pp. 233–244.
93. Gaillard, P., Carrupt, P.-A. and Testa, B., *Use of molecular lipophilicity potential for the prediction of log P*, J. Mol. Graphics, 12 (1994) 73.
94. *Gaillard, P.,Carrupt, P.-A., Testa, B. and Boudon, A., *Molecular lipophilicity potential, a tool in 3D-QSAR: Method and applications*, J. Comput.-Aid. Mol. Design, 8 (1994) 83–96.
95. *Gaillard, P., Carrupt, P.-A., Testa, B. and Schambel, P., *Binding of arylpiperazines, (aryloxy) propanolamines, and tetrahydropyridylindoles to the 5-HT1A receptor: Contribution of the molecular lipophilicity potential to three-dimensional quantitative structure–affinity relationship models*, J. Med. Chem., 39 (1996) 126–134.
96. *Gamper, A.M., Winger, R.H., Liedl, K.R., Sotriffer, C.A., Varga, J.M., Kroemer, R.T. and Rode, B.M., *Comparative molecular field analysis of haptens docked to the multispecific antibody IgE(Lb4)*, J. Med. Chem., 39 (1996) 3882–3888; 40 (1997) 1047–1048.
97. *Gantchev, T.G., Ali, H. and van Lier, J.E., *Quantitative structure–activity relationships/comparative molecular field analysis (QSAR/CoMFA) for receptor-binding properties of halogenated estradiol derivatives*, J. Med. Chem, 37 (1994) 4164–4176.
98. *Glennon, R.A., Herndon, J.L. and Dukat, M., *Epibatidine-aided studies toward definition of a nicotine receptor pharmacophore*, Med. Chem. Res., 4 (1994) 461–473.
99. Good, A.C., So, S.S. and Richards, W.G., *Structure–activity relationships from molecular similarity–matrices*, J. Med. Chem., 36 (1993) 433–438.
100. Good, A.C., Peterson, S.J. and Richards, W.G., *QSAR's from similarity matrices: Technique validation and application in the comparison of different similarity evaluation methods*, J. Med. Chem., 36 (1993) 2929–2937.
101. *Greco, G., Novellino, E., Fiorini, I., Nacci, V., Campiani, G., Ciani, S.M., Garofalo, A., Bernasconi, P. and Mennini, T., *A comparative molecular field analysis model for 6-arylpyrrolo[2,1-d][1,5]benzothiazepines binding selectively to the mitochondrial benzodiazepine receptor*, J. Med. Chem., 37 (1994) 4100–4108.
102. *Greco, G., Novellino, E., Pellecchia, M., Silipo, C. and Vittoria, A., *Effects of variable sampling on CoMFA coefficient contour maps in a set of triazines inhibiting DHFR*, J. Comput.-Aided Mol. Design, 8 (1994) 97–112.
103. Greco, G., Novellino, E., Pellecchia, M., Silipo, C. and Vittoria, A., *Effects of variable section on CoMFA coefficient contour maps*, J. Mol. Graphics, 12 (1994) 67–68.
104. *Greco, G., Novellino, E., Pellecchia, M., Silipo, C. and Vittoria, A., *Use of the hydrophobic substituent constant in a comparative molecular field analysis (CoMFA) on a set of anilides inhibiting the Hill reaction*, SAR QSAR Environ. Res., 1 (1993) 301–334.
105. Green, S.M. and Marshall, G.R., *3D-QSAR: A current perspective*, Trends Pharm. Sci., 16 (1995) 285–291.
106. *Grunewald, G.L., Skjaerbaek, N. and Monn, J.A., *An active site model of phenylethanolamine N-methyltransferase using CoMFA*, In Wermuth, C-G. (Ed.) Trends in QSAR and Molecular Modeling 92, Proceedings of the 9th European Symposium on Structure–Activity Relationships: QSAR and Molecular Modeling, ESCOM, Leiden, The Netherlands, 1993, pp. 513–516.
107. Hahn, M. and Rogers, D. *Receptor surface models: 2. Application to quantitative structure–activity relationships studies*, J. Med. Chem., 38 (1995) 2091–2102.

108. Hahn, M. *Receptor surface models: 1. Definition and construction*, J. Med. Chem., 38 (1995) 2080–2090.
109. *Hannongbua, S., Lawtrakul, L., Sotriffer, C.A. and Rode, B.M., *Comparative molecular field analysis of HIV-1 reverse transcriptase inhibitors in the class of 1[2-hydroxyethoxy)-methyl]-6-(phenylthio)thymine*, Quant. Struct.-Act. Relat., 15 (1996) 389–394.
110. Hansch, C. and Fujita, T., *Status of QSAR at the end of the twentieth century*, In Hansch, C. and Fujita, T. (Eds.) Classical and three-dimensional QSAR in agrochemistry, ACS Symposium series Vol. 606, American Chemical Society, Washington, DC, 1995, pp. 1–12.
111. *Harpalani, A.D., Snyder, S.W., Subramanyam, B., Egorin, M.J. and Callery, P.S., *Alkylamides as inducers of human leukemia cell differentiation: A quantitative structure–activity relationship study using comparative molecular field analysis*, 53 (1993) 766–771.
112. *Heinisch, G., Langer, T. and Lukavsky, P., *Lipophilicity determination of diazine analogs of ridogrel: 2. Application of 3D QSAR for prediction of log k′(w) and log P*, Pharmazie, 51 (1996) 840–842.
113. *Hocart, S.J., Reddy, V., Murphy, W.A. and Copy, D.H., *Three-dimensional quantitative structure–activity relationships of somatostatin analogs: 1. Comparative molecular field analysis of growth hormone release-inhibiting potencies*, J. Med. Chem., 38 (1995) 1974–1989.
114. *Hoffmann, R. and Langer, T., *Use of the CATALYST program as a new alignment tool or 3D QSAR*, In Sanz, F., Giraldo, J. and Manaut, F. (Eds.) QSAR and molecular modeling: Concepts, computational tools and biological applications, Proceedings of the 10th European Symposium on Structure–Activity Relationships: QSAR and Molecular Modeling, Barcelona, Spain, September 4–9, 1994, J.R. Prous Science Publishers, Barcelona, 1995, 466–469.
115. Hopfinger, A., Burke, B.J. and Dunn III, W.J., *A generalized formalism of three-dimensional quantitative structure–property relationship analysis for flexible molecules using tensor representation*, J. Med. Chem., 37 (1994) 3768–3774.
116. Horwell, D.C., Howson, W., Higginbottom, M., Naylor, D., Ratcliffe, G.S. and Williams, S., *Quantitative structure–activity relationships (QSARs) of N-terminal fragments of nkl tachykinin antagonists: A comparison of classical QSARs and 3-dimensional QSAR from similarity-matrices*, J. Med. Chem., 38 (1995) 4454–4462.
117. *Horwitz, J.P., Massova, I., Wiese, T.E., Besler, B.H. and Corbett, T.H., *Comparative molecular field analysis of the antitumor activity of 9H-thioxanthen-9-one derivatives against pancreatic ductal carcinoma 03*, J. Med. Chem., 37 (1994) 781–786, 3196.
118. *Horwitz, J.P., Massova, I., Wiese, T.E., Wozniak, A.J., Corbett, T.H., Seboltleopold, J.S., Capps, D.B. and Leopold, W.R., *Comparative molecular-field analysis of in vitro growth-inhibition of L1210 and HCT-8 cells by some pyrazoloacridines*, J. Med. Chem., 36 (1993) 3511–3516.
119. Itai, A., Tomioka, N., Yamada, M., Inoue, A. and Kato, Y., *Molecular similarity*, In Kubinyi, H. (Ed.) 3D QSAR in drug design: Theory, methods and applications, ESCOM, Leiden, The Netherlands, 1993, pp. 200–225.
120. Jain, A.N., Dietterich, T.G., Lathrop, R.H., Chapman, D., Critchlow, R.E., Bauer, B.E., Webster, T.A. and Lozano-Perez, T., *Compass: A shape-based machine learning tool for drug design*, J. Comput.-Aided Mol. Design, 8 (1994) 635–652.
121. Jain, A.N., Koile, K. and Chapman, D., *Compass: Predicting biological activities from molecular surface properties — performance comparisons on a steroid benchmark*, J. Med. Chem., 37 (1994) 2315–2327.
122. Jiang, H.-L., Chen, K.-X., Wang, H.-W., Tang, Y., Chen, J.-X. and Ji, R.-Y., *3D-QSAR study on ether and ester analogs of artemisinin with comparative molecular field analysis*, Zhongguo Yaoli Xuebao, 15 (1994) 481–487.
123. Jiang, H.-L., Chen, K.X., Chen, J.Z., Tang, Y., Wang, Q.M., Li, Q., Shen, X. and Ji, R.Y., *3D-QSAR study on huperzine A analogs with molecular modeling and comparative molecular field analysis (CoMFA) methods*, Chin. Chem. Lett., 7 (1996) 253–256,
124. Jonathan, P., McCarthy, W.V. and Roberts, A.M.I., *Discriminant analysis with singular covariance matrices: A method incorporating cross-validation and efficient randomized permutation tests*, J. Chemomet., 10 (1996) 189–213.
125. *Jones, J.P., He, M., Trager, W.F. and Rettie, A.E., *Three-dimensional quantitative structure–activity relationships for inhibitors of cytochrome P450 2C9*, Drug Metab. Disp., 24 (1996) 1–6.

126. Kafafi, S.A., Afeefy, H.Y., Ali, A.H., Said, H.K. and Kafafi, A.G., *Binding of polychlorinated biphenyls to the aryl hydrocarbon receptor*, Environ. Health Perspect. 101 (1993) 422–428.
127. Kaminski, J.J., *Computer-assisted drug design and selection*, Advanced Drug Delivery Reviews, 14 (1994) 331–337.
128. *Kellogg, G.E., Kier, L.B., Gaillard, P. and Hall, L.H., *E-state fields: Applications to 3D QSAR*, J. Comput.-Aided Mol. Design, 10 (1996) 513–520.
129. Kenny, P.W., *Prediction of hydrogen bond basicity from computed molecular electrostatic properties: Implications for comparative molecular field analysis*, J. Chem. Soc. Perkin Trans., 2 (1994) 199–202.
130. *Kim, K.H., *3D-quantitative structure–activity relationships: Describing hydrophobic interactions directly from 3D structures using a comparative molecular–field analysis (CoMFA) approach*, Quant. Struct.-Act. Relat., 12 (1993) 232–238.
131. *Kim, K.H. and Kim, D.H., *Description of hydrophobicity parameters of a mixed set from their three-dimensional structures*, Bioorg. Med. Chem., 3 (1995) 1389–1396.
132. *Kim, K.H. and Kim, D.H., *Calculation of the reversed-phase high-performance liquid chromatography (RP-HPLC) capacity factors and octanol–water partition coefficients of substituted benzyl N,N-dimethylcarbamates as a measure of hydrophobicity by comparative molecular field analysis (CoMFA) approach*: In Sanz, F., Giraldo, J. and Manaut, F. (Eds.) QSAR and molecular modeling: Concepts, computational tools and biological appliations, Proceedings of the 10th European Symposium on Structure–Activity Relationships: QSAR and Molecular Modeling, Barcelona, Spain, September 4–9, 1994, J.R. Prous Science Publishers, Barcelona, 1995, 101–106.
133. *Kim, K.H., *Calculation of hydrophobic parameters directly from their three-dimensional structures using comparative molecular field analysis*, J. Comput.-Aid. Mol. Design, 9 (1995) 308–318.
134. Kim, K.H., *Comparative molecular field analysis (CoMFA), In Dean, P.M. (Ed.) Molecular similarity in drug design*, Blackie Academic & Professional, London, 1995, pp. 291–331.
135. Kim, K.H., *Comparison of classical and 3D QSAR*, In Kubinyi, H. (Ed.) 3D QSAR in drug design: Theory, methods and applications, ESCOM, Leiden, The Netherlands, 1993, pp. 619–642.
136. *Kim, K.H., *Comparison of classical QSAR and comparative molecular field analysis: Toward lateral validations*, In Hansch, C. and Fujita, T. (Eds.) Classical and three-dimensional QSAR in agrochemistry, ACS Symposium series Vol. 606, American Chemical Society, Washington, DC, 1995, pp. 302–317.
137. *Kim, K.H., *Description of the reversed-phase high-performance liquid chromatography (RP-HPLC) capacity factors and octanol–water partition coefficients of 2-pyrazine and 2-pyridine analogues directly from the three-dimensional structures using comparative molecular field (CoMFA) approach*, Quant. Struct.-Act. Relat., 14 (1995) 8–18.
138. *Kim, K.H., *Nonlinear dependence in comparative molecular field analysis* (CoMFA), J. Comput.-Aid. Mol. Design, 7 (1993) 71–82.
139. *Kim, K.H., *Separation of electronic, hydrophobic, and steric effects in 3D-quantitative structure–activity relationships with descriptors directly from 3D structures using a comparative molecular field analysis (CoMFA) approach*, Current Topics Med. Chem., 1 (1993) 453–467.
140. *Kim, K.H., *Use of indicator variable in comparative molecular field analysis*, Med. Chem. Res., 3 (1993) 257 -267.
141. *Kim, K.H., *Use of the hydrogen-bond potential function in comparative molecular field analysis (CoMFA): An extension of CoMFA*, In Wermuth, C.-G. (Ed.) Trends in QSAR and molecular modeling 92, Proceedings of the 9th European Symposium on Structure–Activity Relationships: QSAR and Molecular Modeling, ESCOM, Leiden, The Netherlands, 1993, pp. 245–251.
142. *Kim, K.H., Greco, G., Novellino, E., Silipo, C. and Vittoria, A., *Use of the hydrogen bond potential function in a comparative molecular field analysis (CoMFA) on a set of benzodiazepines*, J. Comput.-Aid. Mol. Deign, 7 (1993) 263–280.
143. *Kimura, T., Miyashita, Y., Funatsu, K. and Sasaki, S.-i., *Quantitative structure–activity relationships of the synthetic substrates for elastase enzyme using nonlinear partial least squares regression*, J. Chem. Inf. Comput. Sci., 36 (1996) 185–189.
144. *Kireev, D.B., Chretien, J.R. and Raevsky, O.A., *Molecular modeling and quantitative structure–activity studies of anti-HIV-1 2-heteroarylquinoline-4-amines*, Eur. J. Med. Chem., 30 (1995) 395–402.

145. *Klebe, G. and Abraham, U., *On the prediction of binding properties of drug molecules by comparative molecular field analysis*, J. Med. Chem., 36 (1993) 70–80.
146. Klebe, G., Abraham, U. and Mietzner, T., *Molecular similarity indices in a comparative analysis (CoMSIA) of drug molecules to correlate and predict their biological activity*, J. Med. Chem., 37 (1994) 4130–4146.
147. Klebe, G., Mietzner, T. and Weber, F., *Different approaches towards an automatic structural alignment of drug molecules: Applications to sterol mimics thrombin and thermolysin inhibitors*, J. Comput.-Aided Mol. Design, 8 (1994) 751–778.
148. *Kneubuhler, S., Thull, U., Altomare, C., Carta, V., Gaillard, P., Carrupt, P.-A., Carotti, A. and Testa, B., *Inhibition of monoamine oxidase-B by 5H-Indeno[1,2-c]pyridazines: Biological activities, quantitative structure–activity relationships (QSARs) and 3D-QSARs*, J. Med. Chem., 38 (1995) 3874–3883.
149. *Kopponen, P., Sinkkonen, S., Poso, A., Gynther, J. and Karenlampi, S., *Sulfur analogues of polychlorinated dibenzo-p-dioxins, dibenzofurans and diphenyl ethers as inducers of CYP1A1 in mouse hepatoma cell culture and structure–activity relationships*, Env. Toxicol. Chem., 13 (1994) 1543–1548.
150. *Kroemer, R.T. and Hecht, P., *A new procedure for improving the predictiveness of CoMFA models and its application to a set of dihydrofolate reductase inhibitors*, J. Comput.-Aid. Mol. Design, 9 (1995) 396–406.
151. *Kroemer, R.T. and Hecht, P., *Replacement of steric 6-12 potential-derived interaction energies by atom-based indicator variables in CoMFA leads to models of higher consistency*, J. Comput.-Aid. Mol. Design, 9 (1995) 205–212.
152. *Kroemer, R.T., Ettmayer, P. and Hecht, P., *3D-Quantitative structure–activity relationships of human immunodeficiency virus type-1 proteinase inhibitors: Comparative molecular field analysis of 2-hetero-substituted statine derivatives — implications for the design of novel inhibitors*, J. Med. Chem., 38 (1995) 4917–4928.
153. *Kroemer, R.T., Hecht, P. and Liedl, K.R., *Different electrostatic descriptors in comparative molecular field analysis: A comparison of molecular electrostatic and Coulomb potentials*, J. Comput. Chem., 17 (1996) 1296–1308.
154. *Krystek Jr., S.R., Hunt, J.T., Stein, P.D. and Stouch, T.R., *Three-dimensional quantitative structure–activity relationships of sulfonamide endothelin inhibitors*, J. Med. Chem., 38 (1995) 659–668.
155. Kubinyi, H. and Abraham, U., *Practical problems in PLS analyses*, In Kubinyi, H. (Ed.) 3D QSAR in drug design: Theory, methods and applications, ESCOM, Leiden, The Netherlands, 1993, pp. 717–728.
156. Kubinyi, H. (Ed.), *3D QSAR in drug design: Theory, methods and applications*, ESCOM, Leiden, The Netherlands, 1993 759 pp.
157. Laguerre, M., Dubost, J.-P., Kummer, E. and Carpy, A., *Molecular modeling of 5-HT3 receptor antagonists: Geometrical, electronic, and lipophilic features of the pharmacophore and 3D-QSAR study*, Drug Design Discovery, 11 (1994) 205–222.
158. *Langer, T. and Wermuth, C.G., *Inhibitors of prolyl endopeptidase — characterization of the pharmacophoric pattern using conformational analysis and 3D QSAR*, 7 (1993) 253–262.
159. Langer, T., *Molecular similarity determination of heteroaromatics using CoMFA and multivariate data analysis*, Quant. Struct.-Act. Relat., 13 (1994) 402- 405.
160. *Langlois, M., Bremont, B., Rousselle, D. and Gaudy, F., *Structural analysis by the comparative molecular field anlaysis method of the affinity of beta adrenoceptor blocking agents for 5-HT1A and 5-HT1B receptors*, Eur. J. Pharmacol. 244 (1993) 77–87.
161. Li, H., Xu, L. and Su, Q., *Studies on three-dimensional quantitative structure–activity relationships between the structures of N-nitroso compounds and their carcinogenic activities*, Gaodeng Xuexiao Huaxue Xuebao, 17 (1996) 1450–1453.
162. Lindgren, F. and Wolds, S., *A PLS kernel algorithm for data sets with many variables and few objects: Part 2. Cross-validation, missing data and examples*, J. Chemomet., 9 (1995) 459–470.
163. *Lindgren, F., Geladi, P., Berglund, A., Sjostrom, M. and Wold, S., *Interactive variable selection (IVS) for PLS: Part 2. Chemical applications*, J. Chemomet., 9 (1995) 331–342.
164. Lindgren, F., Geladi, P., Rannar, S. and Wold, S.J., *Interactive variable selection (IVS) for PLS: Part 1. Theory and algorithms*, J. Chemomet., 8 (1994) 349–363.

165. Lindgren, F., Geladi, P. and Wold, S., *The kernel algorithm for PLS*, J. Chemomet., 7, (1993) 45–59.
166. *Liu, R. and Matheson, L.E., *Comparative molecular field analysis combined with physicochemical parameters for rediction of polydimethylsiloxane membrane flux in isopropanol*, Pharmaceu. Res., 11 (1994) 257–266.
167. Llorente, B., Leclerc, F. and Cedergren, R., *Using SAR and QSAR analysis to model the activity and structure of the quinolone–DNA complex*, Bioorg. Med. Chem., 4 (1996) 61–71.
168. *Mabilia, M., Belvisi, L., Bavi, G., Catalano, G. and Scolastico, C., *A PCA/PLS analysis on nonpeptide angiotensin II receptor antagonists*, In Sanz, F., Giraldo, J. and Manaut, F. (Eds.) QSAR and molecular modeling: Concepts, computational tools and biological applications, Proceedings of the 10th European Symposium on Structure–Activity Relationships: QSAR and Molecular Modeling, Barcelona, Spain, September 4–9, 1994, J.R. Prous Science Publishers, Barcelona, 1995, pp. 456–460.
169. Marshall, G.R., Ho, C.M.W., Head, R.D., Smythe, M.L., Oprea, T.I., Waller, C.L. and Green, S.M., 3D QSAR and de novo design: choosing the appropriate tools. In Sanz, F., Giraldo, J. and Manaut, F. (Eds.) QSAR and Molecular Modeling: Concepts, Computational Tools and Biological Applications, Proceedings of the 10th European Symposium on Structure–Activity Relationships: QSAR and Molecular Modeling, Prous Science Pub., Barcelona, Spain, 1995, pp. 623–629.
170. Martin, Y.C. and Lin, C.T., *Three-dimensional quantitative structure–activity relationships: D2 dopamine agonists as an example*, In Wermuth, C.-G. (Ed.), The practice of medicinal chemistry, Academic Press, London, 1996, pp. 459–483.
171. Martin, Y.C., Bures, M.G., Danaher, E.A. and DeLazzer, J., *New strategies that improve the efficiency of the 3D design of bioactive molecules*, In Wermuth, C.-G. (Ed.) Trends in QSAR and molecular modeling 92, Proceedings of the 9th European Symposium on Structure–Activity Relationships: QSAR and Molecular Modeling, ECOM, Leiden, The Netherlands, 1993, pp. 20–26.
172. Martin, Y., *Distance comparisons: A new strategy for examining three-dimensional structure–activity relationships*, In Hansch, C. and Fujita, T., (Eds.) Classical and three-dimensional QSAR in agrochemistry, ACS Symposium series Vol. 606, American Chemical Society, Washington, DC, 1995, pp. 318–329.
173. Martin, Y.C., Kim, K.H. and Lin, C.T., *Comparative molecular field analysis: CoMFA*, In Charton, M. (Ed.) Advances in quantitative structure–property relationships, JAI Press, Greenwich, CT, 1996, Vol. 1, pp. 1–52.
174. *Martin, Y.C., Lin, C.T. and Wu, J., *Application of CoMFA to the design and structural optimization of D1 dopaminergic agonists*, In Kubinyi, H. (Ed.) 3D QSAR in drug design: Theory, methods and applications, ESCOM, Leiden, The Netherlands, 1993, pp. 643–660.
175. Martin, Y.C., Lin, C.T., Hetti, C. and Delazzer, J., *PLS analysis of distance matrices to detect non-linear relationships between biological potency and molecular-properties*, J. Med. Chem., 38 (1995) 3009–3015.
176. *Martinez-Merino, V., Martinez-Gonzalez, A., Gonzalez, A. and Gil, M.J., *3D-QSAR of the diaryl-sulfonylureas as antineoplastic agents*, In Sanz, F., Giraldo, J. and Manaut, F. (Eds.) QSAR and molecular modeling: Concepts, computational tools and biological applications, Proceedings of the 10th European Symposium on Structure–Activity Relationships: QSAR and Molecular Modeling, Barcelona, Spain, September 4–9, 1994, J.R. Prous Science Publishers, Barcelona, 1995, pp. 478–480.
177. *Mascarella, S.W., Bai, X., Williams, W., Sine, B., Bowen, W. and Carroll, F.I., *(+)-cis-N-(para-, meta-, and ortho-substituted benzyl)-N-normetazocines: Synthesis and binding affinity at the [3H]-(+)-pentazocine-labeled (s1) site and quantitative structure–affinity relationship studies*, J. Med. Chem., 38 (1995) 565–569.
178. Mason, K.A., Katz, A.H. and Shen, C.F., *Grid-assisted similarity perception (GRASP): A new method of overlapping molecular structures*, In Wermuth, C.-G. (Ed.) Trends in QSAR and molecular modeling 92, Proceedings of the 9th European Symposium on Structure–Activity Relationships: QSAR and Molecular Modeling, ESCOM, Leiden, The Netherlands, 1993, pp. 394–395.
179. *Masuda, T., Nakamura, K., Jikihara, T., Kasuya, F., Igarashi, K., Fukui, M., Takagi, T. and Fujiwara, H., *3D-quantitative structure–activity relationships for hydrophobic interactions: Comparative molecular field analysis (CoMFA) including molecular lipophilicity potentials as applied to the glycine conjugation of aromatic as well as aliphatic carboxylic acids*, Quant. Struct.-Act. Relat., 15 (1996) 194–200.

180. McGaughey, G.B., *Molecular mechanics parameterization of positively charged nitrogen-containing compounds and its application to comparative molecular field analysis of choline acetyltransferase inhibitors*, University of Georgia, Athens, GA, 1996.
181. *McLay, I.M. and Mason, J.S., *MLR and PLS: A comparison of the techniques applied to the QSAR analysis of a series of structurally diverse biologically active compounds*, In Wermuth, C.-G. (Ed.) Trends in QSAR and molecular modeling 92, Proceedings of the 9th European Symposium on Structure–Activity Relationships: QSAR and Molecular Modeling, ESCOM, Leiden, The NEtherlands, 1993, pp. 519–521.
182. *McNaught, K.S.P., Thull, U., Carrupt, P.-A., Altomare, C., Cellamare, S., Carotti, A., Testa, B., Jenner, P. and Marsden, C.D., *Effects of isoquinoline derivatives structurally related to 1-methyl-4-phenyl-1,2,3,6-tetrahydropyridine (MPTP) on microchondrial respiration*, Biochem. Pharmacol., 51 (1996) 1503–1511.
183. *Medvedev, A.E., Ivanov, A.S., Veselovsky, A.V., Skvortsov, V.S. and Archakov, A.I., *QSAR analysis of indole analogues as monoamine oxidase inhibitors*, J. Chem. Inf. Comput. Sci., 36 (1996) 664–671.
184. *Merz, A. and Folkers, G., *Contribution of electrostatic energies to the CoMFA analysis of herpes simplex virus thimidine kinase inhibitors*, J. Mol. Graphics, 12 (1994) 75–76.
185. *Minor, D.L., Wyrick, S.D., Charifson, P.S., Watts, V.J., Nichols, D.E. and Mailman, R.B., *Synthesis and molecular modeling of 1-phenyl-1,2,3,4-tetrahydroisoquinolines and related 5,6,8,9-tetrahydro-13bH-dibenzo[a,h]quinolizines as D1 dopamine antagonists*, J. Med. Chem., 37 (1994) 4317–4328.
186. Dean, P.M. (Ed.) Molecular similarity in drug design, Blackie Academic & Professional, London, 1995, 342 pp.
187. Montanari, C.A., Tute, M.S., Beezer, A.E. and Mitchell, J.C., *Determination of receptor-bound drug conformations by QSAR using flexible fitting to derive a molecular similarity index*, J. Comput.-Aid. Mol. Design, 10 (1996) 67–73.
188. Mosimann, S., Meleshko, R. and james, M.N.G., *A critical assessment of comparative molecular modeling of tertiary structures of proteins*, Proteins: Struct. Funct. Genet., 23 (1995) 301–317.
189. Muresan, S., Bologa, C., Chiriac, A., Jastorff, B., Kurunczi, L. and Simon, Z., *Comparative structure–affinity relations by MTD for binding of cycloadenosine monophosphate derivatives to protein kinase receptors*, Quant. Struct.-Act. Relat., 13 (1994) 242–248.
190. Muresan, S., Sulea, T., Ciubotariu, D., Kurunczi, L. and Simon, Z., *Van der Waals intersection envelope volumes as a possible basis for steric interaction in CoMFA*, Quant. Struct.-Act. Relat., 15 (1996) 31–32.
191. *Myers, A.M., Charifson, P.S., Owens, C.E., Kula, N.S., McPhail, A.T., Baldessarini, R.J., Booth, R.G. and Wyrick, S.D., *Conformational analysis pharmacophore identification and comparative molecular field analysis of ligands for the neuromodulatory sigma3 receptor*, J. Med. Chem., 37 (1994) 4109–4117.
192. *Nakagawa, Y., Shimizu, B., Oikawa, N., Akamatsu, M., Nishimura, K., Kurihara, N., Ueno, T. and Fujita, T., *Three-dimensional quantitative structure–activity analysis of steroidal and dibenzoyl-hydrazine-type ecdysone agonists*, In Hansch C. and Fujita, T. (Eds.) Classical and three-dimensional QSAR in agrochemistry, ACS Symposium series, Vol. 606, American Chemical Society, Washington, DC, 1995, pp. 288–301.
193. *Navajas, C., Kokkola, T., Poso, A., Honka, N., Gynther, J. and Laitinen, J.T., *A rhodopsin-based model for melatonin recognition at its G protein-coupled receptor*, Eur. J. Pharmacol., 304 (1996) 173–183.
194. *Navajas, C., Poso, A., Tuppurainen, K. and Gynther, J., *Comparative molecular field anlaysis (CoMFA) of MX compounds using different semi-empirical methods: LUMO field and its correlation with mutagenic activity*, Quant. Struct.-Act. Relat., 15 (1996) 189–193.
195. *Nayak, V.R. and Kellogg, G.E., *Cyclodextrin-barbiturate inclusion complexes: A CoMFA/HINT 3-D QSAR study*, Med. Chem. Res., 3 (1994) 491–502.
196. *Nicklaus, M.C., Ford Jr., H.F., Hegedus, L., Milne, G.W.A. and Kelley, J.A., *Comparative molecular field analysis of hydrophobicity descriptors of cytosine nucleosides*, Quant. Struct.-Act. Relat., 14 (1995) 335–343.
197. *Nordvall, G., Sundquist, S., Johansson, G., Glas, G., Nilvebrant, L. and Hacksell, U., *3-(2-benzofuranyl)quinuclidin-2-ene derivatives: Novel muscarinic antagonists*, J. Med. Chem., 39 (1996) 3269–3277.

198. *Norinder, U., *A PLS QSAR anlaysis using 3D generated aromatic descriptors of principal property type: Application to some dopamine D2 benzamide antagonists*, J. Comput.-Aid. Mol. Design, 7 (1993) 671–682.
199. *Norinder, U., *Single and domain model variable selection in 3D QSAR applications*, J. Chemomet., 10 (1996) 95–105.
200. *Norinder, U., *The alignment problem in 3-D QSAR: A combined approach using CATALYST and a 3-D QSAR technique*, In Sanz, F., Giraldo, and J. Manaut, F. (Eds.) QSAR and molecular modeling: Concepts, computational tools and biological applications, Proceedings of the 10th European Symposium on Structure–Activity Relationships: QSAR and Molecular Modeling, Barcelona, Spain, September 4–9, 1994, J.R. Prous Science Publishers, Barcelona, 1995, pp. 433–438.
201. *Novellino, E., Fattorusso, C. and Greco, G., *Use of comparative molecular field analysis and cluster analysis in series design*, Pharm. Acta Helv., 70 (1995) 149–154.
202. *Ohta, M., Koga, H., Sato, H. and Ishizawa, T., *Comparative molecular field analysis of benzopyran-4-carbothioamide potassium channel openers*, Bioorg. Med. Chem. Lett., 4 (1994) 2903–2906.
203. *Oprea, T.I. and Garcia, A.E., *Three-dimensional quantitative structure–activity relationships of steroid aromatase inhibitors*, J. Comput.-Aid. Mol. Design, 10 (1996) 186–200.
204. Oprea, T.I., Ciubotariu, D., Sulea, T.I. and Simon, Z., *Comparison of the minimal steric difference (MTD) and comparative molecular field analysis (CoMFA) methods for anlaysis of binding of steroids to carrier proteins*, Quant. Struct.-Act. Relat., 12 (1993) 21–26.
205. *Oprea, T.I., Head, R.D. and Marshall, G.R., *The basis of cross-reactivity for a series of steroids binding to a monoclonal antibody (DB3) against progesterone: A molecular modeling and QSAR study*, In Sanz, F., Giraldo, J. and Manaut, F. (Eds.) QSAR and molecular modeling: Concepts, computational tools and biological applications, Proceedings of the 10th European Symposium on Structure–Activity Relationships: QSAR and Molecular Modeling, Barcelona, Spain, September 4–9, 1994, J.R. Prous Science Publishers, 1995, Barcelona, pp. 451–455.
206. *Oprea, T.I., Waller, C.L. and Marshall, G.R., *3D-QSAR of human immunodeficiency virus (I) protease inhibitors: 3. Interpretation of CoMFA results*, Drug Des. Discovery, 12 (1994) 29–51.
207. *Oprea, T.I., Waller, C.L. and Marshall, G.R., *Three-dimensional quantitative structure–activity relationship of human immunodeficiency virus (I) protease inhibitors: 2. Predictive power using limited exploration of alternate binding modes*, J. Med. Chem., 37 (1994) 2206–2215.
208. *Ortiz, A.R., Pisabarro, M.T., Gago, F. and Wade, R.C., *Prediction of drug binding affinities by comparative binding energy analysis: Application to human synovial fluidphospholipase A1 inhibitors*, In Sanz, F., Giraldo, J. and Manaut, F. (Ed.) QSAR and molecular modeling: Concepts, computational tools and biological applications, Proceedings of the 10th European Symposium on Structure–Activity Relationships: QSAR and Molecular Modeling, Barcelona, Spain, September 4–9, 1994, J.R. Prous Science Publishers, Barcelona, 1995, pp. 439–443.
209. *Palluotto, F., Carotti, A., Casini, G., Campagna, F., Genchi, G., Rizzo, M. and De Sarro, G.B., *Structure–activity relationships of 2-aryl-2,5-dihydropyridazino[4,3-b]indol-3(3H)-ones at the benzodiazepine receptor*, Bioorg. Med. Chem., 4 (1996) 2091–2104.
210. *Palomer, A., Giolitti, A., Fos, E., Cabre, F., Mauleon, D. and Carganico, G., *Molecular modeling and CoMFA investigations on LTD4 receptor antagonists*, In Sanz, F., Giraldo, J. and Manaut, F. (Eds.) QSAR and molecular modeling: Concepts, computational tools and biological applications, Proceedings of the 10th European Symposium on Structure–Activity Relationships: QSAR and Molecular Modeling, Barcelona, Spain, September 4–9, 1994, J.R. Prous Science Publishers, Barcelona, 1995, pp. 444–450.
211. Pastor, M. and Cruciani, G., *A novel strategy for improving ligand selectivity in receptor-based drug design*, J. Med. Chem., 38 (1995) 4637–4647.
212. Patterson, D.E., Cramer, R.D., Ferguson, A.M., Clark, R.D. and Weinberger, L.E., *Neighborhood behavior: A useful concept for validation of a 'molecular diversity' descriptors*, J. Med. Chem., 39 (1996) 3049–3059.
213. *Pellicciari, R., Natalini, B., Costantino, G., Garzon, A., Luneia, R., Mahmoud, M.R., Marinozzi, M., Roberti, M., Rosato, G.C. and Shiba, S., *Heterocyclic modulators of the NMDA receptor*, Il Farmaco, 48 (1993) 151–157.
214. Phatak, A., Reilly, P.M. and Penlidis, A., *An approach to interval estimation in partial least squares regression*, Anal. Chim. Acta., 227 (1993) 495–501.

215. Poso, A., *Modeling of some bioactive compounds utilizing CoMFA with different field types*, University of Kuopio, 1995 Ph.D. thesis.
216. *Poso, A, Juvonen, R. and Gynther, J., *Comparative molecular field analysis of compounds with CYP2A5 binding affinity*, Quant. Struct.-Act. Relat., 14 (1995) 507–511.
217. *Poso, A., Tuppurainen, K. and Gynther, J., *Modeling of molecular mutagenicity with comparative molecular field analysis (CoMFA): Structural and electronic properties of MX compounds related to TA100 mutagenicity*, J. Mol. Struct. (Theochem), 304 (1994) 255–260.
218. *Poso, A., Tuppurainen, K. and Gynther, J., *Molecular genotoxicity of MX compounds and the correlation with LUMO: Comparative molecular field analysis*, J. Mol. Graphics, 12 (1994) 70.
219. *Poso, A., Tuppurainen, K., Ruuskanen, J. and Gynther, J., *Binding of some dioxins and dibenzofurans to the Ah receptor: A QSAR model based on comparative molecular field analysis (CoMFA)*, J. Mol. Struct. (Theochem), 282 (1993) 259–264.
220. *Prendergast, K., Adams, K., Greenlee, W.J., Nachbar, R.B., Patchett, A.A., and Underwood, D.J., *Derivation of a 3D pharmacophore model for the angiotensin-II site one receptor*, J. Comput.-Aided Mol. Design, 8 (1994) 491- 512.
221. *Raghavan, K., Buolamwini, J.K., Fesen, M.R., Pommier, Y., Kohn, K.W. and Weinstein, J.N., *Three-dimensional quantitative structure–activity relationship (QSAR) of HIV integrase inhibitors: A comparative molecular field analysis (CoMFA) study*, J. Med. Chem., 38 (1995) 890–897.
222. *Ragno, R., Botta, M., Corelli, F., Mai, A., Massa, S., Porretta, G.C. and Artico, M., *Comparative molecular field analysis of new human rhinovirus-14 inhibitors*, In Sanz, F., Giraldo, J. and Manaut, F. (Eds.) QSAR and molecular modeling: Concepts, computational tools and biological applications, Proceedings of the 10th European Symposium on Structure–Activity Relationships: QSAR and Molecular Modeling, Barcelona, Spain, September 4–9, 1994, J.R. Prous Science Publishers, Barcelona, 1995, pp. 488–492.
223. Rannar, S., Lindgren, F., Geladi, P. and Wold, S., *A PLS kernel algorithm for data sets with many variables and fewer objects: Part 1. Theory and algorithm*, J. Chemomet., 8 (1994) 111–125.
224. *Recanatini, M., *Comparative molecular field analysis of non-steroidal aromatase inhibitors related to fadrozole*, J. Comput.-Aid. Mol. Design, 10 (1996) 74–82.
225. Rowberg, K.A., Martin, E.M. and Hopfinger, A.J., *QSAR and molecular shape analyses of three series of 1-(phenylcarbamoyl)-2-pyrazoline I insecticides*, J. Agric. Food Chem., 42 (1994) 374–380.
226. *Said, M., Ziegler, J.C., Magdalou, J., Elass, A. and Vergoten, G., *Inhibition of bilirubin UDP-glucuronosyltransferase: A comparative molecular field analysis (CoMFA)*, Quant. Struct.-Act. Relat., 15 (1996) 382–388.
227. *Sams II, R.L., Compadre, R.L., Castleberry, A., Samokyszyn, V.M., Ronis, M. and Compadre, C.M., *Quantitation of physico-chemical properties affecting the mutagenicity and rates of reduction of nitroaromatics*, In Sanz, F., Giraldo, J. and Manaut, F. (Eds.) QSAR and molecular modeling: Concepts, computational tools and biological applications, Proceedings of the 10th European Symposium on Structure–Activity Relationships: QSAR and Molecular Modeling, Barcelona, Spain, September 4–9, 1994, J.R. Prous Science Publishers, Barcelona, 1995, pp. 484–487.
228. Semus, S.F., *CoMFA: A field of dreams?*, Network Sci., 2 (1996); URL: http;//www.awod.com/netsci/Issues/Jan96/.
229. Seri-Levy, A., Salter, R., West, S. and Richards, W.G., *Shape similarity as a single independent variable in QSAR*, Eur. J. Med. Chem., 29 (1994) 687–694.
230. Seri-Levy, A., West, S. and Richards, W.G., *Molecular similarity, quantitative chirality, and QSAR for chiral drugs*, J. Med. Chem., 37 (1994) 1727–1732.
231. *Seydel, J.K., Czaplinski, K.-H., Wiese, M., Kansy, M. and Hansel, W., *QSAR-CoMFA- and PC-analysis of the inhibitory activity of new benzylpyrimidines against DHFR derived from various species*, In Sanz, F., Giraldo, J. and Manau, F. (Eds.) QSAR and molecular modeling: Concepts, computational tools and biological applications, Proceedings of the 10th European Symposium on Structure–Activity Relationships: QSAR and Molecular Modeling, Barcelona, Spain, September 4–9, 1994, J.R. Prous Science Publishers, Barcelona, 1995, pp. 91–93.
232. *Siddiqi, S.M., Pearlstein, R.A., Sanders, L.H. and Jacobson, K.A., *Comparative molecular field anlaysis of selective A3 adenosine receptor agonists*, Bioorg. Med. Chem., 3 (1995) 1331–1343.
233. Silverman, B.D. and Platt, D.E., *Comparative molecular moment analysis (CoMMA): 3D-QSAR without molecular superposition*, J. Med. Chem., 39 (1996) 2129–2140.

234. Simon, Z., *MTD and hyperstructure approaches*, In Kubinyi, H. (Ed.) 3D QSAR in drug design: Theory, methods and applications, ESCOM, Leiden, The Netherlands, 1993, pp. 307–319.
235. Simon, Z., Chiriac, A., Holban, S., Ciubotariu, D. and Mihals, G.I., *Minimum steric difference: The MTD-method for QSAR studies*, Research Studies press, Letchworth, U.K., 1994.
236. Srivastava, S., Richardson, W.W., Bradley, M.P. and Crippen, G.M., *Three-dimensional receptor modeling using distance geometry and Voronoi polyhedra*, In Kubinyi, H. (Ed.) 3D QSAR in drug design: Theory, methods and applications, ESCOM, Leiden, The Netherlands, 1993, pp. 409–430.
237. *Steinmetz, W.E., *A CoMFA analysis of selected physical properties of amino acids in water*, Quant. Stlruct.-Act. Relat., 14 (1995) 19–23.
238. *Steinmetz, W.E., *A CoMFA model of steric and electronic effects of phosphorus ligands*, Quant. Struct.-Act. Relat., 15 (1996) 1–6.
239. *Tafi, A., Anastassopoulou, J., Theophanides, T., Botta, M., Corelli, F., Massa, S., Artico, M., Costi, R., Santo, R.D. and Ragno, R., *Molecular modeling of azole antifungal agents active against* Candida albicans *:1. A comparative molecular field analysis study*, J. Med. Chem., 39 (1996) 1227–1235.
240. Tafi, A.A.J., Botta, M., Corelli, F. and Theophanides, T., *Azole fungicides: CoMFA study of* Candida albicans *lanosterol 14.alpha.-demethylase azole inhibitors*, In Merlin, J.C.T., Huvenne, S. and Pierre, J. (Eds.) Proceedings of the 6th European Spectroscopy, Biological and Molecular Conference, Kluwer Academic Publishers, Dordrecht, The Netherlands, 1995, pp. 157–.
241. *Tang, Y.C., Jiang, K.X., Jin, H.L., Zhang, G. and Ji, R.Y., *Studies on dopamine receptors and tetrahydroprotoberberines: III. 3D-QSAR study on tetrahydroprotoberberines using CoMFA approach*, Chin. Chem. Lett., 7 (1996) 249–252.
242. Testa, B., Carrupt, P.-A., Gaillard, P., Billois, F. and Weber, P., *Lipophilicity in molecular modeling*, Pharma. Res., 13 (1996) 335–343.
243. Thibaut, U., *Applications of CoMFA and related 3D QSAR approaches*, In Kubinyi, H. (Ed.) 3D QSAR in drug design: Theory, methods and applications, ESCOM, Leiden, The Netherlands, 1993, pp. 661–696.
244. Thibaut, U., Folkers, G., Klebe, G., Kubinyi, H., Merz, A. and Rognan, D., *Recommendations for CoMFA studies and 3D QSAR publications*, Quant. Struct.-Act. Relat. 13 (1994) 1–3.
245. Thibaut, U., Folkers, G., Klebe, G., Kubinyi, H., Merz, A. and Rognan, D., *Recommendations for CoMFA studies and 3D QSAR publications*, In Kubinyi, H. (Ed.) 3D QSAR in drug design: Theory, methods and applications, ESCOM, Leiden, The Netherlands, 1993, pp. 711–716.
246. *Thull, U., Kneubuhler, S., Gaillard, P., Carrupt, P.-A., Testa, B., Altomare, C., Carotti, A., Jenner, P. and McNaught, K.S.P., *Inhibition of monoamine oxidase by isoquinoline derivatives*, Biochem. Pharmacol., 50 (1995) 869–877.
247. Tokarski, J.S. and Hopfinger, A.J., *Three-dimensional molecular shape anlaysis: Quantitative structure–activity relationship of a series of cholecystokinin-A receptor antagonists*, J. Med. Chem., 37 (1994) 3639–3654.
248. *Tomkinson, N.P., Marriott, D.P., Cage, P.A., Cox, D., Davis, A.M., Flower, D.R., Gensmantel, N.P., Humphries, R.G., Ingall, A.H. and Kindon, N.D., *P2T purinoceptor antagonists: A QSAR study of some 2-substituted ATP analogues*, J. Pharm. Pharmacol., 48 (1996) 206–209.
249. *Tong, W., Collantes, E.R., Chen, Y. and Welsh, W.J., *A comparative molecular field analysis study of N-benzylpiperidines as acetylcholinesterase inhibitors*, J. Med. Chem., 39 (1996) 380–387.
250. *Tung, C.-S., Oprea, T.I., Hummer, G. and Garcia, A.E., *Three-dimensional model of a selective theophylline-binding RNA molecule*, J. Mol. Recognition, 9 (1996) 275–286.
251. van de Waterbeemd, H., Carrupt, P.-A., Testa, B. and Kier, L.B., *Multivariate data modeling of new steric, topological and CoMFA-derived substituent parameters*, In Wermuth, C.-G. (Ed.) Trends in QSAR and Molecular Modeling 92, Proceedings of the 9th European Symposium on Structure–Activity Relationships: QSAR and Molecular Modeling, ESCOM, Leiden, The Netherlands, 1993, pp. 69–75.
252. van de Waterbeemd, H., Clementi, S., Costantino, G., Carrupt, P.-A. and Testa, B., *CoMFA-derived substituent descriptors for structure–property correlations*, In Kubinyi, H. (Ed.) 3D QSAR in drug design: Theory, methods and applications, ESCOM, Leiden, The Netherlands, 1993, pp. 697–707.
253. *van Helden, S.P. and Hamersma, H., *3D-QSAR of the receptor binding of steroids: A comparison of multiple regression, neural networks and comparative molecular field analysis*, In Sanz, F., Giraldo, J.

and Manaut, F. (Eds.) QSAR and molecular modeling: Concepts, computational tools and biological applications, Proceedings of the 10th European Symposium on Structure–Activity Relationships: QSAR and Molecular Modeling, Barcelona, Spain, September 4–9, 1994, J.R. Prous Science Publishers, Barcelona, 1995, pp. 481–483.

254. *van Steen, B.J., van Wijngaarden, I., Tulp, M.T.M. and Soudjin, W. *Structure–affinity relationship studies on 5-HT1A receptor ligands: 2. Heterobicyclic phenylpiperazines with N4-aralkyl substituents*, J. Med. Chem., 37 (1994) 2761–2773.
255. Verhaar, H.J.M., Erksson, L., Sjostrom, M., Schuurmann, G., Seinen, W. and Hermens, J.L.M., *Modeling the toxicity of organophosphates: A comparison of the multiple linear regression and PLS regression methods*, Quant. Struct.-Act. Relat., 13 (1994) 133–143.
256. Wade, R.C., *Molecular interaction fields*, In Kubinyi, H. (Ed.) 3D QSAR in drug design: Theory, methods and applications, ESCOM, Leiden, The Netherlands, 1993, pp. 486–506.
257. Wakeling, I.N. and Morris, J.J., *A test of significance for partial least squares regression*, J. Chemomet., 7 (1993) 291–304.
258. *Waller, C.L., *A three-dimensional technique for the calculation of octanol–water partition coefficients*, Quant. Struct.-Act. Relat., 13 (1994) 172–176.
259. Waller, C.L. and Kellogg, G.E., *Adding chemical information to CoMFA models with alternative 3D QSAR fields*, Network Sci., 2 (1996); http://www.awod.com/netsci/Science/Compchem/feature10.html.
260. *Waller, C.L. and marshall, G.R., *3-dimensional quantitative structure–activity relationship of angiotensin-converting enzyme and thermolysin inhibitors: 2. A comparison of CoMFA models incorporating molecular-orbital fields and desolvation free-energies based on active-analog and complementary-receptor field alignment rules.*, J. Med. Chem., 36 (1993) 2390–2403.
261. *Waller, C.L. and McKinney, J.D., *Three-dimensional quantitative structure–activity relationships of dioxins and dioxin-like compounds: Model validation and Ah receptor characterization*, Chem. Res. Toxicol., 8 (1995) 847–858.
262. *Waller, C.L., Evans, M.V. and McKinney, J.D., *Modeling the cytochrome P450-mediated metabolism of chlorinated volatile organic compounds*, Drug Metab. Dispos., 24 (1996) 203–210.
263. *Waller, C.L., Juma, B.W., Gray Jr., L.E. and Kelce, W.R., *Three-dimensional quantitative structure–activity relationships for androgen receptor ligands*, Toxicol. Appl. Pharmacol., 137 (1996) 219–227.
264. *Waller, C.L., Minor, D.L. and McKinney, J.D., *Using three-dimensional quantitative structure–activity relationships to examination of the estrogen-receptor binding affinities of polychlorinated hydroxybiphenyls using three-dimensional quantitative structure–activity relationships*, Environ. Health Perspect., 103 (1995) 702–707.
265. *Waller, C.L., Oprea, T.I., Chae, K., Park, H.-K., Korach, K.S., Laws, S.C., Wiese, T.E., Kelce, W.R. and Gray, Jr., L.E., *Ligand-based identification of environmental estrogens*, Chem. Res. Toxicol., 9 (1996) 1240–1248.
266. *Waller, C.L., Oprea, T.I., Giolitti, A. and Marshall, G.R., *Three-dimensional QSAR of human-immunodeficiency-virus-I protease inhibitors: 1. A CoMFA study employing experimentally determined alignment rules*, J. Med. Chem. 36 (1993) 4152–4160.
267. *Waller, C.L., Wyrick, S.D., Kemp, W.E., Park, H.M. and Smith, F.T., *Conformational-analysis, molecular modeling, and quantitative structure–activity relationship studies of agents for the inhibition of astrocytic chloride transport*, Pharm. Res., 11 (1994) 47–53.
268. Walters, D.E. and Hinds, R.M., *Genetically evolved receptor models: A computational approach to construction of receptor models*, J. Med. Chem., 37 (1994) 2527–2536.
269. *Wang, M.-M., Huang, N., Yang, G.-Z. and Guo, Z.-R., *Study on 3D-QSAR of retinoids: 3D-interaction between retinoids and their receptor*, J. Chinese Pharm. Sci., 5 (1996) 57–62.
270. *Watson K., Michell, E.P., Johnson, L.N., Cruciani, G., Son, J.C., Bichard, C.J.F., Fleet, G.W.J., Oikonomakos, N.G., Kontou, M. and Zographos, S.E., *Glucose analogue inhibitors of glycogen phosphorylase: From crystallographic analysis to drug prediction using GRID force-field and GOLPE variable selection*, Acta Cryst., D51 (1995) 458–472.
271. *Welch, W., Ahmad, S., Airey, J.A., Gerzon, K., Humerickhouse, R.A, Besch Jr., H.R., Ruest, L., Deslongchamps, P. and Sutko, J.L., *Structural determinants of high-affinity binding of ryanoids to the vertebrate skeletal muscle ryanodine receptor: A comaprative molecular field analysis*, Biochem., 33 (1994) 6074–6085.

272. *Welsh, W.J., Tong, W., Collantes, E.R, Chickos, J.S. and Gagarin, S.G., *Enthalpies of sublimation and formation of polycyclic aromatic hydrocarbons (PAHs) derived from comparative molecular field anlaysis (CoMFA): Application of moment of inertia for molecular alignment*, Thermochim. Acta, 290 (1996) 55–64.
273. Wiese, M., *The hypothetical active-site lattice*, In Kubinyi, H. (Ed.) 3D QSAR in drug design: Theory, methods and applications, ESCOM, Leiden, The Netherlands, 1993, pp. 431–442.
274. Wold, S., Johansson, E. and Cocchi, M., *PLS — partial least-squares projections to latent structures*, In Kubinyi, H. (Ed.) 3D QSAR in drug design: Theory, methods and applications, In Kubinyi, H. (Ed.) 3D QSAR in drug design: Theory, methods and applications, ESCOM, Leiden, The Netherlands, 1993, pp. 523- 550.
275. *Wong, G., Koehler, K.F., Skolnick, P., Gu, Z.Q., Ananthan, S., Schonholzer, P., Hunkeler, W., Zhang, W.J. and Cook, J.M., *Synthetic and computer-assisted analysis of the structural requirements for selective, high-affinity ligand-binding to diazepam-insensitive benzodiazepine receptors*, J. Med. Chem., 36 (1993) 1820–1830.
276. *Xia, Q., Li, Z.-x., Zhou, J.-g., Li, R.-I., Feng, J., Pang, S.-h., Zhou, J. and Wu, J., *Molecular design of lipophilic antifolates by the aid of Hansch analysis and CoMFA method*, Fourth China–Japan Joint Development Paper, Symposium on Drug Design, October 4–7, 1995.
277. *Yamakawa, M., Ezumi, K., Takeda, K., Suzuki, T., Horibe, I., Kato, G. and Fujita, T. (Eds.) *Classical and three-dimensional quantitative structure–activity analyses of steroid hormones: Structure–receptor binding patterns of anti-hormonal drug candidates*, In Fujita, T. (Ed.) QSAR and drug design: New developments and applications, Elsevier, Amsterdam, The Netherlands, 1995, pp.125–150.
278. *Yoo, S.-e. and Cha, O.J., *Correlation between the reactant complex or transition state conformations and the reactivity of 4-nitrophenyl benzoate and its sulfur analogues with anionic nucleophiles by comparative molecular field analysis (CoMFA)*, Bull. Korean Chem. Soc., 17 (1996) 653–655.
279. *Yoo, S.-e. and Cha, O.J., *Theoretical study on the [3,3]-sigmatropic rearrangement of allylic esters by comparative molecular field analysis (CoMFA)*, Bull. Korean Chem. Soc., 15 (1994) 889–890.
280. *Yoo, S.-e. and Shin, Y.A., *Prediction of lipophilicity of orthopramides by comparative molecular field analysis (CoMFA)*, Bull. Korean Chem. Soc., 16 (1995) 1189–1193.
281. *Yoo, S.U. and Cha, O.J., *Prediction of LUMO energy and rate constant by comparative molecular field analysis (CoMFA)*, J. Comput. Chem., 16 (1995) 449–453.
282. Yoo, S.U. and Shin, Y.A., *A new 3D-QSAR method for developing new medicine: Comparative molecular field analysis (CoMFA)*, Hwahak Sekye, 234 (1994) 423–425.
283. *Yoshii, F. and Hirono, S. *Construction of a quantitative three-dimensional model for odor quality using comparative molecular field analysis (CoMFA)*, Chem. Senses, 21 (1996) 201–210.
284. *Zhang, W.e.a., *Synthesis of 5-thenyl- and 5-furyl-substituted benzodiazepines: Probes of the pharmacophore for benzodiazepine receptor agonists*, Eur. J. Med. Chem., 30 (1995) 483–496.
285. *Zhu, L., Yu, q., Chen, K. and Lin, R., *Study on quantitative structure–activity relationship of 1-cyclopropyl-7-(4-methylpiperazinyl)-6-fluoro-1,4-dihydro-4-oxo-3-quinolinecarboxylic acid by comparative molecular field analysis*, Chinese J. Med. Chem. (Zhongguo Yaowu Huaxue Zazhi), 5 (1995) 187–191.
286. *Zhu, L., Yu, q., Chen, K. and Lin, R., *Study on quantitative structure–activity relationship of N1 position of quinolone*, Acta Physico-Chimica Sinica (Wuli Huaxue Xuebao), 11 (1995) 925–928.
287. *Zhu, L.-G., Yu, Q.-S., Chen, K.-X., Lin, R.-S. and Cai, G.-Q., *Studies on the quantitative structure–activity relationship of 1-cyclopropyl-5,7,8-substituted 6-fluoro-1,4-dihydro-4-oxo-3-quinoline acid by comparative molecular field analysis*, Chem. J. Chinese Univ. (Gaodeng Xuexiao Huaxue Xuebao), 16 (1995) 1592–1596.

288a. Navajas, C., Poso, A. and Gynther, J., *CoMFA of flavonoids with antimutagenic activity against 2-amino-3-methylimidazol[4,5-F]quinoline (IQ)*, Elect. J. Theo. Chem., 1, (1996) 45–51.

287b. Wold, S., Kettaneh, N., Tjessem, K., *Hierarchical multiblock PLS and PC models for easier model interpretation and as an alternative to variable selection*, J. Chemomet., 10, (1996) 463–482.

(b) List of CoMFA References, 1997

288. Abrahamian, E., and Hurst, T., *Automated GA/CoMFA — a genetic algorithm driver for producing CoMFA models*, Book of Abstracts, 214th ACS National Meeting, Las Vegas, 1997, COMP-036.
289. *Akamatsu, M., Ozoe, Y., Ueno, T., Fujita, T., Mochida, K., Nakamura, T. and Matsumura, F., *Sites of action of noncompetitive GABA antagonists in houseflies and rats: Three-dimensional QSAR analysis*, Pesticide Sci., 49, (1997) 319–332.
290. Anderson, C.Y., Kellogg, G.E., and Freer, R.J., *C5aR ligand peptide 3D QSAR study performed with an applied linear conformation*, J. Peptide Res. 49, (1997) 476–483.
291. Azzaoui, K., Diazperez, M. J., Price, G.B., Wainer, I.W., *3D-QSAR study of steroids involving in DNA-replication*, Book of Abstracts, 214th ACS National Meeting, Las Vegas, 1997, COMP-109.
292. *Berglund, A. and Wold, S., *INLR (Implicit Non-linear Latent Variable Regression). II. Blockscaling of Expanded Terms with QSAR Examples.*, In: Computer-Assisted Lead Finding and Optimization. Current Tools for Medicinal Chemistry, van de Waterbeemd H., Testa, B., Folkers, G., ed. Verlag Helvetica Chimica Acta: Basel, (1997) in press.
293. *Berglund, A. and Wold, S. *INLR, Implicit Non-linear Latent Variable Regression*, J. Chemomet., in press.
294. Beusen, D.D., Takeuchi, Y., Shands, E.F.B., and Marshall, G. R., *Derivation of a 3D pharmacophore model of substance P antagonists at the neurokinin-1 receptor*, Book of Abstracts, 213th ACS National Meeting, San Francisco, 1997, COMP-010.
295. Bravi, G., Gancia, E., Mascagni, P., Pegna, M., Todeschini, R. and Zaliani, A., *MS-WHIM, new 3D theoretical descriptors derived from molecular surface properties: A comparative 3D QSAR study in a series of steroids*, J. Comput.-Aided Mol. Design, 11, (1997) 79–92.
296. Bures, M.G., *Designing combinatorial libraries using automated docking methods and 3D-QSAR*, Book of Abstracts, 213th ACS National Meeting, San Francisco, 1997, CINF-008.
297. Caldwell, T.M., Criscione, K.R., Dahanukar, V.H., Jalluri, R.K., Slavica, M., and Grunewald, G.L., *Highly selective inhibitors of phenylethanolamine N-methyltransferase*, Book of Abstracts, 213th ACS National Meeting, San Francisco, 1997, MEDI-019.
298. Carrieri, A., Brasili, L., Leonetti, F., Pigini, M., Giannella, M., Bousquet, P., and Carotti, A., *2-D and 3-D modeling of imidazoline receptor ligands: Insights into pharmacophore*, Bioorg. Med. Chem. 5, (1997) 843–856.
299. *Carrigan, S.W., Fox, P.C., Wall, M.E., Wani, M.C. and Bowen, J.P., *Comparative molecular field analysis and molecular modeling studies of 20-(S)-camptothecin analogs as inhibitors of DNA topoisomerase I and anticancer/antitumor agents*, J. Comput.-Aided Mol. Design, 11, (1997) 71–78.
300. Chen, H., Zhou, J., Xie, G., and Pang, S., *Studies on pharmacophore model of K+ channel opener*, Wuli Huaxue Xuebao 13, (1997) 101–105.
301. Collantes, E.R., Tong, W., and Welsh, W.J., *Predicting the chromatographic retention and thermodynamic properties of polycyclic aromatic hydrocarbons (PAHs) based on 3D-QSAR models*, Book of Abstracts, 214th ACS National Meeting, Las Vegas, NV, 1997, ENVR-103.
302. *Corelli, F., Manetti, F., Tafi, A., Campiani, G., Nacci, V. and Botta, M., *Diltiazem-like Calcium Entry Blockers: A Hypothesis of the Receptor-Binding Site Based on a Comparative Molecular Field Analysis Model*, J. Med. Chem., 40 (1997) 125–131.
303. Crippen, G.M., *Validation of EGSITE2, A mixed-integer program for deducing objective site models from experimental binding data*, J. Med. Chem., 40, (1997) 3161–3172.
304. Cruciani, G., Pastor, M. and Clementi, S., *Region Selection in 3D-QSAR*, In: Computer-Assisted Lead Finding and Optimization. Current Tools for Medicinal Chemistry, van de Waterbeemd H, Testa, B., Folkers, G., ed. Verlag Helvetica Chimica Acta: Basel, (1997) in press.
305. Dove, S., and Buschauer, A., *Stepwise leave-one-isomer-out free-Wilson approaches as preprocessing tools in QSAR analysis of racemates*, Quant. Struct.-Act. Relat. 16, (1997) 11–19.
306. Doweyko, A.M., *Predictive 3D-pharmacophores developed from HASL models*, Book of Abstracts, 213th ACS National Meeting, San Francisco, 1997, COMP–306.
307. Easmon, J., Heinisch, G., Hofmann, J., Langer, T., Grunicke, H.H., Fink, J., and Purstinger, G., *Thiazolyl and benzothiazolyl hydrazones derived from a-(N)-acetylpyridines and diazines: Synthesis, antiproliferative activity and CoMFA studies*, Eur. J. Med. Chem. 32, (1997) 397–408.

308. Ettorre, A., Biava, M., Fioravanti, R., Porretta, G.C., *The antifungal agent 1-[2-(4-chlorobenzylamino)-benzyl]-1H-imidazole*, Acta Cryst. Sect. C. Cryst. Struct. Comm., 53, (1997) 761–762.
309. Ferguson, A.M., Heritage, T., Jonathon, P., Pack, S.E., Phillips, L., Rogan, J., and Snaith, P.J., EVA: *A new theoretically based molecular descriptor for use in QSAR/QSPR analysis*, J. Comput.-Aided Mol. Design 11, (1997) 143–152.
310. Fleischer, R., Wiese, M., Troschutz, R., and Zink, M., *3D-QSAR analysis and molecular modeling investigations of piritrexim and analogues*, J. Mol. Model. 3, (1997) 338–346.
311. Gamper, A.M., Winger, R.H., Liedl, K.R., Sotriffer, C.A., Varga, J.M., Kroemer, R.T., and Rode, B.M., *Comparative molecular field analysis of haptens docked to the multispecific antibody IgE(Lb4)*, J. Med. Chem. 40, (1997) 1047–1048.
312. Ginn, C.M.R., Turner, D.B., Willett, P., Ferguson, A.M., and Heritage, T.W., *Similarity searching in files of three-dimensional chemical structures: Evaluation of the EVA descriptor and combination of rankings using data fusion*, J. Chem. Inf. Comput. Sci. 37, (1997) 23–37.
313. Greco, G., Novellino, E. and Martin, Y.C. *Approaches to Three-Dimensional Quantitative Structure-Activity Relationships*, (in press).
314. Hahn, M., *Three-dimensional shape-based searching of conformationally flexible compounds*, J. Chem. Inf. Comput. Sci. 37, (1997) 80–86.
315. Hasegawa, K., Kimura, T., and Funatsu, K., *Nonlinear CoMFA using QPLS as a novel 3D-QSAR approach*, Quant. Struct.-Act. Relat. 16, (1997) 219–223.
316. He, M., Li, T.H., Cong, P.S., *Nonlinear pls improved by numeric genetic algorithm for QSAR modeling*, Chem. J. Chinese Universities, 18, (1997) 854–859.
317. Heritage, T.W., and Hurst, T., *HQSAR — a highly predictive QSAR technique based on molecular holograms*, Book of Abstracts, 214th ACS National Meeting, Las Vegas, 1997, COMP-080.
318. Hinds, T.A., Drake, R.R., and Compadre, C.M., *Analysis of the binding modes of substrates and inhibitors of the herpes simplex virus type I thymidine kinase (HSV-I TK) using 3D QSAR and molecular surface properties*, Book of Abstracts, 213th ACS National Meeting, San Francisco, 1997, MEDI-264.
319. Hurst, T., *HQSAR — a highly predictive QSAR technique based on molecular holograms*, Book of Abstracts, 213th ACS National Meeting, San Francisco, 1997, CINF-019.
320. Jiang, H.L., Chen, K.X., Tang, Y., Chen, J.Z., Li, Q., Wang, Q.M., and Ji, R.Y., *Molecular modeling and 3D-QSAR studies on the interaction mechanism of tripeptidyl thrombin inhibitors with human a-thrombin*, J. Med Chem. 40, (1997) 3085–3090.
321. Kaminski, J.J. and Doweyko, A.M. *Antiulcer Agents. 6. Analysis of the in vitro Biochemical and Pyridines and Related Analogs using Comparative Molecular Field Analysis and Hypothetical Active-Site Lattice Methodologies*, J. Med. Chem., 40, (1997) 427–436.
322. Kim, K.H., Brusniak, M.-Y. K., Pearlman, R. ., *Union dot surface-based comparative molecular field analysis. 1. Toward obtaining consistent results*, in "Rational Molecular Design in Drug Delivery," Alfred Benzon Symposium No. 42, Munksgaard, Copenhagen, Denmark, in press.
323. *Kim, K.H., *Description of an electrostatic nonlinear relationship in comparative molecular field analysis*, Med. Chem. Res., 7 (1997) 45–52.
324. *Kim, K.H., *Electrostatic nonlinear relationships in comparative molecular field analysis derived from the PLS analysis of distance matrices*, (unpublished).
325. Klebe, G., *Structural Alignment of Molecules*, In Kubinyi, H. (Ed.) 3D QSAR in Drug Design. Theory Methods and Applications. ESCOM: Leiden, The Netherlands, 1993, pp. 173–199.
326. Kubinyi, H., *A general view on similarity and QSAR studies, Computer-Assisted Lead Finding and Optimization* (11th Eur. Symp. Quant. Struct.-Act. Relat., Lausanne, 1996), Editors Van de Waterbeemd, H., Testa, B., and Folkers, G., Verlag Helvetica Chimica Acta, Basel, Switzerland, 1997, pp. 7–28.
327. Laguerre, M., Saux, M., Dubost, J.P., and Carpy, A., *MLPP: a program for the calculation of molecular lipophilicity potential in proteins,* Pharm. Sci. 3, (1997) 217–222.
328. Li, H., Xu, L., Su, Q., and Guo, M., *Three-dimensional quantitative structure–activity relationship studies of some steroids and their antiinflammatory activities*, Jisuanji Yu Yingyong Huaxue 14, (1997) 27–30.
329. Li, Y.L., MacKerell, A.D., Egorin, M.J., Ballesteros, M.F., Rosen, D.M., Wu, Y.Y., Blamble, D.A., and Callery, P.S., *Comparative molecular field analysis-based predictive model of structure–function relationships of polyamine transport inhibitors in L1210 cells*, Cancer Res. 57, (1997) 234–239.

330. Liu, J., Wang, X., Ma, Y., Li, Z.M., Lai, C.M., Jia, G.F., and Wang, L.X., *Comparative molecular field analysis on a set of new herbicidal sulfonylurea compounds*, Chin. Chem. Lett. 8, (1997) 503–504.
331. Lopez-Rodriguez, M.L., Rosado, M.L., Benhamu, B., Morcillo, M.J., Fernandez, E., amd Schaper, K.J., *Synthesis and structure–activity relationships of a new model of arylpiperazines. 2. Three-dimensional quantitative structure–activity relationships of hydantoin-phenylpiperazine derivatives with affinity for 5-HT1A and α(1) receptors. A comparison of CoMFA models*, J. Med. Chem. 40, (1997) 1648–1656.
332. Luo, Q., Darsey, J.A., Compadre, R.L., Marles, R.J., and Compadre, C.M., *Structure–activity relationships of sesquiterpene lactones with potential antimigraine activity*, Book of Abstracts, 213th ACS National Meeting, San Francisco, 1997, MEDI-057.
333. Matter, H., *Selecting optimally diverse compounds from structure databases: A validation study of two-dimensional and three-dimensional molecular descriptors*, J. Med. Chem. Vol. 40, (1997) 1219–1229.
334. Mestres, J., Rohrer, D. C., and Maggiora, G.M., MIMIC: *A molecular-field matching program. Exploiting applicability of molecular similarity approaches*, J. Comput. Chem. 18, (1997) 934–954.
335. Meyer, C., *Sweetness pharmacophore elucidation*, Book of Abstracts, 213th ACS National Meeting, San Francisco, 1997, COMP-036.
336. Morita, H., Gonda, A., Wei, L., Takeya, K., Itokawa, H., *3D QSAR analysis of taxoids from taxus-cuspidata var. nana by Comparative molecular-field approach*, Bioorg. Med. Chem. Lett., 7, (1997) 2387–2392.
337. Nilsson, J., Wikstrom, H., Smilde, A., Glase, S., Pugsley, T., Cruciani, G., Pastor, M. and Clementi, S. *GRID/GOLPE 3D Quantitative Structure–Activity Relationship Study on a Set of Benzamides and Naphthamides, with Affinity for the Dopamine D3 Receptor Subtype*, J. Med. Chem., 40, (1997) 833–840.
338. Norinder, U. *3D-QSAR investigation of the tripos benchmark steroids and some protein-tyrosine kinase inhibitors of styrene type using the TDQ approach*, J. Chemom., 11, (1997) in press.
339. *Oprea, T.I., Kurunczi, L. and Timofei, S. *QSAR Studies of Disperse Azo Dyes — Towards the Negation of the Pharmacophore Theory of Dye-Fiber Interaction*, Dyes Pigments, 33, (1997) 41–64.
340. Ortiz, A.R., Pastor, M., Palomer, A., Cruciani, G., Gago, F., and Wade, R.C., *Reliability of Comparative Molecular Field Analysis Models: Effects of Data Scaling and Variable Selection Using a Set of Human Synovial Fluid Phospholipase A_2 Inhibitor*, J. Med. Chem. 40 (1997), 1136–1148.
341. Pajeva, I.K., and Wiese, M., *QSAR and molecular modeling of catamphiphilic drugs able to modulate multidrug resistance in tumors*, Quant. Struct.-Act. Relat. 16, (1997) 1–10.
342. Parretti, M.F., Kroemer, R.T., Rothman, J.H., and Richards, W.G., *Alignment of molecules by the Monte Carlo optimization of molecular similarity indices,* J. Comput. Chem. 18, (1997) 1344–1353.
343. *Pastor, M. and Cruciani, G., *The Role of Water in Receptor–Ligand Interactions. A 3D-QSAR Approach*, In: Computer-Assisted Lead Finding and Optimization. Current Tools for Medicinal Chemistry, van de Waterbeemd H., Testa, B., Folkers, G., ed. Verlag Helvetica Chimica Acta: Basel, (1997) in press.
344. Pastor, M., Cruciani, G. and Clementi, S., *Smart Region Definition SRD: a new way to improve the predictive ability and interpretability of 3D QSAR models*, J. Med. Chem. 40, (1997) 1455–1464.
345. Polanski, J., *The receptor-like neural network for modeling corticosteroid and testosterone binding globulins*, J. Chem. Inf. Comput. Sci. 37, (1997) 553–561.
346. *Poso, A., von Wright, A. and Gynther, J., *An empirical and theoretical study on mechanisms of mutagenic activity of hydrazine compounds*, Mutation Res., in press.
347. *Rong, S.B., Zhu, Y.C., Jiang, H.L., Wang, Q.M., Zhao, S.R., Chen, K.X. and Ji, R.Y., *Interaction models of 3-methylfentanyl derivatives with mu-opioid receptors*, Acta Pharmacol. Sinica, 18, (1997) 128–132.
348. Schmetzer, S., Greenidge, P., Kovar, K.A., Schulze-Alexandru, M., and Folkers, G., Structure–activity relationships of cannabinoids: A joint CoMFA and pseudoreceptor modeling study, J. Comput.-Aided Mol. Design 11, (1997) 278–292.
349. Schnitker, J., Gopalaswamy, R., and Crippen, G.M., *Objective models for steroid binding sites of human globulins*, J. Comput.-Aided Mol. Design 11, (1997) 93–110.
350. Shim, J.-Y., Collantes, E.R., Welsh, W.J., Berglund, B., and Howlett, A.C., *Rational drug design of potent agonists and antagonists for the CB1 cannabinoid receptor*, Book of Abstracts, 214th ACS National Meeting, Las Vegas, 1997, COMP-077.

351. *Sicsic, S., Serraz, I., Andrieux, J., Bremont, B., Matheallainmat, M., Poncet, A., Shen, S. and Langlois, M., *3-Dimensional quantitative structure–activity relationship of melatonin receptor ligands — A comparative molecular-field analysis study*, J. Med. Chem., 40, (1997) 739–748.
352. Singh, S., Basmadjian, G.P., Avor, K.S., Pouw, B., Searle, T.W., *Synthesis and lignd-binding studies of 4′-iodobenzoyl esters of tropanes and piperidines at the dopamine transporter*, J. Med. Chem., 40, (1997) 2474–2481.
353. Sotomatsuniwa, T., Ogino, A., *Evaluation of the hydrophobic parameters of the amino-acid side-chains of peptides and their application in QSAR and conformational studies*, THEOCHEM, (1997).
354. T. Sulea, T.I., Oprea, S.L. Chan and S. Muresan, *A Different Method for Steric Field Evaluation in CoMFA Improves Model Robustness*, J. Chem. Inf. Comput. Sci., accepted.
355. T.I. Oprea and C.L. Waller, *Theoretical and Practical Aspects of 3D-QSAR*, in Reviews in Computational Chemistry, vol 11, D. Boyd and K. Lipkowitz (Eds), VCH Publishers, New York, NY, 1997, in press.
356. T.I. Oprea, R.D. Head and G.R. Marshall, *The basis of cross-reactivity for a series of steroids binding to a monoclonal antibody against progesterone (DB3). A molecular modeling and QSAR study*, in QSAR and Molecular Modeling: Concepts, Computational Tools and Biological Applications, F. Sanz, J. Giraldo and F. Manaut (Eds.), JR Prous Publishers, Barcelona, 1995, pp. 451–455.
357. Thorner, D.A., Willett, P., Wright, P.M., and Taylor, R., *Similarity searching in files of three-dimensional chemical structures: Representation and searching of molecular electrostatic potentials using field-graphs*, J. Comput.-Aided Mol. Design 11, (1997) 163–174.
358. Todeschini, R., Moro, G., Boggia, R., Bonati, L., Cosentino, U., Lasagni, M., and Pitea, D., *Modeling and prediction of molecular properties. Theory of grid-weighted holistic invariant molecular (G-WHIM) descriptors*, Chemom. Intell. Lab. Systems 36, (1997) 65–73.
359. Todeschini, R.; Gramatica, P., *3D-Modeling and prediction by WHIM descriptors. 5. Theory, development and chemical meaning of WHIM descriptors*, Quant. Struct.-Act. Relat., 16, (1997) 113–119.
360. Todeschini, R., Gramatica, P., *3D-Modeling and prediction by WHIM descriptors. 6. Application of WHIM descriptors in QSAR studies*, Quant. Struct.-Act. Relat., 16, (1997) 120–125.
361. Tokarski, J.S., Hopfinger, A.J., *Prediction of ligand-receptor binding thermodynamics by free-energy force-field (FEFF) 3D-QSAR analysis — Application to a set of peptidometic renin inhibitors*, J. Chem. Inform. Comput. Sci., 37, (1997) 792–811.
362. Tong, W.D., Perkins, R., Xing, L., Welsh, W.J., and Sheehan, D.M., *QSAR models for binding of estrogenic compounds to estrogen receptor a and b subtypes*, Endocrinology 138, (1997) 4022–4025.
363. Tong, W., Collantes, E.R., Shim, J.-Y., Welsh, W.J., Berglund, B., and Howlett, A., *Pharmacophoric mapping of the CB1 cannabinoid receptor*, Book of Abstracts, 213th ACS National Meeting, San Francisco, 1997, COMP-012.
364. Tong, W., Perkins, R., Chen, Y., Shvets, V., Xing, L., Welsh, W., and Sheehan, D.M., *QSAR models for estrogen binding to estrogen receptors α and β*, Book of Abstracts, 214th ACS National Meeting, Las Vegas, 1997, ENVR-102.
365. Tong, W., Perkins, R., Collantes, E.R., Welsh, W.J., Branham, W.S., and Sheehan, D. M., *Quantitative structure–activity relationships (QSARS) for estrogen binding to the estrogen receptor: Predictions across species*, Book of Abstracts, 214th ACS National Meeting, Las Vegas, NV, 1997, ENVR-101.
366. Tong, W., Perkins, R., Sheehan, D.M., Welsh, W.J., Lowis, D.R., Heritage, T., and Goddette, D.W., *Application of the holographic QSAR (HQSAR) method to predict the biological activity of environmental estrogens*, Book of Abstracts, 214th ACS National Meeting, Las Vegas, 1997, COMP-081.
367. Tong, W., Perkins, R., Strelitz, R., Collantes, E.R., Welsh, W.J., and Sheehan, D.M., *QSAR studies of estrogen receptor binding affinity*, Book of Abstracts, 213th ACS National Meeting, San Francisco, 1997, COMP-037.
368. Turner, D.B., Willett, P., Ferguson, A.M., and Heritage, T., *Evaluation of a novel infrared range vibration-based descriptor (EVA) for QSAR studies. 1. General application*, J. Comput.-Aided Mol. Design 11, 409–422 (1997).
369. Turner, D.B., Willett, P., Ferguson, A.M., and Heritage, T., *Development and validation of the EVA descriptor for QSAR studies*, Book of Abstracts, 214th ACS National Meeting, Las Vegas, NV, 1997, COMP-158.

370. Turner, D.B., Willett, P., Ferguson, A.M., Heritage, T., *Evaluation of a novel infrared range vibration-based descriptor (EVA) for QSAR studies. 1. General application*, J. Comput.-Aided Mol. Design, 11, (1997) 409–422.
371. Ungwitayatorn, J., Pickert, M., and Frahm, A.W., *Quantitative structure–activity relationship (QSAR) study of polyhydroxyxanthones*, Pharm. Acta Helv. 72, (1997) 23–29.
372. Vaz, R.J., *Use of electron-densities in comparative molecular-field analysis (CoMFA) — A quantitative structure–activity relationship (QSAR) for electronic effects of groups*, Quant. Struct.-Act. Relat., 16, (1997) 303–308.
373. *Welch, W., Williams, A.J., Tinker, A., Mitchell, K.E., Deslongchamps, P., Lamothe, J., Gerzon, K., Bidasee, K.R., Besch, H.R., Airey, J.A., Sutko, J.L., Ruest, L., *Structural components of ryanodine responsible for modulation of sarcoplasmic-reticulum calcium-channel function*, Biochem., 36, (1997) 2939–2950.
374. Welsh, W.J., Tong, W.D., Collantes, E.R., Chickos, J.S., and Gagarin, S.G., *Enthalpies of sublimation and formation of polycyclic aromatic hydrocarbons (PAHs) derived from comparative molecular field analysis (CoMFA): Application of moment of inertia for molecular alignment*, Thermochim. Acta 290, (1997) 55–64.
375. Woolfrey, J.R., Avery, M.A., *The design and synthesis of potential selective progesterone receptor antagonists*, Book of Abstracts, 213th ACS National Meeting, San Francisco, 1997, MEDI-015.
376. Xie, G., Qang, D., Feng, J. and Zhou, J., *QSAR Study of a-Oxocyclododecylsulphonamides Series Compounds by CoMFA*, Science Bulletin, in press.
377. Zefirov, N.S., Palyulin, V.A., and Radchenko, E.V., *Molecular field topology analysis (MFTA) technique in QSAR studies of organic compounds*, Doklady Akademii Nauk 352, (1997) 630–633.
378. Haraldson, C.A., Karle, J.M., Freeman, S.G., Duvadie, R.K., Avery, M.A., *The synthesis of 8,8-disubstituted tricyclic analogs of artemisinin*, Bioorg. Med. Chem. Lett., 7, (1997) 2357–2362.
379. Hopfinger, A.J., Wang, S., Tokarski, J.S., Jin, B.Q., Albuquerque, M., Madhav, P.J., Duraiswami, C., *Construction of 3D-QSAR models using the 4D-QSAR analysis formalism*, J. Am. Chem. Soc., 119, (1997) 10509–10524.
380. Kubinyi, H., *QSAR and 3D QSAR in drug design. 1. Methodology*, Drug Discovery Today, 2, (1997) 457–467.
381. Shimizu, B., Nakagawa, Y., Hattori, K., Nishimura, K., Kurihara, N., Ueno, T., *Molting hormonal and larvicidal activities of aliphatic acyl analogs of dibenzoylhydrazine insecticides*, Steroids, 62, (1997) 638–642.
382. Teitler, M., Scheick, C., Howard, P., Sullivan, J.E., Iwamura, T., Glennon, R.A., *5-HT5a serotonin receptor-binding. A preliminary structure affinity investigation*, Med. Chem. Res., 7, (1997) 207–218.
383. Wiese, T.E., Polin, L.A., Palomino, E., Brooks, S.C., *Induction of the estrogen specific mitogenic response of MCF-7 cells by selected analogs of estradiol-17-beta-A 3D QSAR study*, J. Med. Chem., 40, (1997) 3659–3669.

Author Index

H. Kubinyi et al. (eds.), 3D QSAR in Drug Design, Volume 3. 339.

Subject Index

H. Kubinyi et al. (eds.), 3D QSAR in Drug Design, Volume 3. 341–352.

QSAR = Three-Dimensional Quantitative Structure Activity Relationships

1. H. Kubinyi (ed.): *3D QSAR in Drug Design.* Theory Methods and Applications 1997 ISBN 90-72199-14-6
2. H. Kubinyi, G. Folkers and Y.C. Martin (eds.): *3D QSAR in Drug Design. Volume 2* Ligand-Protein Interactions and Molecular Similarity. 1998 ISBN 0-7923-4790-0
3. H. Kubinyi, G. Folkers and Y.C. Martin (eds.): *3D QSAR in Drug Design. Volume 3* Recent Advances. 1998 ISBN 0-7923-4791-9

KLUWER ACADEMIC PUBLISHERS – DORDRECHT / BOSTON / LONDON